Modulation
of
Protein Function

Academic Press Rapid Manuscript Reproduction

Proceedings of the
1979 ICN–UCLA Symposia on
Molecular and Cellular Biology
Held in Keystone, Colorado
February 25–March 2, 1979

*ICN–UCLA Symposia on Molecular and Cellular Biology
Volume XIII, 1979*

MODULATION
OF
PROTEIN FUNCTION

edited by

DANIEL E. ATKINSON
*Department of Chemistry
and Molecular Biology Institute
University of California, Los Angeles
Los Angeles, California*

C. FRED FOX
*Department of Microbiology
and Molecular Biology Institute
University of California, Los Angeles
Los Angeles, California*

ACADEMIC PRESS   1979
A Subsidiary of Harcourt Brace Jovanovich, Publishers
New York   London   Toronto   Sydney   San Francisco

COPYRIGHT © 1979, BY ACADEMIC PRESS, INC.
ALL RIGHTS RESERVED.
NO PART OF THIS PUBLICATION MAY BE REPRODUCED OR
TRANSMITTED IN ANY FORM OR BY ANY MEANS, ELECTRONIC
OR MECHANICAL, INCLUDING PHOTOCOPY, RECORDING, OR ANY
INFORMATION STORAGE AND RETRIEVAL SYSTEM, WITHOUT
PERMISSION IN WRITING FROM THE PUBLISHER.

ACADEMIC PRESS, INC.
111 Fifth Avenue, New York, New York 10003

United Kingdom Edition published by
ACADEMIC PRESS, INC. (LONDON) LTD.
24/28 Oval Road, London NW1 7DX

Library of Congress Cataloging in Publication Data

Main entry under title:

Modulation of protein function.

    (ICN-UCLA symposia on molecular and cellular
biology ; v. 13)
    Proceedings of a conference held in Keystone,
Colo., Feb. 25-Mar. 2, 1979.
    Includes bibliographical references.
    1. Enzymes—Congresses.   2. Cellular control
mechanisms—Congresses.   I. Atkinson, Daniel E.
II. Fox, C. Fred.   III. Series: ICN-UCLA symposia
on molecular & cellular biology ; v. 13.
QP601.M565   574.1'33   79-21565
ISBN 0-12-066250-7

PRINTED IN THE UNITED STATES OF AMERICA

79 80 81 82   9 8 7 6 5 4 3 2 1

CONTENTS

Contributors ix
Preface xv

I. MODULATION OF ENZYMES OF INTERMEDIARY METABOLISM

1. Molecular Properties of Phosphofructokinase (PFK) Relevant to Modulation of Its Function   1

    Tag E. Mansour, Glenda Choate, and Litai Weng

2. Adenine Nucleotide Pool Maintenance during Bacterial Growth and Starvation   13

    Christopher J. Knowles

3. Multimodulation of Enzyme Activity. Physiological Significance and Evolutionary Origin   27

    Alberto Sols

II. MODULATION AND INACTIVATION

4. Poly(ADP-Ribose) and ADP-Ribosylation of Proteins   47

    Kunihiro Ueda, Osamu Hayaishi, Masashi Kawaichi, Norio Ogata, Kouichi Ikai, Jun Oka, and Hiroto Okayama

5. Selective Inactivation and Degradation of Enzymes in Sporulating Bacteria   65

    Robert L. Switzer, Michael R. Maurizi, Joseph Y. Wong, and Kerry J. Flom

6. Endogenous Proteolytic Modulation of Yeast Enzymes   81

    Heidrun Matern and Helmut Holzer

III. PHOTOSYNTHESIS AND STORAGE POLYSACCHARIDES

7. Thioredoxin and Enzyme Regulation in Photosynthesis   93
   Bob B. Buchanan

8. Regulation of the Photosynthetic Carbon Cycle, Phosphorylation, and Electron Transport in Illuminated Intact Chloroplasts   113
   Ulrich Heber, Ulrike Enser, Engelbert Weis, Ursula Ziem, and Christoph Giersch

9. Regulation of Photosynthetic Carbon Metabolism and Partitioning of Photosynthate   139
   James A. Bassham

10. Comparative Regulation of $\alpha 1,4$-Glucan Synthesis in Photosynthetic and Nonphotosynthetic Systems   161
    Jack Preiss, William K. Kappel, and Elaine Greenberg

IV. CASCADE SYSTEMS

11. Covalently Interconvertible Enzyme Cascade and Metabolic Regulation   185
    P. B. Chock and E. R. Stadtman

12. Metabolite Control of the Glutamine Synthetase Cascade   203
    E. R. Stadtman, P. B. Chock, and S. G. Rhee

13. The Molecular Dynamics and Biochemistry of Complement   219
    Hans J. Müller-Eberhard

V. PROTEIN PHOSPHORYLATION

14. Cyclic Nucleotide-Independent Protein Kinases from Rabbit Reticulocytes and Phosphorylation of Translational Components   233
    Jolinda A. Traugh, Gary M. Hathaway, Polygena T. Tuazon, Stanley M. Tahara, Georgia A. Floyd, Robert W. Del Grande, and Tina S. Lundak

15. An Approach to the Study of Phosphoprotein and Cyclic Nucleotide Metabolism in Cultured Cell Lines with Differentiated Properties   247
    Ora M. Rosen, Chen K. Chou, Jeanne Piscitello, Barry R. Bloom, Charles Smith, Peter J. Wejksnora, Rameshwar Sidhu, and Charles S. Rubin

16. Glycogen Synthase Kinase-2 and Phosphorylase Kinase Are the Same Enzyme ... 257
   Noor Embi, Dennis B. Rylatt, and Philip Cohen

VI. METHYLATION IN CHEMOTAXIS

17. Methylation and Demethylation in the Bacterial Chemotactic System ... 273
   Sharon M. Panasenko and Daniel E. Koshland, Jr.

18. Requirement of Transmethylation Reactions for Eukaryotic Cell Chemotaxis ... 285
   Ralph Snyderman and Marilyn C. Pike

19. Role for Methylation in Leukocyte Chemotaxis ... 299
   E. Schiffmann, R. F. O'Dea, P. K. Chiang, K. Venkatasubramanian, B. Corcoran, F. Hirata, and J. Axelrod

VII. CYCLIC GMP AND CYCLIC CMP

20. Enzymatic Formation of Cyclic CMP by Mammalian Tissues ... 315
   Louis J. Ignarro and Stella Y. Cech

21. Cyclic CMP Phosphodiesterase: Biological Involvement and Its Regulation by Agents ... 335
   J. F. Kuo, Mamoru Shoji, David M. Helfman, and Nancy L. Brackett

VIII. PROTEIN SYNTHESIS

22. Regulation of Phosphoenolpyruvate Carboxykinase (GTP) Synthesis ... 357
   Richard W. Hanson, Michele A. Cimbala, J. Garcia-Ruiz, Kathelyn Nelson, and Dimitris Kioussis

23. Peptide-Chain Initiation in Heart and Skeletal Muscle ... 369
   Leonard S. Jefferson, Kathryn E. Flaim, and Howard E. Morgan

24. Effect of Phosphorylation of eIF-2 ... 391
   William C. Merrick

25. Modulation of Protein Synthesis and Eukaryotic Initiation Factor 2 (eIF-2) Function during Nutrient Deprival in the Ehrlich Ascites Tumor Cell   407

    Edgar C. Henshaw, Walter Mastropaolo, and A. R. Subramanian

IX. CLINICAL IMPLICATIONS

26. Genetic Defects of the Human Red Blood Cell and Hemolytic Anemia   423

    William N. Valentine and Donald E. Paglia

27. Enzyme Replacement Therapy   449

    Ernest Beutler and George L. Dale

Index   *463*

CONTRIBUTORS

Numbers in parentheses indicate the pages on which the author's contributions begin.

J. AXELROD (299), National Institute of Mental Health, National Institutes of Health, Bethesda, Maryland 20205

JAMES A. BASSHAM (139), Lawrence Berkeley Laboratory, University of California, Berkeley, California 94720

ERNEST BEUTLER (449), Department of Clinical Research, Scripps Clinic and Research Foundation, La Jolla, California 92037

BARRY R. BLOOM (247), Departments of Cell Biology, Microbiology, and Immunology, Albert Einstein College of Medicine, Bronx, New York 10461

NANCY L. BRACKETT (335), Department of Pharmacology, Emory University School of Medicine, Atlanta, Georgia 30322

BOB B. BUCHANAN, (93), Department of Cell Physiology, University of California, Berkeley, California 94720

STELLA Y. CECH (315), Department of Pharmacology, Tulane University School of Medicine, New Orleans, Louisiana 70112

P. K. CHIANG (299), National Institute of Mental Health, National Institutes of Health, Bethesda, Maryland 20205

GLENDA CHOATE (1), Department of Pharmacology, Stanford University School of Medicine, Stanford, California 94305

P. B. CHOCK (185, 203), Building 3, Room 202, National Institutes of Health, Bethesda, Maryland 20205

CHEN K. CHOU (247) Departments of Molecular Pharmacology and Molecular Biology, Albert Einstein College of Medicine, Bronx, New York 10461

MICHELE A. CIMBALA (357), Department of Biochemistry, University School of Medicine, Cleveland, Ohio 44106

PHILIP COHEN (257), Department of Biochemistry, University of Dundee, Dundee, DD1 4HN, Scotland

B. CORCORAN (299), Laboratory of Developmental Biology and Anomalies, National Institutes of Health, Bethesda, Maryland 20205

GEORGE L. DALE (449), Division of Medicine, City of Hope National Medical Center, Duarte, California 91010

ROBERT W. DEL GRANDE (233), Department of Biochemistry, University of California, Riverside, California 92521

NOOR EMBI (257), Department of Biochemistry, University of Dundee, Dundee, 4HN, Scotland

ULRIKE ENSER (113), der Universitat Dusseldorf, Universitatstrabe 1, 4000 Düsseldorf, West Germany

KATHRYN E. FLAIM (369), Department of Physiology, Pennsylvania State University, Hershey, Pennsylvania 17033

KERRY J. FLOM (65), 318 Roger Adams Laboratory, Biochemistry Department, University of Illinois, Urbana, Illinois 61801

GEORGIA A. FLOYD (233), Department of Biochemistry, University of California, Riverside, California 92521

J. GARCIA-RUIZ (357), Departments of Biochemistry and Molecular Biology, Autónoma University of Madrid, Madrid, Spain

CHRISTOPH GIERSCH (113) der Universität Düsseldorf, Universitatstrabe 1, 4000 Düsseldorf, West Germany

ELAINE GREENBERG (161), Department of Biochemistry and Biophysics, University of California, Davis, California 95616

RICHARD W. HANSON (357), Department of Biochemistry, Case Western Reserve, University School of Medicine, Cleveland, Ohio 44106

GARY M. HATHAWAY (233), Department of Biochemistry, University of California, Riverside, California 92521

OSAMU HAYAISHI (47), Department of Medical Chemistry, Kyoto University Faculty of Medicine, Sakyo-ku, Kyoto 606, Japan

ULRICH HEBER (113), der Universität Düsseldorf, Universitatstrabe 1, 4000 Düsseldorf, West Germany

DAVID M. HELFMAN (335), Department of Pharmacology, Emory University School of Medicine, Atlanta, Georgia 30322

EDGAR C. HENSHAW (407), University of Rochester Cancer Center, Rochester, New York 14642

F. HIRATA (299), National Institute of Mental Health, National Institutes of Health, Bethesda, Maryland 20205

CONTRIBUTORS   xi

HELMUT HOLZER (81), Biochemisches Institut, der Universität, Herman-Herder-Strasse 7, D-7800 Freiburg in Breisgau, West Germany

LOUIS J. IGNARRO (315), Department of Pharmacology, Tulane University School of Medicine, New Orleans, Louisiana 70112

KOUICHI IKAI (47), Department of Medical Chemistry, Kyoto University Faculty of Medicine, Sakyo-ku, Kyoto 606, Japan

LEONARD S. JEFFERSON (369), Department of Physiology, Pennsylvania State University, Hershey, Pennsylvania 17033

WILLIAM K. KAPPEL (161), Departments of Biochemistry and Biophysics, University of California, Davis, California 95616

MASASHI KAWAICHI (47), Department of Medical Chemistry, Kyoto University Faculty of Medicine, Sakyo-ku, Kyoto 606, Japan

DIMITRIS KIOUSSIS (357), Department of Biochemistry, University School of Medicine, Cleveland, Ohio 44106

CHRISTOPHER J. KNOWLES (13), Biological Laboratory, University of Kent, Canterbury, CT2 7NJ, England

DANIEL E. KOSHLAND, JR. (273), Department of Biochemistry, University of California, Berkeley, California 94720

J. F. KUO (335), Department of Pharmacology, Emory University School of Medicine, Atlanta, Georgia 30322

TINA S. LUNDAK (233), Department of Biochemistry, University of California, Riverside, California 92521

TAG E. MANSOUR (1), Department of Pharmacology, Stanford University School of Medicine, Stanford, California 94305

WALTER MASTROPAOLO (407), University of Rochester Cancer Center, Rochester, New York 14642

HEIDRUN MATERN (81), Biochemisches Institut, der Universität, Hermann-Herder-Strasse 7, D-7800 Freiburg im Breisgau, West Germany

MICHAEL R. MAURIZI (65), Laboratory of Biochemistry, NHLBI, Bethesda, Maryland 20205

WILLIAM C. MERRICK (391), Department of Biochemistry, Case Western Reserve, University School of Medicine, Cleveland, Ohio 44106

HOWARD E. MORGAN (369), Department of Physiology, Pennsylvania State University, Hershey, Pennsylvania 17033

HANS J. MÜLLER-EBERHARD (219), Department of Molecular Immunology, Research Institute of Scripps Clinic, La Jolla, California 92037

KATHELYN NELSON (357), Department of Biochemistry, University School of Medicine, Cleveland, Ohio 44106

R. F. O'DEA (299), University of Minnesota Medical School, Minneapolis, Minnesota 55455

NORIO OGATA (47), Department of Medical Chemistry, Kyoto University Faculty of Medicine, Sakyo-ku, Kyoto 606, Japan

JUN OKA (47), Department of Medical Chemistry, Kyoto University Faculty of Medicine, Sakyo-ku, Kyoto 606, Japan

HIROTO OKAYAMA (47), Department of Medical Chemistry, Kyoto University Faculty of Medicine, Sakyo-ku, Kyoto 606, Japan

DONALD E. PAGLIA (423), Division of Surgical Pathology, University of California, Los Angeles, California 90024

SHARON M. PANASENKO (273), Department of Biochemistry, University of California, Berkeley, California 94720

MARILYN C. PIKE (285), Box 3892, Duke University Medical Center, Durham, North Carolina 27710

JEANNE PISCITELLO (247), Department of Molecular Pharmacology, Albert Einstein College of Medicine, Bronx, New York 10461

JACK PREISS (161), Departments of Biochemistry and Biophysics, University of California, Davis, California 95616

S. G. RHEE (203), Laboratory of Biochemistry, National Institutes of Health, Bethesda, Maryland 20205

ORA M. ROSEN (247), Departments of Molecular Pharmaology and Medicine, Albert Einstein College of Medicine, Bronx, New York 10461

CHARLES S. RUBIN (247), Departments of Molecular Pharmacology and Neuroscience, Albert Einstein College of Medicine, Bronx, New York 10461

DENNIS B. RYLATT (257), Department of Biochemistry, University of Dundee, Dundee DD1 4HN, Scotland

E. SCHIFFMANN (299), Building 30, Room 410, National Institutes of Health, Bethesda, Maryland 20205

MAMORU SHOJI (335), Department of Pharmacology, Emory University School of Medicine, Atlanta, Georgia 30322

RAMESHWAR SIDHU (247), Department of Molecular Pharmacology, Albert Einstein College of Medicine, Bronx, New York 10461

CHARLES SMITH (247), Department of Molecular Pharmacology, Albert Einstein College of Medicine, Bronx, New York 10461

RALPH SNYDERMAN (285), Box 3892, Duke University Medical Center, Durham, North Carolina 27710

ALBERTO SOLS (27), Instituto de Enzimología y Patología Molecular, Universidad Autónoma, Madrid 34, Spain

E. R. STADTMAN (185, 203), Building 3, Room 202, National Institutes of Health, Bethesda, Maryland 20205

A. R. SUBRAMANIAN (407), University of Rochester Cancer Center, Rochester, New York 14642

ROBERT L. SWITZER (65), 318 Roger Adams Laboratory, Biochemistry Department, University of Illinois, Urbana, Illinois 61801

STANLEY M. TAHARA (233), Department of Biochemistry, University of California, Riverside, California 92521

JOLINDA A. TRAUGH (233), Department of Biochemistry, University of California, Riverside, California 92521

POLYGENA T. TUAZON (233), Department of Biochemistry, University of California, Riverside, California 92521

KUNIHIRO UEDA (47), Department of Medical Chemistry, Kyoto University Faculty of Medicine, Sakyo-ku, Kyoto 606, Japan

WILLIAM N. VALENTINE (423), Department of Medicine, Center for the Health Sciences, University of California, Los Angeles, California 90024

K. VENKATASUBRAMANIAN (299), Building 30, Room 410, National Institutes of Health, Bethesda, Maryland 20205

ENGELBERT WEIS (113), Botanisches Institut, der Universität Düsseldorf, Universitatstrabe 1, 4000 Düsseldorf, West Germany

PETER J. WEJKSNORA (247), Department of Biochemistry, Albert Einstein College of Medicine, Bronx, New York 10461

LITAI WENG (1), Department of Pharmacology, Stanford University School of Medicine, Stanford, California 94305

JOSEPH Y. WONG (65), 318 Roger Adams Laboratory, Biochemistry Department, University of Illinois, Urbana, Illinois 61801

URSULA ZIEM (113), Botanisches Institut, der Universität Düsseldorf, Universitatstrabe 1, 4000 Düsseldorf, West Germany

PREFACE

During the past two decades it has come to be recognized that biological functions of many, if not all, types are controlled to a very large extent through modulation of the functions of individual proteins or of multimolecular protein systems that result from interaction with metabolites or with specialized messenger compounds of low molecular weight. It was the aim of this symposium to bring together workers from several fields, all of which deal with such modulation of protein function. Discussion of representative metabolic control systems, ranging from single-enzyme responses to complex regulatory cascades, and the control of photosynthesis and of protein synthesis and enzyme inactivation dealt with the general topic at perhaps its most fundamental cellular level. Modulations and conformational changes in proteins that underlie higher-level interactions, such as those involved in cyclic nucleotide function, sensing and chemotactic response to foreign materials, and the complement system, were described. Two talks dealt with potential clinical relevance of phenomena of the types described by other participants. The common thread of functionally significant consequences of protein–small-molecule interaction led to extensive interaction among participants who work on widely diverse systems, and the editors hope that common thread will similarly unify this published record of the symposium.

We wish to thank the symposium speakers and poster session contributors for providing the basis of the program. We also wish to acknowledge the continuing support that the Life Sciences Division of ICN Pharmaceuticals, Inc., endows for the general support of this conference series, and, finally, we cite the generous contribution made by The National Foundation in partial support of the present meeting.

Daniel E. Atkinson

MOLECULAR PROPERTIES OF PHOSPHOFRUCTOKINASE (PFK) RELEVANT TO MODULATION OF ITS FUNCTION[1]

Tag E. Mansour, Glenda Choate[2], and Litai Weng
Department of Pharmacology
Stanford University School of Medicine
Stanford, California 94305

Studies during the past twenty years on the molecular properties of phosphofructokinase have contributed immensely to our understanding of its role as an important regulatory enzyme in glycolysis. Both covalent and non-covalent changes in enzyme structure have been reported. Evidence has been accumulating showing variation in phosphofructokinase activity in connection with different physiological conditions. In many cases the changes in enzyme activity is implied from indirect evidence and on the basis of what we already know of the properties of the enzyme. We wish to summarize briefly our current knowledge of some of the most important molecular properties of the enzyme. We will then report on some recent experiments on its allosteric sites and the nature of inhibition by vanadate. Finally, we will discuss briefly the relationship between these properties and the regulatory function of phosphofructokinase.

Molecular structure. Information based on data from our own laboratory on heart phosphofructokinase as well as laboratories of Lardy (1,2) and others indicates that the smallest fully active phosphofructokinase is a tetramer with a molecular weight of 360,000 and an $S_{20,w}$ value of 13. High enzyme concentration or the presence of fructose-1,6-P_2 or fructose-6-P favor the formation of high aggregates of the enzyme with an $S_{20,w}$ value as high as 54, while the presence of ATP or low enzyme concentration favor the low molecular form. The tetrameric form of the enzyme can be dissociated to dimers which are inactive. Enzyme protomers can be obtained in the presence of 4mM of SDS. Each protomer can be dissociated to 4 subunits with a molecular weight of 24,000 in the presence of 5M guanidine HCl.

Kinetics. Studies on the kinetics of phosphofructokinase have indicated that pH determines the nature of these kinetics (1). At pH 8.2, which is the optimal pH for enzyme activity, it exhibited Michaelis-Menten type of kinetics. At pH 6.9 typical allosteric kinetics are seen. The curve for ATP is hyperbolic until the activity

---

[1] Some of this work was supported by Public Health Service Research Grant HL17976.
[2] Recipient of a U.S. Public Health Service Research Fellowship: Arthritis, Metabolic and Digestive Diseases Institute, #1 F32 AM05663-01.

TABLE 1

SOME OF THE IMPORTANT EFFECTORS
OF PHOSPHOFRUCTOKINASE

Inhibitors	Deinhibitors of ATP or Activators
ATP	3',5'-cyclic AMP
Citrate	5'-AMP
P-creatine	ADP
3-P-glycerate	Fructose 6-P
2-P-glycerate	Fructose-1,6-P_2
2,3-P_2-glycerate	Glucose-1,6-P_2
P-enolpyruvate	NH_4^+, Pi

is maximal, followed by a steep inhibition curve as the ATP concentration is increased. In the presence of an activator the catalytic part of the curve is not changed while the inhibitory curve reaches almost a plateau after maximal activity. Thus activators exert their effect by relieving ATP inhibition, i.e. by "de-inhibition". The saturation curve for fructose-6-P at pH 6.9 is sigmoidal. Inhibitors will increase the sigmoidicity while activators will convert the sigmoidal kinetics to hyperbolic kinetics.

The list of allosteric effectors of mammalian phosphofructokinase (Table 1) is long and more agents are being added to it. Among the activators listed, AMP and Pi are of special interest since their levels are increased after anoxia. Furthermore, cyclic 3',5'-AMP, whose level is increased following administration of several hormones, also is one of the activators.

Nature of Allosteric Sites. Our approach to study the molecular properties of allosteric sites of PFK is largely through chemical modification and through identification of the structures that have been modified. Previously we have used photo-oxidation (3,4) and ethoxyformic anhydride (4) to modify sheep heart PFK. The modified enzyme became less sensitive to ATP inhibition in connection with the loss of ATP inhibitory binding sites; its sigmoidal kinetics for fructose-6-P was also abolished, while the catalytic effect was only slightly decreased. Reaction of ethoxyformic anhydride with PFK specifically modified four histidine residues per protomer; thus those histidine residues presumably served as cationic binding sites for inhibitory ATP.

The use of affinity label reagents is a more effective way of selectively binding to the site in the enzyme prior to its covalent reaction. Recently we have used two reagents that react covalently

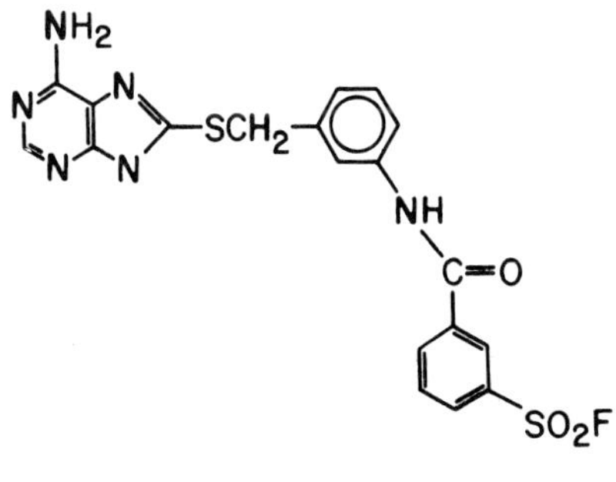

8-[m-(m-FLUOROSULFONYLBENZA-MIDO)BENZYLTHIO]ADENINE
(FSB-ADENINE)

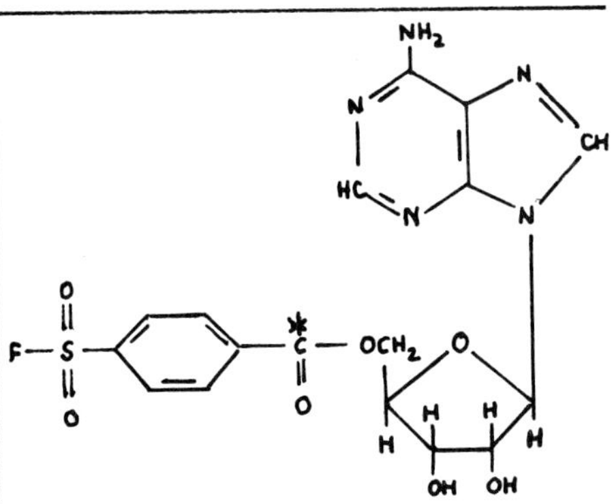

5'-p-FLUOROSULFONYLBENZOYL ADENOSINE
(FSB-ADENOSINE)

FIGURE 1.

with sheep heart phosphofructokinase making it much less sensitive to inhibition by ATP (5,6). Modified enzyme becomes insensitive to activation by AMP, ADP and cyclic AMP.

Fig 1 shows the structure of these two compounds. The first compound used is 8-[m-(m-fluorosulfonylbenzamido)benzylthio] adenine which will be referred to as FSB-adenine (7). The other reagent is an adenosine derivative, 5'-p-fluorosulfonylbenzoyl adenosine which will be referred to as FSB-adenosine (8). Both affinity label reagents have been used to label several other enzymes. For example, FSB-adenine was first used by Graves to label the AMP site on glycogen phosphorylase (7). The adenosine reagent was used as an affinity label of the inhibitory DPNH site of bovine liver glutamate dehydrogenase by Colman (8). The catalytic sites of rabbit muscle pyruvate kinase (9) and mitochondrial ATPase (10) were also labelled with the same reagent following inactivation of the enzymes.

Our recent studies show that the adenosine reagent is a specific reagent for the AMP-ADP sites of phosphofructokinase and protects against ATP inhibition better than the adenine derivatives (Figure 2). This may be due to the presence of the ribose moiety which is important for the right orientation of the reagent molecule. Conditions were first established to abolish allosteric kinetics without affecting maximal enzyme activity at pH 8.2. This was achieved when the enzyme covalently binds approximately 1 mole of the reagent per protomer. The modified enzyme completely lost its sensitivity to inhibition by ATP at moderately low levels. Inhibition can only be produced at concentrations as high as 700μM.

The results summarized in Fig. 3 show the sensitivity of the enzyme to AMP activation when inhibited by ATP at a concentration that causes 60% inhibition. The results show that while the native enzyme is sensitive to activation by AMP at concentration as low as

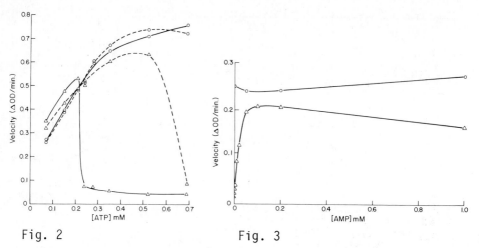

Fig. 2. Fig. 3

Fig. 2. Sensitivity of native (▲) and modified (O) phosphofructokinase to inhibition by ATP. Enzyme modified by 5-FSBO$_2$B$_Z$-adenosine and native enzyme were prepared as reported. Initial velocity of phosphofructokinase was measured at different ATP concentrations in the regular reaction mixture at pH 6.9 without AMP (straight line) or with 1 mM AMP (dashed lines). The concentration of fructose-6-P was fixed at 0.5 mM and the MgCl$_2$ at 1 mM.

Fig. 3. Sensitivity of native (▲) and modified (O) phosphofructokinase to activation by AMP. Enzyme activity was measured at pH 6.9 in the presence of 0.25 mM ATP, 0.5 mM fructose -6-P and 1 mM MgCl$_2$.

10 μM, the modified enzyme is completely insensitive to AMP activation. Titration curves for the second substrate fructose 6-P showed no sigmoidal kinetics for the modified enzyme. Kinetics of the enzyme at pH 8.2, on the other hand, was not significantly influenced.

The effect of enzyme modification on nucleotide binding to PFK showed that the binding of AMP, cAMP and ADP was abolished following enzyme modification. In contrast, the modified and native enzyme did not differ significantly to the maximal amount of binding of the ATP imidoanalog, App(NH)p. Affinity of the modified enzyme to App(NH)p was reduced.

This reagent therefore appears to be affecting the allosteric sites specifically and does not appear to involve the catalytic sites of PFK. The fact that this reagent abolished the binding of the activators AMP, cAMP and ADP but only interfered with the affinity for the ATP analog App(NH)p indicates that occupation of the AMP site does not eliminate ATP binding. Thus, the desensitization to ATP inhibition by FSB-adenosine modification must result from a change in the interaction of PFK with ATP subsequent to ATP binding.

Our success in labelling the allosteric site with a specific affinity label prompted us to investigate the chemical nature of the AMP-ADP site. Following modification of PFK with [^{14}C]-FSB-adenosine, we attempted to isolate the labelled peptide and to identify the residue which is modified. For these experiments we used the [^{14}C] reagent that is labelled at the carbonyl moiety instead of the adenosine (10); since the ester bond between the benzoyl and adenosine is hydrolyzed slowly under slightly alkaline conditions. Following modification with the reagent, PFK was subjected to reduction and carboxymethylation. This was followed by citraconylation in order to solubilize the modified enzyme and to prevent cleavage at the lysine sites. The enzyme was then treated with trypsin and chromatographed on Sephadex G-50 column. The major radioactive peptide was identified in one peak. We are currently isolating the labelled peptide for sequence analysis.

We have also attempted to identify the amino acid residue that is modified by the adenosine reagent. The side chains of serine, tyrosine, lysine and histidine residues in protein are capable of reacting with sulfonyl halides. Since sulfonylated serine and histidine are base-labile (10) they cannot be expected to isolate under such an isolation step; thus the sulfonylated derivatives of lysine and tyrosine are the reasonable candidates. Carboxybenzensulfonyl (CBS) derivatives of lysine and tyrosine (CBS-Lys and CBS-Tyr) can be synthesized (10) as standards for the identification. A sample of the radioactive fraction was treated with alkali and hydrolyzed with 6NHCl for 22 hrs. at 110°C. The acid hydrolysate was subjected to high voltage paper electrophoresis (HVPE) at pH 3.5 along with samples of CBS-Lys and CBS-Tyr, which had been treated under similar conditions. The spot with the major radioactivity moved at the same velocity as CBS-lysine. The results therefore identify lysine as the amino acid involved in enzyme modification. It is possible that this lysine residue serves as a cationic binding site for AMP, ADP and cAMP.

<u>Vanadate as a Potent PFK Inhibitor.</u> Ortho vanadate has been reported to inhibit several enzymes, including alkaline phosphatase (11), Na, K-ATPase (12), and dynein ATPase (13,14). Some of the chemical properties of vanadate are summarized in Figure 4. Since the tetrahedral structure of vanadate is so similar to that of phosphate, vanadate may inhibit enzymes by competing at phosphate-binding sites. Other studies have shown that vanadate can form a trigonal bipyramidal structure in solution and may inhibit as a transition-state analog. Enzymes catalyzing phosphate-transfer reactions have been proposed to go through this transition state. Vanadate can also exist in several polymeric states. The degree of polymerization is dependent upon the pH and concentration of vanadate in solution. At a basic pH, monomeric vanadate is

PROPERTIES OF VANADATE IN SOLUTION

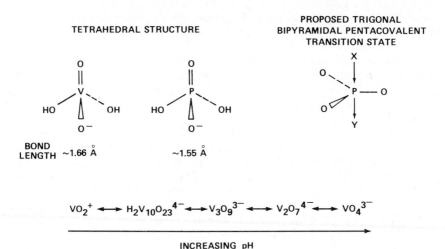

Fig. 4

predominant, and at an acidic pH, dimeric, trimeric and decameric species are found. Vanadate was chosen for this study as a potential inhibitor of PFK, since PFK is a phospho-transfering enzyme, binds several phosphate-containing ligands, and is inhibited by several polyanionic effectors.

Preliminary experiments showed that vanadate, when prepared under conditions that favored decavanadate, was a potent inhibitor of PFK (15). The enzyme was not inhibited by monomeric vanadate. When inhibitory curves for different samples of vanadate were normalized on the basis of decavanadate concentration, there was a good correlation between decavanadate concentration and the inhibitory effect on the enzyme.

Unlike the known inhibitors of PFK, vanadate was shown to inhibit both at pH 8.2 and at pH 6.9. Vanadate also can inhibit in the presence of 50mM of phosphate. Thus, the inhibitory effect of vanadate is not because of its structural similarity to phosphate but because of the effect of decavanadate as an allosteric modifier of the enzyme.

Further kinetic data showed that vanadate was a more potent inhibitor at pH 6.9 than at pH 8.2. The concentration of vanadate to inhibit PFK to the extent of 50% of its activity (I_{50} value) at pH 6.9 was 0.45 µM and at pH 8.2 was 5.5 µM. The usual allosteric activators of PFK such as cyclic AMP and glucose-1,6-P_2 were also activators of the vanadate inhibited enzyme (Table 2). In many ways, vanadate inhibition behaves like other inhibitors that act synergistically with ATP. Citrate is a good representative of this group.

TABLE 2

INHIBITION OF PFK BY VANADATE AND
DE-INHIBITION BY ALLOSTERIC EFFECTORS

PFK was assayed at pH 6.9 in 50mM morpholinopropane sulfonic acid buffer or at pH 8.2 in 50mM glycylglycine buffer. The reaction mix also contained 20mM KCl, 3mM $MgCl_2$, 1mM DTT, 0.1mg/ml BSA, 0.2mg/ml NADH, aldolase, triose-P-isomerase, α-glycero-P-dehydrogenase, and where indicated, vanadate, glucose-1,6-diphosphate, and cyclic AMP. FSB-adenosine modified PFK was prepared as described previously (6). The concentrations of vanadate reported below represent the concentration required for maximal inhibition at each condition in the absence of positive effectors.

Preparation	pH	[Vanadate] (µM)	Additions	Activity (% control)
Native	6.9	1.5	none	2
		1.5	5µM Glu-1,6-P_2	41
		1.5	20µM cAMP	38
	8.2	15	None	1
		15	20µM Glu-1,6-P_2	57
		15	50µM cAMP	41
Modified	6.9	13	None	6
		13	20µM Glu-1,6-P_2	44
	8.2	125	None	10
		125	500µM Glu-1,6-P_2	38

Vanadate, like citrate, acts synergistically with ATP; however, unlike other inhibitors, vanadate remains effective at pH 8.2. As well as inhibiting synergistically, vanadate affects the response of PFK to fructose-6-P. At pH 8.2, vanadate converts the response of PFK with respect to fructose-6-P from hyperbolic to sigmoidal kinetics.

The question arose whether PFK modified with FSB-adenosine reagent changes the sensitivity of the enzyme to vanadate inhibition. The results indicate that the modified PFK, although completely desensitized to ATP inhibition, was still sensitive to vanadate inhibition. Modified enzyme however was less sensitive to the inhibition than the native enzyme (Table 2), yielding I_{50} values over six times greater than that observed for the native enzyme. Vanadate inhibition was still sensitive to deinhibition by glucose-1,6-P_2, but was uneffected by cAMP, consistent with the observation that the modification blocks the AMP site.

The results obtained with the vanadate studies have shown that PFK possesses a polyanionic site for a potent synergistic inhibitor. Since decavanadate depolymerizes rapidly in solution and has not been detected in tissues, it is probably not an inhibitor of PFK in vivo. The physiological inhibitor may be a polyphosphate or a polyanion yet to be determined. At pH 6.9, poly $(P_i)_{14}$ was found to inhibit PFK in vitro at concentrations less than 1µM and was deinhibited by glucose 1,6-P_2 and cAMP.

The vanadate inhibition studies have also increased our understanding of allosterism in PFK. Some workers have proposed that allosteric kinetics is coupled to the protonation of active tetrameric PFK and eventual dissociation to inactive dimers. This proposal is not supported by the present studies showing allosterism at pH 8.2, where PFK is unlikely to dissociate. Inactivation studies provided further evidence against allosterism coupling to dissociation. At pH 6.5, dilute concentrations of PFK inactivate rapidly due to dissociation. Rather than enhancing dissociation, decavanadate actually was found to protect PFK against inactivation. PFK can no longer be defined as an allosteric enzyme below pH 7.5 and as a Michaelis-Menten enzyme above it and is certainly not regulated by a simple two-state concerted process. Rather, the activity and allosteric sensitivity of PFK varies over a wide range and reflects numerous conformations elicited by substrates and effectors.

Activation of PFK by hormones and by changes in physiological condition of the cell. A change in enzyme activity is often assumed on the basis of a decrease or an increase of enzyme modifiers. In other cases a change in enzyme activity is based on more direct evidence; that is, an increase or a decrease in assayable enzyme activity in tissue extracts. A summary of these changes in enzyme activity is included in Table 3.

Activation of PFK in aerobic cells as a result of anoxia is assumed on the basis of an increase in the levels of activators such as AMP, ADP and Pi (21). An increase in enzyme activity in this case is kinetic since the agents act as allosteric deinhibitors.

In the case of serotonin activation of PFK in the liver fluke, the evidence seems to indicate that the effect is more direct. Enzyme activation is more stable and there is an increase in the specific activity of the enzyme. This means that when the enzyme is diluted for the assay and any effector ligand present is also diluted, there is still an increase in enzyme activity. We previously showed that cyclic AMP is necessary for such an effect. We ascribe such an activation to either covalent modification or tightly bound activators.

TABLE 3

INFLUENCE OF HORMONES, OTHER CHEMICAL AGENTS AND
PHYSIOLOGICAL CONDITIONS ON PHOSPHOFRUCTOKINASE ACTIVITY

Agent or Condition	Tissue or Organism	Mechanism	Ref.
Serotonin	Liver flukes	Covalent Activation ? or Tightly Bound Activator ?	(16)
Epinephrine	Rabbit Muscle	Covalent Activation ? or Tightly Bound Activator ?	(17)
Epinephrine or Dibutyryl cAMP	Rat Diaphragm	Cyclic AMP ↑, Fru-di-P ↑, Glu-di-P ↑	(18)
Serum or Epidermal G.F. or Insulin	Quiescent 3T3 Cells	No New Protein Synthesis Covalent Activation ? Tightly Bound Activator ?	(19)
Insulin	Liver (PFK-L_2)	Synthesis is Enhanced Degradation is Greatly Diminished	(20)
Anoxia	All Aerobic Cells	Levels of Deinhibitors Increased	(21)

Injection of epinephrine into a rabbit and rapid assay of muscle extracts at pH 6.9 showed that the enzyme is modified to a more active form. Sensitivity to ATP inhibition and sigmoidal kinetics were decreased. Thus, there can be an apparent increase in enzyme activity when the enzyme was assayed at pH 6.9. No significant change in enzyme activity was observed at pH 8.2. Again the molecular explanation may be due to tightly bound activator or an as yet unknown covalent enzyme modification. The enzyme, after considerable dilution, may still bind to those ligands that it was exposed to in the cell. Hence, the enzyme activation persists after dilution.

Recently Beitner and coworkers (18) reported an increase in PFK activity following incubation of rat diaphragm with epinephrine. They ascribe this effect to the combined increase in the levels of glucose-1,6-P_2, fructose-1,-6-P_2 and cyclic AMP. The use of dibutyryl cAMP had the same effect as incubation with epinephrine. Schneider et al.(19) reported last year that serum, epidermal growth factor, or insulin, when added to quiescent cultures of 3T3 cells, markedly enhanced the specific activity of PFK. Such an increase is not due to an increase in enzyme synthesis or an increase in the glycolytic flux but appears to result from the activation of pre-existing inactive molecules of PFK. Dunaway et al. (20) reported recently that the activity of liver (PFK-L_2) was decreased in diabetic rats and increased to normal or super-normal amounts following insulin treatment. These changes appear to be related to the rate of degradation of the enzyme, which is controlled by changes in the levels of a peptide stabilizing factor.

Conclusion. The above examples reveal that regulation of PFK can occur by many mechanisms. Several studies have confirmed that PFK is affected in vivo by many metabolites which are known to influence the in vitro activity. In addition to changes in levels of soluble metabolites, PFK may also be regulated by rates of protein synthesis and degradation, covalent modification, such as phosphorylation, and aggregation state of the enzyme. Thus, the cell has a variety of immediate and longer-acting regulatory mechanisms available with which to respond to changes in energy-demand.

REFERENCES

1. Mansour, T.E. in Current Topics in Cellular Regulation (Horecker, B.L., and Stadtman, E.R., eds.), (1972). Vol. 5, pp. 1-46. New York, Academic Press.
2. Paetkau, V.H., Younathan, E.S. and Lardy, H.A. (1968). J.Mol.-Biol. 33, 721-736.
3. Ahlfors, C.E. and Mansour, T.E. (1969). J.Biol.Chem. 244,1247-1251.
4. Setlow, B. and Mansour, T.E. (1970). J.Biol.Chem. 253, 5524-5533.
5. Mansour, T.E. and Martensen, T.M. (1978). J.Biol.Chem. 253, 3628-3634.
6. Mansour, T.E. and Colman, R.F. (1978). Biochem.Biophys.Res. Comm. 81, 1370-1376.
7. Anderson, R.A. and Graves, D.T. (1973). Biochemistry 12, 1895-1906.
8. Pal, P.K., Wechter, W.J. and Colman, R.F. (1975) J.Biol.Chem. 250, 8140-8147.
9. Wyatt, J.L. and Colman, R.F. (1977). Biochemistry 16, 1333-1342.
10. Esch, F.S. and Allison, W.S. (1978). J.Biol.Chem. 253, 6100-6106.

11. Lopez, V., Stevens, T. and Lindquist, R.N. (1976). Arch.-Biochem.Biophys., 175, 31-38.
12. Cantley, L.C., Jr., Josephson, L., Warner, R., Yanagisawa, M., Lechene, C. and Guidotti, G. (1977). J.Biol.Chem. 252, 7421-7423.
13. Kobayashi, T., Martensen, T., Nath, J. and Flavin, M. (1978). Biochem.Biophys.Res.Comm. 81, 1313-1318.
14. Gibbons, I.R., Cosson, M.P., Evans, J.A., Gibbons, B.H., Houck, B., Martinson, K.H., Sale, W.S. and Tang, W.-J.Y. (1978) Proc.Nat.Acad.Sci.USA 75, 2220-2224.
15. Choate, G.L. and Mansour, T.E. (1978). Fed.Proc. 37, 1433.
16. Mansour, T.E. and Mansour, J.M. (1962). J.Biol.Chem. 237, 629-634.
17. Mansour, T.E. (1972). J.Biol.Chem. 247, 6059-6066.
18. Beitner, R., Haberman, S., Nordenberg, J. (1978). Mol.&Cell.-Endocrinology 10, 135-147.
19. Schneider, J.A., Diamond, I. and Rozengurt, E. (1978). J.Biol.Chem. 253, 872-877.
20. Dunaway, G.A., Leung, G.L.-Y., Thrasher, J.R., Cooper, M.D. (1978). J.Biol.Chem. 253, 7460-7463.
21. Passonneau, J.V. and Lowry, O.H. (1962) Biochem.Biophys.Res.-Comm. 7, 10-15.

ADENINE NUCLEOTIDE POOL MAINTENANCE DURING BACTERIAL GROWTH AND STARVATION

Christopher J. Knowles

Biological Laboratory, University of Kent,
Canterbury, CT2 7NJ, Great Britain

ABSTRACT Adenine nucleotides (ATP, ADP and AMP) have a uniquely crucial role in the metabolism of living cells, and in the regulation of that metabolic activity. Their role might be expected to be somewhat different in eukaryotes and prokaryotes because of compartmentation of the adenylate pools in mitochondria and the cytosol of the former organisms and the extra metabolic flexibility and control this confers on the cell. However, within bacteria and in eukaryotes it is perhaps *a priori* reasonable to suppose that the adenine nucleotide pools would be maintained in a similar manner. In exponentially growing bacteria this appears to be the case, with relatively similar adenine nucleotide contents and ATP:ADP and ATP:AMP ratios, as shown by similar energy charge values, in a wide variety of species. Exponentially growing cultures are time-averaged and these apparent similarities in the maintenance of the adenine nucleotide pools could disguise differences between cultures. That important differences do occur in the maintenance of the adenine nucleotide pools of bacteria has been highlighted by studies on two relatively similar gram-negative bacteria (*Escherichia coli* and *Beneckea natriegens*). During the transition from exponential growth to the stationary phase, when dramatic changes occur in the demand for adenine nucleotides for micromolecular synthesis and for turnover of anhydride-bound phosphate of ATP, these two bacteria exhibit surprisingly different changes in adenine nucleotide content and energy charge.

INTRODUCTION

In 1941 Lipman first elucidated, in a classic review, the metabolic importance of ATP (1). Since then it has become overwhelmingly clear that the adenine nucleotides (ATP, ADP and AMP) have a unique and crucial role in the functioning of the cell. However, it was not until 1961 that the first measurements were made of the ATP content of a growing bacterium (2). Moreover, only during the last decade have detailed studies been made of the adenine nucleotide pools of growing cells, especially in microorganisms (3, 4, 5).

It is my aim to discuss in this article the concentrations and ratios of the adenine nucleotides observed during growth and starvation of bacteria with particular emphasis on *Escherichia coli* (3, 6-10) and *Beneckea natriegens* (11-13). The function of cyclic AMP will not be discussed: its role in the regulation of gene expression by bacteria has recently been reviewed (14, 15).

ADENINE NUCLEOTIDE METABOLISM

<u>Metabolic Intermediates</u>. Adenine nucleotides are utilised as metabolic "building-blocks" for synthesis of nucleic acids, proteins (part of the purine ring of ATP is incorporated into the imidazole ring of histidine), several cofactors (NAD, NADP, FAD and coenzyme A) and cyclic AMP. In addition, AMP may be catabolised to adenine, adenosine or IMP. Thus, there is net consumption of adenine nucleotides during growth, which must be replenished by *de novo* synthesis from phosphoribose pyrophosphate or by scavenging AMP released during catabolism of RNA. Although mRNA has a rapid turnover during growth, utilising ATP and regenerating AMP, only a small part of this turnover, proportional to the total increase in mRNA concentration, represents net consumption of ATP.

The proportion of the total adenine nucleotide consumption involved in synthesis of RNA is strongly dependent on the growth rate. Chapman & Atkinson (4) have calculated that in two enteric bacteria RNA synthesis consumes from 47 to 75% of the total adenylates utilised as the growth rate increases from a generation time (t_d) of 300 min. to 24 min. DNA and protein synthesis represent 7-13% and 18-36% of net utilisation, whereas formation of cyclic AMP, NAD, NADP, FAD and coenzyme A probably account for only 1 or 2% of the total adenine nucleotide consumption. These figures exclude any catabolism of AMP; the regulatory properties of the enzymes involved in AMP degradation suggest that they are active only under conditions of metabolic stress (16-19). The turnover time of the adenylate pool varies with the growth rate. In *E. coli* it is 45 sec at a t_d of 24 min and 210 sec at a t_d of 110 min (4).

<u>Energy Transduction</u>. ATP is the key molecule involved in energy transduction by living cells. Biosynthetic processes, which are otherwise endergonic, are driven by the

removal of the δ- or the δ-plus β-phosphate groups of ATP, to form ADP and AMP respectively (ATP-utilising or U-type pathways). Catabolic pathways regenerate ATP from ADP by oxidative phosphorylation, photophosphorylation or substrate level phosphorylation (ATP-regenerating or R-type pathways). The three adenine nucleotides are interconverted by the action of adenylate kinase (ATP + AMP ⇆ 2ADP) (ref. 20, 21).

Unlike utilisation of adenine nucleotides for synthesis of nucleic acids, proteins, etc., there is no net consumption of the adenine nucleotides during recycling of the phosphoanhydride groups of ATP. The rapid turnover of mRNA during growth contributes to this recycling of the phospho-anhydride groups of ATP. However, any change in the steady-state concentration of RNA will affect the rate of turnover of ATP if it alters the size of the adenine nucleotide pool. Turnover of anhydride-bound phosphate also occurs during formation of cyclic AMP from ATP and its breakdown to 5'-AMP.

The rate of turnover of anhydride-bound phosphate residues of the adenylate pool is extremely high in growing bacteria, probably several times per second (4-6, 11, 22, 23). In non-growing bacteria, since the net rate of nucleic acid and protein synthesis decreases dramatically, the turnover of anhydride-bound phosphate is much slower, but is still required for maintenance purposes (24, 25)

Metabolic Regulation. The activities of specific metabolic pathways are controlled by the end product(s) of the pathway via allosteric regulation of the first enzyme or a branch point enzyme of the pathway. Activity of the pathway must also be integrated with the activities of other metabolic sequences and the prevailing overall metabolic activity of the cell. The key role of adenine nucleotides in the stoichiometric coupling of virtually every metabolic pathway, discussed in the preceding section, suggests that they are ideally placed to act as an overall metabolic regulatory system. Thus the activities of many regulatory enzymes are affected by ATP, ADP or AMP or by the ATP : ADP or ATP : AMP ratios, where the adenine nucleotides are substrates and products and/or allosteric effectors of the enzymes. In general, catabolic pathways contain regulatory enzymes that are activated by ADP or AMP and inhibited by ATP, whereas the converse occurs for biosynthetic pathways.

The effect of the individual adenine nucleotides and the ratios of ATP : ADP or ATP : AMP on the activities of a wide range of isolated enzymes have been investigated. These studies are only meaningful if they reflect the responses of the enzymes to the changes in concentration of each of the adenine nucleotides that occur in the intracellular environment. It is important to realise that in the cell a change in concentration of any one of the adenine nucleotides causes a change in concentration of *both* the other adenylates.

In order to take this into account, Atkinson & Walton, (26) introduced the energy charge concept, which was later expounded in greater detail (27). Adenylate energy charge (E.C.) is $\{ATP\} + \frac{1}{2}\{ADP\}/\{ATP\} + \{ADP\} + \{AMP\}$. It is a measure on a linear scale from 0 (all AMP) to 1 (all ATP) of the amount of energy, metabolically available in the total adenine nucleotide pool, as anhydride-bound phosphate of high free energy of hydrolysis. The theory pre-supposes that the adenine nucleotide pool is equilibrated by adenylate kinase, and at intermediate E.C. values the concentration of each of the adenine nucleotides depends on the value of the equilibrium constant of the adenylate kinase reaction.

According to the hypothesis, catabolic or ATP-generating (R-type) pathways are inactivated by high E.C. values, and biosynthetic or other ATP-utilising (U-type) pathways are activated at high E.C. values. The greatest change in enzyme activity occurs in the E.C. region of 0.6 - 1.0, and there is a cross-over at 0.8 - 0.95. Decreases in E.C. from this region will inactivate U-type and activate R-type pathways, resulting in a re-establishment of the higher E.C. value, and hence metabolic stabilisation.

The responses of a wide range of enzymes to E.C. have been studied, and shown to have appropriate R- and U-type curves (see ref. 4, 5, 28 for references). The E.C. hypothesis is therefore an attractive explanation of metabolic energy regulation. However, there are several drawbacks, which have recently been the subject of some controversy (29-36), suggesting a refinement or modification of the theory is required to take into account various factors. I have discussed these problems in detail elsewhere (5). Briefly, they include the following :-

(a) The hypothesis requires the adenine nucleotide pool to be equilibrated by adenylate kinase. This enzyme occurs ubiquitously in living cells (20) and, although it is associated with mitochondria in eukaryotes, it is functionally part of the cytosol (37). The adenylate pools of mitochondria and the cytosol are separate, with stoichiometric exchange of ATP and ADP by adenine nucleotide translocase (38). Mitochondrial AMP is converted to ADP by a transphosphorylase (37). Regulation of enzyme activity by energy charge, due to the absence of adenylate kinase from the mitochondrial matrix, is only directly applicable to the cytosol. In controlled mitochondria the ATP : ADP ratio is much lower than the extramitochondrial ratio (39).

(b) Most enzymes respond to $MgATP^{2-}$ and $MgADP^{-}$ rather than ATP^{4-} and ADP^{3-}. Indeed, enzymes that are activated by the chelated form of the nucleotides may be inhibited by the free nucleotides. The relative intracellular ratio of magnesium ion to total adenine nucleotide concentration is therefore important. In mammalian cells available Mg^{2+} may be limiting, whereas in bacteria there is probably excess Mg^{2+} (c.f. ref. 5, 32).

(c) As E.C. varies, the magnesium ion level will vary because ATP forms a stronger complex with it than do ADP and AMP. Adenylate kinase reacts according to $MgATP^{2-} + AMP^{2-} \rightleftharpoons MgADP^{-} + ADP^{3-}$ (20, 21). Thus, the equilibrium position alters depending on the availability of Mg^{2+} for complexing. This shift will in turn affect the relative ratios of the adenine nucleotides and hence the ATP : ADP and ATP : AMP ratios, thereby altering the activity of adenylate-sensitive enzymes.

(d) As E.C. varies the intracellular free orthophosphate level changes. The activities of many enzymes are affected by orthophosphate. This effect is not taken into account in the E.C. hypothesis. In addition it has been shown that respiratory activity in isolated mitochondria (40) depends on the $\{ATP\} / \{ADP\}\{P_i\}$ ratio (in bacteria the mechanism of regulation of respiratory activity is not known).

(e) *In vivo* most non-adenylate substrates and products are not saturating (41), and E.C. responses act by affecting the affinity of enzymes for their substrates. In contrast, most *in vitro* experiments have been carried out using saturating substrate and Mg^{2+} concentrations. More studies are required to demonstrate that isolated enzymes exhibit the appropriate responses to E.C. at non-saturating or saturating substrate, product and Mg^{2+} levels, etc. that relate to their intracellular concentrations.

(f) E.C. is a unitless parameter and a knowledge of its value does not indicate the adenine nucleotide pool size, nor the rate of turnover of ATP. Therefore, metabolic regulation by E.C. does not take into account the effect of variations of the adenine nucleotide pool content on the activity of those enzymes regulated by only one of the nucleotides, for example the AMP-stimulated citrate synthase of aerobic gram-negative bacteria (42). However, if the affinity of a particular adenine nucleotide for the regulatory site of an enzyme is considerably lower than the intracellular adenylate content, then wide variations of the adenylate pool size will not affect the activity of the enzyme. With enzymes affected by two of the adenine nucleotides, the ratio of the adenylates (which may be expressed in terms of the energy charge) will in any case be more important than the absolute concentrations of the adenylates.

MEASUREMENTS OF ADENINE NUCLEOTIDES

Because of the extremely rapid turnover during growth of the anhydride-bound phosphate residues of ATP, sampling methods must involve rapid quenching of metabolic activity prior to assay of the adenine nucleotides. This poses considerable technical problems when studying multicellular and especially differentiated organisms. In eukaryotes, due to the separate mitochondrial and cytosolic adenine nucleotide pools, rapid quenching of metabolic activity must be accompanied by separation of the two pools with no intermixing of the adenylates. This formidable problem has yet to be resolved, although imaginative attempts have been made (43).

In unicellular bacteria neither of the latter problems occur, and simple rapid sampling plus quenching techniques have been developed (11, 22). Some workers appear to prefer pipetting samples from growing cultures into a quenching reagent, assuming that the growth conditions of pH, substrate concentration, oxygen tension, temperature, etc. are unaltered during the 10 or 15 sec. sampling period. In the authors opinion it is dangerous to rely on data from this type of experiment, unless comprehensive control experiments are presented to show that the adenine nucleotide pools are not perturbed during sampling. Earlier reports of the adenine nucleotide composition of growing microorganisms using filtration or centrifugation methods should be discounted.

ADENINE NUCLEOTIDE POOLS OF GROWING BACTERIA

In order to fully comprehend the metabolic role of the adenine nucleotides, it is clearly important that we have an understanding of their concentrations in growing cells, their net rates of synthesis and utilisation, the rate of phosphate-bond turnover, the responses of the adenylate pool to metabolic stress, etc. It is therefore surprising how little is known about the adenine nucleotide content of growing organisms, possibly due to the problems outlined in the previous section. The remainder of this article will be devoted to a discussion of the adenylate pools of vegetative cultures of bacteria. The adenylate pools of eukaryotes will not be discussed, because virtually all measurements of them to date have yielded only average values of their cellular content rather than their distribution and ratios within mitochondria and the cytosol (c.f. ref. 4, 5).

The E.C. values of a wide range of growing bacteria have been shown to be in the 0.75 - 0.95 range, in good agreement with the predictions of the E.C. concept (4, 5). The adenylate pool sizes of growing bacteria appear to be similar, in the range of 2 - 10 µmol per g dry wt, which is about 1-5mM and 1 - 5% of all the small metabolites of the cell. There have been only a limited number of studies on the effect of growth rate on adenine nucleotide content of bacteria (4, 5) but continuous culture experiments on *Klebsiella aerogenes* suggest that their concentrations are independent of growth rate (22).

To date all measurements of adenine nucleotide content and E.C. values have been made using exponentially growing batch or continuous cultures of bacteria, where the cultures are "time-averaged". During the cell cycle there is probably a discontinuous demand for phosphate-bond energy and for adenylates as building blocks for macromolecular synthesis. Thus, E.C. and adenylate content could fluctuate during the cell cycle. Indeed, experiments with synchronised cultures of eukaryotic microorganisms have already shown such variations (44 - 46). Moreover, the respiration rate (and hence the rate of ATP synthesis) of bacteria varies widely during the cell cycle (47,48).

An alternative approach to examining the effect of changes in turnover requirements for high energy phosphate and adenylates for macromolecular synthesis is during the transition into the stationary phase and in starved cells, including extended starvation and cell death. When the stationary phase commences the demand for adenylates for DNA synthesis stops, and decreases for RNA and protein (histidine) synthesis, coupled with a decrease in demand for ATP turnover of phosphate-bond energy supply for biosynthesis and mRNA turnover, although energy is still required for maintenance purposes. It is possible that a lowered E.C. in the stationary phase causes turn-off of growth related processes, yet permits maintenance reactions to continue.

The stationary phase of heterotrophic bacteria may occur due to depletion of the carbon source from the medium, when survival depends on the ability to utilise endogenous materials as a source of energy. These include storage polymers (glycogen, polyphosphate or poly-β-hydroxybutyrate) and catabolism of mRNA or non-essential proteins (24,25). In most cases, storage polymers are unlikely to be an important source of energy, as they are usually only deposited in appreciable quantities when growth terminates due to depletion of a nutrient other than the carbon source from the medium (e.g. nitrogen). When this happens the excess carbon may be used to form an energy reserve polymer during the carbon-rich period at the start of the stationary phase (25). The reserve polymer is later used to maintain viability during extended periods of starvation.

The effects of transitions into the stationary phase of batch cultures due to depletion of carbon or ammonia from the medium, which I will refer to as "carbon-limited" and "nitrogen-limited" cultures, have been examined in *E. coli* (6-10) and *B. natriegens* (11-13). Baumann & Baumann (50) have proposed that bacteria of the genus *Beneckea* are "marine enterobacteria" and hence closely related to *E. coli*, which is a terrestial enteric bacterium. The basis for this suggestion is the similar pattern of regulation of key enzymes of biosynthetic pathways (50,51). It is therefore of interest to determine whether the adenine nucleotide pools of *B. natriegens* and *E. coli* are maintained in a similar or different manner. It should be added that there has been some controversy as to whether there is in fact a genus *Beneckea*, because of the similarity of these organisms to the genus *Vibrio* (52).

Chapman *et al* (3) reported that the size of the adenine nucleotide pool of glucose-limited cultures of *E. coli* strain B decreases by about 50% at the start of the stationary phase. The E.C. falls from a growth value of 0.8 to 0.6. Later experiments (9), using an improved assay system (8), gave somewhat higher E.C. values (0.9 - 0.93 and 0.9 - 0.8, respectively). There is a doubling of viable cell numbers after the maximal culture absorbance has been obtained, indicating a round of cell division after growth stops. These data contrast with earlier studies on the ATP pool of *E. coli* strain ML308 (6), where the ATP content was constant during a second phase of respiratory activity that followed depletion of glucose from the medium. This was due to utilisation of acetate that had accumulated during the growth on glucose. *E. coli* strain B also accumulates acetate during growth on glucose (9). The E.C. of 0.9 - 0.8, noted immediately after termination of growth of strain B, is maintained for 5 to 7 hours, followed by a gradual decrease to 0.4, and it is probable that the former E.C. is preserved only until acetate is completely utilised.

Carbon-limitation of cultures grown on succinate or glycerol causes total starvation of exogenous substrates. In *E. coli* strain B/r there is a drop of E.C. from 0.85 to 0.6 at the start of the stationary phase in succinate-limited cultures (10). In *E. coli* strain ML308 the ATP pool and respiratory activity decreases immediately after glycerol-limitation (6).

Unlike carbon-limited cultures of E. coli, depletion of glucose from cultures of *B. natriegens* does not cause a measurable decrease in E.C., which remains at about 0.9 (11). *B. natriegens* also excretes volatile acids (mainly pyruvate) when grown on glucose, which are oxidised without further growth after glucose depletion from the medium. The E.C. stays at 0.9 for many hours following complete oxidation of the volatile acids (13). Also in contrast to *E. coli*, the adenine nucleotide pool content increases slightly after glucose depletion from the medium. Total cell numbers of *B. natriegens* in glucose-limited cultures increase by 20 - 30%

after cell growth has finished (13). When grown in succinate-limited cultures, which do not exhibit a secondary, post-growth burst of respiratory activity, *B. natriegens* again has a high E.C. in the stationary phase and the adenine nucleotide pool is constant (11).

The apparent maintenance of the E.C. of *B. natriegens* at the growth value of about 0.9 in carbon-limited cultures is surprising in comparison to the drop in E.C. observed in *E. coli* at the start of the stationary phase. It should be remembered that a decrease in E.C. of about 0.05 or less is difficult to detect yet represents a considerable change in the ATP : ADP and ATP : AMP ratios, but which have a major effect on the activity of adenylate-sensitive enzymes. Therefore, there may be only a qualitative difference between *E. coli* and *B. natriegens*. If the E.C. in *B. natriegens* does in fact not shift, then it is primarily the rate of catabolism as determined by substrate supply that is important in affecting metabolic activity. The maintenance of an E.C. at 0.8 or 0.9 represents a balance of metabolic activity regulated at an ATP : ADP ratio of 10^7 or 10^8 times the equilibrium value for the H_2O + ATP $\rightleftharpoons$ ADP + phosphate reaction (when reasonable assumptions are made about the physiological phosphate concentration). This balance is all the more remarkable if it is remembered that it is conserved at a time when there is a switch from one exogenous substrate to another (from glucose to acetate) or from an exogenous substrate (succinate or acetate) to endogenous substrates, and when the rate of metabolic activity shifts dramatically (as seen by the changes in respiratory activity).

At the point of change-over to non-growing conditions the demand for adenylates as building-blocks for macromolecular synthesis also decreases. Because the size of the adenine nucleotide pool depends on the balance between the rates of *de novo* synthesis of AMP and the rate of utilisation of adenylates there must be an extremely sensitive regulation of the AMP biosynthetic pathway. Phospho-ribose-pyrophosphate (PRPP) synthase activity has been shown to be affected by E.C. (52). A complicating factor here is catabolism of RNA which, if it occurs, releases AMP, or IMP, hypoxanthine or adenosine that may be converted back to AMP. This would alter the size of the adenylate pool. In addition, AMP and the other nucleotides or nucleosides may be further catabolised as endogenous sources of energy. Unfortunately the size of the RNA pool is large compared to the adenine nucleotide pool (4). Only a small change in RNA concentration which is experimentally difficult to measure sufficiently accurately, would dramatically alter the size of the adenylate pool unless the rates of *de novo* synthesis of AMP and AMP catabolism are coordinately regulated. Competition for PRPP for *de novo* synthesis of AMP and recycling of AMP from the products of RNA catabolism appears to be the crucial factor (see ref. 4, p. 258-259 for a more detailed discussion).

The pathway of AMP catabolism is also probably important in maintaining the E.C. When substrate supply is limited or no substrate is available, E.C. will tend to fall. This activates AMP nucleosidase in bacteria (AMP $\rightarrow$ adenine + ribose-5-phosphate) or AMP deaminase in eukaryotes (AMP $\rightarrow$ IMP + NH) (ref. 16-19). Removal of AMP buffers E.C. at the expense of the adenine nucleotide pool size. A total drain of the adenylates does not occur due to the build-up of phosphate, which inhibits the enzymes. Activity of AMP nucleosidase may explain the decrease in adenine nucleotide concentration of *E. coli* on carbon-limitation (3,9). We have been unable to detect this enzyme in *B. natriegens* (53,54) and the bacterium does not exhibit a decrease in adenine nucleotide content on glucose or succinate-limitation (11,13). We earlier suggested, by indirect assays, that AMP deaminase is present in *B. natriegens*, but further attempts to detect and purify the enzyme have been unsuccessful (54).

Excretion of AMP does not occur in either *E. coli* or *B. natriegens* as a method of regulating E.C. on carbon (or nitrogen)-limitation. However, measurements of changes in intracellular phosphate content, phosphorylation potential or proton motive force on carbon-limitation have not been made. A shift in phosphate content and phosphorylation potential in response to changes in E.C. charge, adenine nucleotide concentration and RNA content is potentially important in regulating metabolic activity. Changes in the rate of transport of phosphate into or out of the cell may also be significant.

Limitation of growth of both *E. coli* (3,9) and *B. natriegens* (11,13) due to ammonia depletion from the medium causes some of the excess glucose in the medium to be deposited as glycogen and the rest to be oxidised as an energy source. After complete utilisation of glucose from the medium glycogen is degraded as a source of energy, thereby extending the period of viability of the starving bacteria (25).

Unlike carbon-limited cultures, the E.C. of *E. coli* strain B remains at the growth value (0.8 - 0.9) after nitrogen-limitation (3), and it has been claimed that the E.C. of *E. coli* strain W4597(K) rises from the rather low value of 0.74 observed during exponential growth to 0.87 during the period of glycogen deposition (7). The total adenine nucleotide content is unchanged during the transition to the stationary phase of both strains. A possible reason for the constant adenylate pool could be the inactivity of AMP nucleosidase because of the high E.C. (17). Two enzymes are involved in glycogen synthesis by *E. coli*, ADPglucose pyrophosphorylase and ADPglucose glucosyl transferase. The former enzyme shows an extremely sharp activation with increases in E.C. (55). It has been proposed that it is the increase in E.C. which occurs on nitrogen-limitation of glucose-grown cultures of *E. coli* W4597(K) that causes an increase in the rate of formation of glycogen.

An interesting effect occurs on nitrogen-limitation of glucose- or succinate-containing cultures of *B. natriegens*. Immediately after growth terminates the adenine nucleotide pool increases until there is complete utilisation of the excess carbon from the medium. Depending on the initial concentration of the carbon source, the concentration of the adenine nucleotides may increase up to 4-fold of the exponential growth content (11,13). The E.C. remains at 0.9 during the phase of increasing adenylate content. With glucose as the substrate, about 10% of the residual glucose is deposited as glycogen.

Little change in RNA content occurs as the adenine nucleotide content of *B. natriegens* increases. It is therefore likely that the increase is caused by an imbalance in the rate of *de novo* synthesis and the decreased demand for adenine nucleotides for protein and RNA synthesis, with little salvage of AMP from RNA degradation.

During starvation of bacteria energy is not required for growth but it is needed for essential maintenance processes, including turnover of macromolecules that are essential for survival. A low basal level of metabolic activity occurs, as indicated by a low rate of respiration (25). Cell death must inevitably follow total depletion of endogenous energy supplies, unless an inert form of the bacterium occurs, i.e. spore formation. The inability to regenerate ATP will be seen as a decrease in E.C. This occurs during extended starvation of glucose-grown *E. coli* (nitrogen-limited). A gradual decrease in E.C. to 0.5 occurs with little loss of activity until 60-80 hours after inoculation (3). This is followed by a rapid decrease in viability coincident with a further fall in E.C. Throughout the period from 10 to 60-80 hours after inoculation there is a gradual decrease in intracellular adenine nucleotide content and excretion of AMP into the medium. The latter effect may be a mechanism for maintaining E.C. at the expense of the adenine nucleotide pool size.

E. coli grown under these conditions accumulates considerable glycogen reserves, which are used to extend the period of viability (24,25,56). Unfortunately there have been no measurements made on the correlation between viability and adenine nucleotide content, and glycogen levels during starvation of nitrogen-limited cultures. Glucose-limited cultures form little glycogen and utilise only RNA to maintain

viability during starvation. It remains to be seen whether
a similar relationship to that discussed above holds between
E.C., adenine nucleotide content and viability of starving
E. coli that had been grown under glucose-limited conditions.
Peptococcus prévotii, which does not form reserve polymers,
uses RNA as the sole endogenous substrate (57). During
starvation the adenine nucleotide content initially falls
rapidly followed by a steep increase, presumably due to AMP
salvage from RNA. Finally, there is a second decrease in
adenine nucleotide concentration. The E.C. remains at 0.5
and there is high viability until the intracellular RNA content
reaches a minimum, when both E.C. and viability decrease.

REFERENCES

1. Lipmann, F. (1941). *Adv. Enzymology* 1, 99.
2. Franzen, J. S., and Binkley, S. B. (1961). *J. Biol. Chem.* 236, 515.
3. Chapman, A. G., Fall, L. and Atkinson, D. E. (1971). *J. Bacteriol.* 108, 1072.
4. Chapman, A. G. and Atkinson, D. E. (1977). *Adv. Microbiol. Physiol.* 15, 253.
5. Knowles, C. J. (1977). In *Microbial Energetics*, ed. B. Haddock and W. A. Hamilton, *Soc. Gen. Microbiol. Symp.* 27, 241. London: Cambridge University Press.
6. Holms, W. H., Hamilton, I. D. and Robertson, A. G. (1972). *Arch. Mikrobiol.* 83, 95.
7. Dietzler, D. N., Lais, C. J. and Leckie, M. P. (1974). *Arch. Biochem. Biophys.* 160, 14.
8. Swedes, J. S., Sedo, R. J. and Atkinson, D. E. (1975). *J. Biol. Chem.* 250, 6930.
9. Walker-Simmons, M. and Atkinson, D. E. (1977). *J. Bacteriol.* 130, 676.
10. Anderson, K. B. and von Meyenburg, K. (1977). *J. Biol. Chem.* 252, 4151.
11. Niven, D. F., Collins, P. A. and Knowles, C. J. (1977). *J. Gen. Microbiol.* 98, 95.
12. Niven, D. F., Collins, P. A. and Knowles, C. J. (1977). *J. Gen. Microbiol.* 103, 141.
13. Nazly, N., Carter, I. C. and Knowles, C. J. (1979) unpublished observations.
14. Pastan, I. (1976). *Bacteriol Rev.* 40, 527.
15. Peterkovsky, A. (1976). *Adv. Cyclic Nucleotide Res.* 7, 1.
16. Chapman, A. G. and Atkinson, D. E. (1973) *J. Biol. Chem.* 248, 8309.
17. Schramm, V. L. and Leung, H. (1973). *J. Biol. Chem.* 248, 8313.
18. Schramm, V. L. and Lazorik, F. C. (1975). *J. Biol. Chem.* 250, 1801.
19. Schramm, V. L. and Leung, H. (1978). *Arch. Biochem.*

Biophys. 190, 46.
20. Noda, L. (1973). In *The Enzymes* ed. P. D. Boyer, 3rd edn, vol. 8, 279. New York and London : Academic Press.
21. Blair, J. McD. (1970). *Eur. J. Biochem.* 13, 384.
22. Harrison, D. E. F. and Maitra, P. K. (1969). *Biochem. J.* 112, 647.
23. Miovic, M. L. and Gibson, J. (1971). *J. Bacteriol.* 108, 954.
24. Dawes, E. A. and Senior, B. J. (1973). *Adv. Microbial Physiol.* 10, 136.
25. Dawes, E. A. (1976). In *The Survival of Vegetative Microbes,* ed. T. R. G. Gray and J. R. Postgate, *Soc. Gen. Microbiol. Symp.* 26, 19. London : Cambridge University Press.
26. Atkinson, D. E. and Walton, G. M. (1967). *J. Biol. Chem.* 242, 3239.
27. Atkinson, D. E. (1968). *Biochemistry* 7, 4030.
28. Atkinson, D. E. (1969). *Ann. Rev. Microbiol.* 23, 47.
29. Atkinson, D. E. (1977). *Trends in Biochem. Sci.* 2, N198.
30. Purich, D. L. and Fromm, H. J. (1972). *J. Biol. Chem.* 247, 249.
31. Purich, D. L. and Fromm, H. J. (1973). *J. Biol. Chem.* 248, 461.
32. Fromm, H. J. (1977). *Trends in Biochem. Sci.* 2, N198.
33. Purich, D. L. (1978). *Trends in Biochem. Sci.* 3, N39.
34. Pradet, A. and Raymond, P. (1978). *Trends in Biochem. Sci.* 3, N40.
35. Shargoal, P. D. (1978). *Trends in Biochem. Sci.* 3, N40.
36. Dietzler, D. N. and Leckie, M. P. (1978). *Trends in Biochem. Sci.* 3, N41.
37. Heldt, H. W. and Schwalbach, K. (1967). *Eur. J. Biochem.* 1, 199.
38. Klingenberg, M. (1970). In *Essays in Biochemistry,* ed. P. N. Campbell and F. Dickens, vol. 6, 119. New York and London : Academic Press.
39. Heldt, H. W., Klingenberg, M. and Milovancev, M. (1972). *Eur. J. Biochem.* 30, 434.
40. Holian, A., Owen, C. S. and Wilson, D. F. (1977). *Arch. Biochem. Biophys.* 181, 164.
41. Cornish-Bowden, A. (1976). *J. Mol. Biol.* 101, 1.
42. Weitzman, P. D. J. and Danson, M. J. (1976). *Cur. Top. Cell. Reg.* 10, 161.
43. Siess, E. A. and Wieland, O. H. (1976). *Biochem. J.* 156, 91.
44. Lloyd, D., Phillips, C. A. and Statham, M. (1978). *J. Gen. Microbiol.* 106, 19.
45. Poole, R. K. and Salmon, I. (1978). *J. Gen. Microbiol.* 106, 153.
46. Edwards, S. W. and Lloyd, D. (1978). *J. Gen. Microbiol.* 108, 197.

47. Poole, R. K. (1977). *J. Gen. Microbiol.* 99, 369.
48. Edwards, C. and Jones, C. W. (1977). *J. Gen. Microbiol.* 99, 383.
49. Baumann, P. and Baumann, L. (1977). *Ann. Rev. Microbiol.* 31, 39.
50. Baumann, L. and Baumann, P. (1973). *Arch. Mikrobiol.* 90, 171.
51. Austin, B., Zachary, A. and Colwell, R. R. (1978). *Int. J. System. Bacteriol.* 28, 315.
52. Klungsoyr, L., Hagemen, J. H., Fall, L. and Atkinson, D. E. (1968). *Biochemistry* 7, 4035.
53. Niven, D. F., Collins, P. A. and Knowles, C. J. (1977). *J. Gen. Microbiol.* 100, 5.
54. Pickard, M. A. and Knowles, C. J. (1979), unpublished observations.
55. Preiss, J. (1973). In *The Enzymes,* ed. P. D. Boyer, 3rd edn, vol. 8, 73. New York and London : Academic Press.
56. Dawes, E. A. and Ribbons, D. W. (1965). *Biochem. J.* 95, 332.
57. Montague, M. D. and Dawes, E. A. (1974). *J. Gen. Microbiol.* 80, 291.

MULTIMODULATION OF ENZYME ACTIVITY. PHYSIOLOGICAL SIGNIFICANCE AND EVOLUTIONARY ORIGIN[1]

Alberto Sols

Instituto de Enzimología y Patología Molecular del CSIC, Facultad de Medicina, Universidad Autónoma, Madrid 34 - Spain

ABSTRACT Knowledge of specific mechanisms of regulation of enzyme activity has increased within less than 25 years from essentially zero to a rich variety, that in some cases reaches a bewildering multiplicity. Such multiplicity of presumptive regulatory effects makes particularly pressing the question of the physiological significance of multimodulation of enzyme activity, trying to sort out facts from artifacts, and biologically relevant mechanisms from *in vitro* curiosities. The basic type of mechanism is heterotropic allosteric regulation, frequently accompanied by one or more of other types of regulatory mechanisms: homotropic cooperativity or metabolic interconversion. Multiple heterotropic effects may range from essentially independent to strongly concerted or conditioned, with physiological activation being frequently based on counteraction of an inhibition. A critical examination of the multimodulation of phosphofructokinases indicates that out of over 20 different reported effectors, about 8, involving as many different sites, seem to be significant for the regulation of the enzyme in most eukaryotic organisms. Pyruvate kinase L has 5 physiologically significant mechanisms for the modulation of its activity. Some other multimodulated enzymes are briefly discussed. An analogy principle in allosteric regulation is formulated on the basis of comparative biochemistry. In general, multimodulated enzymes are highly sophisticated integrators of metabolic signals. The fact that most regulatory enzymes have a single kind of subunit, that some of them have about half a dozen regulatory sites, and that in these cases the subunit size is not greater than that of the average non-regulatory enzymes, suggests that many, perhaps even most of the regulatory sites, arose from mutational development of new specific sites. It is argued that the more basic regulatory mechanisms must be very old: over 1 billion years.

[1]This work was supported by grants from the Comisión Asesora para la Investigación Científica y Técnica.

INTRODUCTION

By the nineteen forties enzymology had reached a golden age. Hundreds of enzymes were already known, with new ones added every month or so, quite a few had already been crystallized as pure proteins, and unraveling of metabolic pathways at the enzyme level was becoming routine work. It was becoming evident that each cell would probably contain about a thousand or more different enzymes. With the realization of this complexity, a need for regulation of enzyme activity was beginning to be felt by some, as exemplified by Carl F. Cori in a famous 1946 lecture (1). But it was only in the fifties when the first specific mechanisms of regulation of enzyme activity were discovered, as feedback mechanisms (2,3,4) or as modulation by reversible covalent modification (5). And it was in the early sixties when a theoretical framework for modulation of enzyme activity was developed, centered in the allostery concepts formulated by the Pasteur group (7). Later in the sixties covalent modulation of enzyme activity transcended from the field of glycogen metabolism to become a general mechanism that has been expanding ever since (with a regular series of international symposia which started in 1970, the fourth one published in 1976 (8)), largely independently of the non-covalent modulation, although necessary links between both major types of modulation of enzyme activity were occasionally emphasized (9). This Symposium comes to fill a gap by attempting an integration between "covalent and non-covalent modulation of protein function".

By now it can be stated confidently that heterotropic allosteric regulation of enzyme activity is the central and more basic mechanism of physiological modulation of enzyme activity. Its discovery and conceptualization opened a 3rd dimension in physiological enzymology. This crucial mechanism is frequently accompanied by other types of regulatory mechanisms: homotropic cooperativity and covalent modulation. And these specific mechanisms of modulation of enzyme activity are in the cells superimposed on coarser possibilities of metabolic control based on certain intrinsic catalytic properties, particularly the magnitude of the enzyme-substrate affinity, and the regulation of enzyme levels, obviously through control of enzyme synthesis, and sometimes also through selective degradation, which is a form of irreversible covalent modification.

The basic mechanisms of modulation of enzyme activity are a marvel of biological engineering that greatly contribute to the high efficiency of the present forms of life. And many particular instances are obviously suited for sophisticated metabolic control. But there are also many claims in particular instances that are more confusing than enlightening. Particularly bewildering is the multiplicity of allosteric effectors reported for some enzymes, up to more than 20 in a few cases! Such multiplicity of presumptive regulatory effects makes particularly pressing the question of the physiological significance of multimodulation of enzyme activity. It is important to try to sort out facts from artifacts, and biologically relevant mechanisms from *in vitro* curiosities. Protein chemistry, ligand specificity, metabolic behaviour, comparative biochemistry, molecular pathology, and molecular evolution, if critically used, could contribute to the formulation of tentative principles and working hypothesis. To begin with it is convenient to precise a framework of definitions and general rules.

Mechanisms of modulation of enzyme activity.
1. Allosteric regulation of enzyme activity involves regulatory sites, catalytic interacting sites, or both.
2. Any compound can be an heterotropic effector for an enzyme, no matter how unrelated to substrate and product, but also even if it is a substrate or product. This definition includes different sites even if the physiological effector is the same.
3. Heterotropic effectors can be either positive or negative. They can act on the molecular activity, the affinity for a substrate or cofactor, the degree of cooperativity, the ratio of affinities for substrate and product, the relative specificity, or any combination of them.
4. Homotropic effects can be either positive or negative, and the commoner positive ones can have an oligomeric or mnemonical basis.
5. If there is more than one kind of allosteric effect in an enzyme:
   a) The enzyme can integrate (essentially) simultaneous signals.
   b) Physiological activation frequently involves counteraction of an inhibition (oligomeric decrease in affinity for a substrate or heterotropic).
   c) Multiple effects may range from essentially independent (contributing algebraically) to strongly concerted.
6. Covalent modification of an enzyme can affect any of the above parameters, either catalytic or allosteric.
7. Metabolic interconversion of enzymes is usually regulated by some allosteric effect on either the metabolic enzyme or a converter enzyme.

It would be too complex to attempt to describe the basis for each of these definitions, and to give individual credits. I will single out a few comments on the need for clarification around the term allosteric. The farsighted concept of chemical transduction proposed by Monod and collaborators borrowing from Kosland's induced fit crystallized with the term allosteric, a catching word that soon became a mixed blessing (10) because of confusions between two frequently linked but essentially independent concepts: specifically regulatory sites, and multiplicity of equal sites in oligomeric proteins. The common mixture of these two concepts can be exemplified by a recent description by Perutz: "Monod, Changeux and Jacob proposed the term allostery for *enzymes that possess two, or at least two, stereospecifically different, nonoverlaping receptor sites* (1963). One of these, the active site, binds the substrate, while the other, or allosteric site, binds the effector. *Such enzymes contain more than one site of each kind, and these act cooperatively*" (11) (italics added). To compound the tendency to confusion, oligomerism is not only not necessarily linked to heterotropic allosteric effects but is not even the only basis for positive cooperativity, since the identification of the mnemonical mechanism to account for sigmoidal kinetics in monomeric enzymes, as proposed by Ricard (12) and developed by Storer and Cornish-Bowden (13).

The clarification that genuinely heterotropic allosteric effects can involve specific regulatory sites for a primary product or a substrate has the following well authenticated prototypes: 1), inhibition of animal hexokinases by glucose-6-P (2,10): 2) inhibition of eukaryotic phosphofructokinases by ATP (14,10): 3) activation of *E. coli* phosphofructokinase by ADP (15,16), and 4) activation of the NAD-dependent isocitrate dehydrogenase by isocitrate (17,18).

Finally, the generalization that many cases of allosteric physiological activation of an enzyme involve the counteraction of a prior inhibition, something easier than primary activation, stems not only from certain antagonistic, frequently called "deinhibitory" allosteric effects, as will be illustrated in the next section, but also from the basic fact that the so called positive cooperativity for substrate binding in oligomeric proteins actually involves an initial hindrance by quaternary constraint (7) for the binding of the substrate in the oligomer respect to the corresponding monomer. Two outstanding examples are aspartate transcarbamylase (19) and hemoglobin (20). In these two proteins, the affinity for the substrate can rise steeply within the physiological range, but in fact it is never greater than that of the monomer at any given substrate concentration.

In summary, allosteric enzyme is not equivalent, nor necessary linked, to cooperativity in oligomeric proteins. Allosteric enzymes are those that can be modulated through conformational changes involving site-site interactions, wether homotropic (involving equal sites, or, by extension, the same site along time), heterotropic (involving different kinds of sites), or both.

CRITICAL ANALYSIS OF SOME OUTSTANDING CASES OF MULTIMODULATED ENZYMES

Modulation of enzyme activity was considered initially as synonymous with feedback inhibition, typically by the end-product. With time, some enzymes appeared to have more than one mechanism of modulation of activity. Eventually, plurality moved for certain enzymes from more than one to many modulation mechanisms, even compounding (or merely piling up) non-covalent and covalent ones. Here I will examine in detail two multimodulated enzymes of which I have long experience, and comment briefly on four others.

Phosphofructokinases. Phosphofructokinase has long been known as a difficult enzyme. When I first met it, while studying hexokinase in Cori's laboratory in 1951 I heard comments about its being an "unreliable enzyme". A decade later, also in St. Louis, Passonneau and Lowry (21) reported that muscle phosphofructokinase activity could be markedly affected by a variety of metabolites. Which led Carl Cori to suggest to me that perhaps it was an "all-hysteric enzyme". With time the number of presumptive regulatory effects reported for phosphofructokinase has grown until reaching the fantastic number of 23 different effectors listed in a recent review (Table 1).

TABLE 1

EFFECTORS OF PHOSPHOFRUCTOKINASE
as listed by Tejwani (22)

Inhibitors	Activators	Deinhibitors of ATP citrate, or M_g^{2+}
ATP	NH_4^+	cAMP
Citrate		AMP
M_g^{2+}	K^+	ADP
Ca^{2+}	Pi	Fru-6-P
P-creatine		Pi
Glycerate-3-P	AMP	
P-enolpyruvate	cAMP	Fru-1,6-P_2
Glycerate-2-P	ADP	Glc-1,6-P_2
Glycerate-2,3-P_2	Fru-1,6-P_2	Man-1,6-P_2
Oleate or palmitate	Peptide stabilizing factor	
Fructosebisphosphatase		
cGMP		

Out of this jungle I have made a critical selection of the known modulator mechanisms that seem really involved in the physiological regulation of phosphofructokinases, particularly from eukaryotic organisms (Table 2). I will comment briefly upon each of them.

TABLE 2

MULTIMODULATION OF PHOSPHOFRUCTOKINASES

Regulatory mechanisms	
Most eukaryotic organisms	E. coli
Cooperativity for Fru-6-P	Cooperativity for Fru-6-P
(-) ATP	(-) P-enolpyruvate
(-) Citrate (& glycerate-3-P, P-enolpyruvate)	(+) ADP
	(-) ATP
(+) AMP (& cAMP)	
(+) Pi	
(+) NH_4^+ (beyond K^+ dependence)	
(-) H^+ in animals	
(-) P-creatine in muscle isozyme	
[o,m] in liver isozyme	
(+) Fru-1,6-P_2 (& Glc-1,6-P_2, Man-1,6-P_2)	

Cooperativity for fructose-6-P is a widespread fact in native phosphofructokinase of most origins. It fits well an oligomeric enzyme which is the first irreversible one in the glycolytic pathway below the glucose-6-P crossroads. Changes in the oligomeric state of phosphofructokinase markedly affect its kinetic behaviour and sensitivity to allosteric effectors (23,24). The dissociation constant, around the micromolar range, is markedly affected by the pH (25).

Inhibition by ATP was ambiguous, since it is a substrate of the enzyme. We showed in 1963, on the basis of differential specificity with the yeast enzyme, that it is an allosteric effect, and pointed out that ATP is an endproduct of glycolysis (14). Additional complication arised when Atkinson found that the inhibition by ATP could be counteracted by AMP, presumably by isosteric displacement of the inhibitory ATP (26). Eventually I obtained evidence, on the basis of desensitization by trypsin in the absence and presence of nucleotides, that two different regulatory sites were involved (27). We have recently arrived to the same conclusion for the ascites tumor enzyme on the basis of clearcut differential specificity for three adenylnucleotide sites (Table 3). The complexity regarding the number of different sites for adenylnucleotides is increased by the report of cAMP as an activator of animal phosphofructokinase, initiated by Mansour and Mansour as early as 1962. Nevertheless, at least in higher animals, cAMP is merely an analogue of AMP acceptable *in vitro* but never able to act *in vivo*, because it can operate only in the concentration range physiological for AMP (28) which cAMP can never approach for thermodynamic limitation. For this reason there never was selective evolutionary pressure to have an AMP site discriminating against cAMP. Just the opposite to a site for regulation by cAMP, which needs both high affinity for it and sharp discrimination respect to the much more abundant AMP.

TABLE 3

PHOSPHOFRUCTOKINASE OF ASCITES TUMOR

Differential specificity of the three adenyl nucleotides sites

---

NTPs as substrates: GTP $\simeq$ ATP > CTP $\simeq$ UTP $\simeq$ ITP

NTPs as inhibitors: ATP $\gg$ CTP > GTP $\simeq$ ITP > UTP

NMPs as activators: AMP $\ggg$ IMP, GMP, CMP, UMP.

NTP and NMP stand for nuceloside tri- and monophosphates respectively.
(J. G. Castaño and A. Sols, unpublished results).

The combination of an inhibitory site for ATP and an activator site for AMP seems to make phosphofructokinase precisely fitted to buffer the energy charge of the cell, adjusting overall energy movilization to energy spenditure, maintaining a stable supply of appropiately charged nucleotides to perhaps hundreds of reactions. Phosphofructokinase would thus be a major factor in energy charge homeostasis which in turn was probably a major acquisition for metabolic regulation.

Phosphofructokinases are also feedback inhibited by a carbon endproduct, citrate in most eukaryotic organisms (29, 30) and P-enolpyruvate in *E. coli* (31,16). NH_4^+ ions can activate phosphofructokinases within their physiological concentration ranges and even in the presence of the physiologically ever present K^+ ions in the 0.1 M range, allowing a high rate of glycolysis when there is demand for carbon skeletons derived from glycolysis (32). Pi, closely related to energy metabolism, is a strong activator of phosphofructokinase, synergistically with AMP (33) and NH_4^+ ions. H^+ ions dramatically inhibit animal phosphofructokinase within the physiological pH range (34), apparently by markedly increasing the affinity of the ATP inhibitory site. This inhibition by a decreasing pH presumably allows for a feedback protective slowdown of glycolysis in anaerobic tissues.

P-creatine is a potent inhibitor of the muscle isozyme (where it is an important quick energy store) but does not affect those of liver and brain (35). Glycerate-3-P and P-enolpyruvate can inhibit the muscle enzyme, but apparently they do so at the citrate inhibitory site (36). Fructose-1,6-P_2, a product of the reaction can, under certain conditions, be a potent activator of phosphofructokinase of various origins (37,38,39); the activation takes place in the micromolar range, and can be mimicked by glucose-1,6-P_2 or mannose-1,6-P_2.

The last addition to the already impressive list of modulatory mechanisms for the physiological regulation of phosphofructokinases concerns the hormonally mediated regulation of the liver isozyme. After several reports from the laboratories of Söling and Hofer showing phosphorylation of liver and muscle phosphofructokinase hard to relate to physiological regulation, J. G. Castaño, A. Nieto and J. E. Felíu in my laboratory have found that treatment of hepatocytes with either glucagon or cAMP leads to a form of the enzyme with higher $S_{0.5}$ and nH for fructose-6-P, which result in a marked inactivation of the enzyme when assayed in near physiological conditions.

From the above considerations, the likely types of sites involved in the physiological multimodulation of animal phosphofructokinases are listed in Table 4.

TABLE 4

SPECIFIC SITES IN ANIMAL PHOSPHOFRUCTOKINASES

Active site (2 subsites: Fru-6-P and nucleosidetriphosphate)		
ATP inhibitory site	AMP activatory site	
H^+ " "	P_i " "	
Citrate " "	NH_4^+ " "	
P-creatine " " (muscle)	Fru-1,6-P_2	
Phosphorylable regulatory site (liver)		
Oligomerization site (s)		

The kinetic behaviour of phosphofructokinases is so complex and variable that Mansour ended a review in 1972 (40) saying that "the reader may well come to the conclusion that phosphofructokinase is beyond control" and that "future advances in our knowledge of the regulation of phosphofructokinase will have to await studies on the nature of the enzyme in the resting cell and its response to different physiological conditions". This goal has been partly accomplished through the *in situ* approach for the kinetic study of intracellular enzymes in permeabilized cells, with the confirmation that at least most of the allosteric effectors studied *in vitro* do operate in cells, with kinetic parameters similar or better suited for physiological regulation of metabolism than those obtained *in vitro*. This holds for the enzymes in *E. coli* (16), yeast (41), erythrocytes and ascites tumor cells (42).

Pyruvate Kinase L. In contrast to phosphofructokinase, pyruvate kinases have not been subject to inflationary pressure with respect to metabolic regulation. Indeed, the first regulatory mechanism, allosteric activation by fructose-1,6-P_2, was discovered in 1966 by Hess (43) almost as a surprise. This property was interpreted as related to the potential shift from glycolysis to gluconeogenesis (44). After this came the study of the liver isozyme, pyruvate kinase L, which has proved to be endowed with an elaborate array of regulatory mechanisms, summarized in Table 5.

TABLE 5

MULTIMODULATION OF PYRUVATE KINASE L

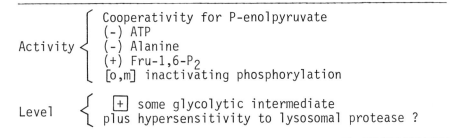

The cooperativity for P-enolpyruvate depends on its oligomeric structure, to the point that lowering the temperature to 0° reversibly shifts the kinetics from markedly sigmoidal to hyperbolic, with increase in the affinity for the substrate (40). Of the two inhibitors, ATP is a product of the enzyme and alanine is structurally related to the other product, pyruvate. We obtained clear evidence, mainly on the basis of differential specificity, that both are allosteric effectors involving two regulatory sites (46). Cysteine is also a strong inhibitor and proline a weaker one, but we found no need to postulate additional sites for them. Alanine is very likely a physiological effector because of the importance in higher animals of the alanine cycle involving muscle and liver. Fructose-1,6-bis-P_2 is a powerful activator that can overcame all the allosteric inhibitions (oligomeric, ATP, and alanine) within its physiologically available range (47). The covalent modulation by phosphorylation leading to a form of the enzyme less active at physiological concentrations of substrate and effectors is solidly stablished both *in vitro* (48) and *in vivo* (49), in the later case favoring glucogeneogenesis in response to glucagon (50). In addition to this series of five regulatory mechanisms for the modulation of the activity of pyruvate kinase L the level of the enzyme is regulated by metabolite induction (51) and perhaps by selective degradation (52).

<u>Miscellaneous Enzymes</u>. I will briefly refer to four other outstanding cases, for various reasons, of multimodulation of enzyme activity.

Phosphorylase has been first in many aspects of metabolic regulation at the enzyme level. It has marked cooperativity for the substrate Pi, is inhibited by glucose-6-P, and can be activated by either non-covalent (AMP) or covalent (phosphorylation) modulation. The two activations make excellent physiological sense, AMP as energy-need signal, and the phosphorylation because it is made by a converter system ultimately under hormonal control. In addition, it has recently been observed that it can be similarly activated by the presence of 20% ethanol. In all three cases activation consists in a shift from sigmoidal to hyperbolic kinetics. It seems that a quaternary structure of the enzyme was first selected that had a convenient initial hindrance for the binding of Pi, leading to low activity at physiological concentrations of the later. Superimposed on this fact, an allosteric effect to activate by inducing the more active conformation was evolved in the form of specific binding of AMP. Eventually, a way of "freezing" the enzyme in this more active conformation was achieved through a specific phosphorylation system. That neither AMP nor covalently bound phosphate have anything very special by themselves other than the induction of the appropiate conformation is suggested by the fact that an organic solvent can lead to the same functional result. Which in a way brings the covalent modulation by metabolic interconversion into the realm of allosteric effects as a third type: covalent interconversion.

A qualitatively complex case of multimodulation is the ribonucleotide reductase, as shown by the work in Reichard's laboratory (53, 54). The *E. coli* enzyme reduces the four common ribonucleoside diphosphates and is allosterically affected by the nucleoside triphosphates in its general activity and in its relative specificity. Two regulatory sites with overlapping specificities have been clearly identified: one that binds dATP, dTTP, dGTP, or ATP, and one that binds only dATP or ATP. The precise working of this multimodulation of the enzyme relative specificity is a fascinating problem still pending solution. Incidentally, ribonucleotide reductase is one of the well authenticated cases of heterotropic allosterism in a monomeric enzyme with hyperbolic kinetics.

Glutamine synthetase in *E. coli* is the key enzyme in nitrogen metabolism and the subject of a highly elaborate system of mechanisms of modulation of its activity, as discovered in Stadtman and Holzer laboratories. Apparently it combines an allosteric multimodulation comparable to that of animal phosphofructokinases and a system for covalent modulation at least as complex as that involved in muscle phosphorylase. I look forward to the presentation of Drs. Stadtman and Chock later in this Symposium and hope that the criteria discussed in this article could help to clarify how many of the observed effects are likely to involve different physiologically significant mechanisms.

Finally I wish to mention what is probably the most complex enzyme of all, the RNA polymerase, that requires modulation of its specificity for the control of gene expression. This enzyme, particularly in eukaryots, is too complex and insufficiently known (55,56) to attempt even a tentative definition of sites and effects. It may take much ingenuity and work to unravel this problem, but the outcome is likely to be of really outstanding value.

SOME LESSONS FROM COMPARATIVE BIOCHEMISTRY

Feedback inhibition of the first enzyme (or the first irreversible enzyme) of a biosynthetic pathway by the endproduct of the pathway (such as the prototypes treonine deaminase inhibition by isoleucine or aspartate transcarbamylase inhibition by CTP) make obvious and indeed almost inescapable sense. However, many other modulatory effects of enzyme activity are less or no obvious at all, from a physiological point of view. Some of the claims run even contrary to physiological sense. Much of this stuff is plain non-sense. In any case, it is unfortunate that many spurious claims have been allowed to creep into the circulating literature of reviews and even text books. Ultimately, modulation of enzyme activity has to integrate with physiology (or viceversa!) or be restricted to the realm of *in vitro* curiosity, perhaps even very important for basic enzymology but irrelevant for metabolic regulation. To use an illustration without risk of anybody being hurt I will mention the dramatic activation by lyxose of the ATPase activity of yeast hexokinase (57), which was of considerable interest in support of Koshland's induced fit theory and for the elucidation of the mechanism of the hexokinase reaction, but has absolutely nothing to do with enzymatic regulation of me-

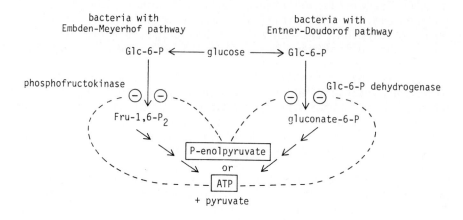

FIGURE 1. A doble analogy in the feedback inhibition of two different but "parallel" key enzymes in the regulation of glucose degradation in prokaryotic organisms.

TABLE 6

ANALOGIES IN ALLOSTERIC REGULATION OF METABOLISM

Energy-need activation:
  of glycogenolysis
    AMP   phosphorylase b
  of glycolysis
    AMP   phosphofructokinase of most organisms
    ADP   "   "   "   " $E.\ coli$
  of the Krebs cycle
    AMP   isocitrate dehydrogenase of yeasts
    ADP   " "   "   "   " animals
Precursor activation of glycogen synthesis:
    Glc-6-P   glycogen synthetase of most organisms
    Glc-1-P   "   "   "   " insects
    Fru-1,6-P_2 ADP-pyrophosphorylase of prokaryots
Feedback inhibition of glycolysis by a carbon endproduct:
    Citrate   phosphofructokinase of eukaryots
    P-enolpyruvate   "   "   "   " prokaryots
Activation of anaplerotic enzymes by acetyl-CoA:
     pyruvate carboxylase of eukaryots
     P-enolpyruvate carboxylase of prokaryots

---

tabolism.

The ascertaining of physiological significance for some presumptive regulatory mechanisms in certain enzymes can be aided by observations from comparative biochemistry. The generalizations summarized in Figure 1 and Table 6, independently of wether multimodulation is involved, indicate what could be referred to as the analogy principle in allosteric regulation of metabolism, involving effectors and enzymes.

CONSIDERATIONS ON THE EVOLUTIONARY ORIGIN OF THE MODULATION OF ENZYME ACTIVITY

As an opening to the highly speculative question of evolutionary origin of regulatory mechanisms I will reproduce a recent statement of Mayr (58): "The understanding of the evolutionary process... led to the realization that every biological problem poses an evolutionary question, that it is legitimate to ask with respect to any biological structure, function or process: Why it is there? What was its selective advantage when it was acquired?". Difficult and risky as it is I will tackle the question.

There is fairly strong presumptive evidence that as catalysts most enzymes have evolved up to nearly maximal capacity (59,60), i.e. they have approached catalytic perfection. Somewhere along this long evolution came for enzymes at metabolic crossroads the additional evolution of acquisition and improvement of modulatory mechanisms with enough physiological advantage to succeed in being selected in the long run. Probably, regulatory efficiency was in part accomplished at the expense of some catalytic potency. On the other hand, there is experimental evidence that evolution of enzymes is not easy. Under strong selective pressure to improve the ability of a microorganism to utilize a marginal nutritional substrate, it has repeatedly been observed significant success through gene duplication leading to more copies of an enzyme with neither greater molecular activity nor affinity for the marginal substrate (61).

Three basic questions on the evolutionary origin of the modulation, and multimodulation, of enzyme activity are in short: why, how, and when?

The *why*, of course has to be a biological advantage significant enough for natural selection. This was simple for the early cases of feedback inhibition by the endproduct of a pathway depending on the inhibitable enzyme. It is much less clear in many other cases. And it is specially difficult to understand in the cases of the more complex multimodulated enzymes. Unless perhaps in terms of coping with alternative contingences, or as back up systems. In general, this difficulty reminisces at the molecular level Darwin's honest perplexity when formulating the evolutionary theory on the basis of natural selection. We must allow for changing conditions in nature as opposed to the artificially standardized ones common in laboratory practice. Perhaps the justificating advantages were related to contingencies of which we are unaware. When the modulatory effect is unmistakable but the advantage unapparent, the search could go in reverse: a little of sensible teleonomy could help in imagining possible conditions in which a given modulatory property could be physiologically useful.

The *how* is more amenable to concretion. Setting out from the initial emphasis on feedback inhibition of biosynthetic pathways Horowitz (62) proposed an explanation for the origin of regulatory sites: evolution by retroevolution, with the first enzyme in the sequence carrying a memory of its origin. This hypothesis is of little help in the presence of the variety and frequent multiplicity of regulatory sites. A similar limitation holds for the regulatory subunits hypothesis, formulated by Gerhart and Schachman (63) when they discovered a regulatory subunit in aspartate transcarbamylase, and supported by Monod et al (7) as a likely general mechanism. It is now known that regulatory subunits occur in rather rare instances, instead of being common in regulatory enzymes. With newer knowledge on the evolution of genes, Engel (64) proposed partial gene duplication as a basic mechanism for regulatory sites for compounds identical or structurally related to the substrate. This seems likely to be the case in liver glutamate dehydrogenase that has an active site for a nucleotide (NAD) and a regulatory site for another nucleotide (GTP), while aminoacid sequences make unlikely a convergent evolution of sites. Koshland (65) has recently proposed that in flexible molecules like the relatively large proteins that most enzymes are, a binding site for an efector might arise by random probability and could be refined by mutational selection. A new site in an enzyme could arise by gene fusion or by crossing over of part of another gene for which the effector was a substrate or product. Such mechanisms would tend to make the subunits of regulatory enzymes, particularly the multimodulated ones, larger than those of nonregulatory enzymes. And this is not the case. Among highly multimodulated enzymes, mammalian phosphofructokinases, pyruvate kinase L, and glutamate synthetase of *E. coli* have identical subunits of 80.000 daltons the former and 50.000 the other two, values within the range of subunits in general (9). Accordingly, it seems that many regulatory sites, particularly in multimodulated enzymes, probably arose by mutational development of a new specific site, able to induce a useful conformational change, out of the large unspecialized area outside the active site. Nevertheless, new site acquisition by gene fusion, followed by size reduction through RNA splicing is also a possibility to be considered.

As for the *when* of the appearance of regulatory sites. I propose that many of them, perhaps the majority, are very very old. The accumulation of a prebiotic organic soup is widely accepted as a precursor to life. But the origin of real life is still a very hard to unravel mystery. It is generally recognized that efficient catalysis and an elaborate reproduction system are essential characteristics of life. I would like to suggest a third essential characteristic for efficient life: a sophisticated network of regulatory mechanisms to modulate key enzymes. Perhaps for a long time there was a "preliminary" life, with poor catalysis, primitive reproduction and no regulation, that was eventually superseded by the life we know. In this picture the advance of the development of the more basic regulatory mechanisms probably took place somewhere between 3 and 2 billion years ago. Only after the transition from protoorganisms to regulated life, could take place the classical organismic evolution to higher forms of life. (Unless study of key enzymes of the methanogenic bacteria, the presumed older "urkingdom" or "archaebacteria" (66) could suggest otherwise).

Let us know examine a few cases. Pyruvate kinases of widely diverse origin are sensitive to allosteric activation by fructose-1,6 -P_2 superimposed on an oligomeric hindrance. This holds from *E. coli* (67) and yeast (68) to mammals. There seems to be no compelling reason for a precursor activation by fructose-1,6-P_2 instead of, say, dihydroxyacetone-P. It seems probable that oligomeric cooperativity and activation by fructose-1,6-P_2 in pyruvate kinase appeared in primitive ancestors, say around 2 billion years ago. Sequence analysis for the fructose-1,6-P_2 site in widely separated organism could support or disprove this hypothesis. As for the specific oligomer, the aldolase tetramer has just been elegantly shown to be that old (69). An extension of this hypothesis is the suggestion that of the three mammalian isozymes, the widely distributed allosteric pyruvate kinase A is probably the ancestor out of which could have arisen, after gene duplication, the more sophisticated L isozyme of the liver, while the non-regulated M isozyme would be merely the result of (partial?) loss of modulatory properties irrelevant in muscle, where there is instead pressure for high activity. Another enzyme with an allosteric property well recognized as of wide occurence is fructosebisphosphatase with its inhibition by AMP, known to occur from *E. coli* to mammals (including yeasts and fishes). One exception that deserves comment is its absence in the bumble-bee, which seems to be a useful loss to facilitate the thermogenic "futile" cycle that confers to this humble creature the ability to have an active life in cold weather by heating itself (70).

Within the context of this discussion, living beings can be classified in three groups: 1) man, 2) the microbes in flasks and the animals in cages, and 3) all others. And natural selection tends to eliminate errors in enzyme regulation in the third group but not in the other two[2] ?. precisely the favorite ones of most biochemists. There is a popular saying that "if anything can go wrong, it will". To loss a modulatory mechanism is much easier that to gain it. Surely it happens a great many times. And indeed many microbial mutants with loss of, or markedly unfavorable quantitative change in an allosteric mechanism have been selected in the laboratory in appropiate conditons, choosen specifically to offer advantage to what in wild life would be a hindrance. At the other end of the scale are humans that go along despite the loss or impairment of a modulatory mechanism. It can be safely predicted that with appropiate assays of tissue enzymes a great many "disallosterisms" and other types of protein disregulation defects will be identified in the not distant future. This will be an important chapter of future molecular pathology.

CONCLUDING REMARKS

After the critical selection of a series of likely regulatory mechanisms for certain enzymes it can be concluded that multimodulation of enzyme activity in metabolic regulation is a quite significant and rather widespread fact. This fact, together with the variety of interacting possibilities between different modulators for a give enzyme, has an heuristic value. Exploratory research on possible regulatory effectors for an enzyme should be carried out in assay conditions with good sensitivity for the detection of effects on the affinity for the substrate (by working below saturation, preferably at concentrations around the $S_{0.5}$), or of modulation of other effectors (by adding them at concentrations around their Ka or Ki) . With this particular condition superimposed to the general rule of avoiding any marked deviation from physiological condition in pH, ionic environment (K^+ !), temperature, and, if possible, enzyme concentration.

Finally, as a consequence of the fact that multimodulation is common, and sometimes complex, in regulatory enzymes,

---

[2] With the predictable result in man of a progressive impairment of the quality of the genetic endowment of the species (hopefully correctable sometime in the future).

care should be taken not to be overconfident in trying to account for the metabolic regulation of an enzyme on the basis of one (or even several !) modulatory effects, even if they are likely to be of physiological significance. There is always the possibility of some additional as yet undiscovered modulatory effect that could be not only important, but even more important for metabolic regulation than what was already known. While multiheaded enzymes are a very rare exception (and almost always limited to two different active sites), multimodulation is rather common and can be quite complex.

ACKNOWLEDGMENTS

I am sure several colleagues have contributed through personal contacts to the early development of some of the ideas presented in this article, although it would be too difficult to try to track them in particular. But I can explicitly acknowledge the contribution to the actual preparation of the manuscript with discussions and criticisms of coworkers in my own laboratory: C. Asensio, G. DelaFuente, J.G. Castaño, J.E. Felíu, C. Gancedo, J.M. Gancedo, R. Lagunas, P.A. Lazo, R. Marco and J. Sebastian and of Renée Clarys, for her very efficient technical assistance.

REFERENCES

1. Cori, C.F. (1947).The Harvey Lectures, Ser. 41 (1945-1946), Academic Press, Inc., New-York, p. 253.
2. Crane, R.K., and Sols, A. (1954). J. Biol. Chem. 210, 597.
3. Umbarger, H.E. (1956). Science 123, 848.
4. Yates, R.A., and Pardee, A.B. (1956). J. Biol. Chem. 221, 757.
5. Krebs, E.G., and Fisher, E.H. (1956). Biochem. Biophys. Acta 20, 150.
6. Monod, J., Changeux, J.-P., and Jacob, F. (1963). J. Mol. Biol. 6, 306.
7. Monod, J., Wyman, J., and Changeux, J.-P. (1956). J. Mol. Biol. 12, 88.
8. Metabolic Interconversion of Enzymes 1975. (1976) Ed. by Shatiel, S., Springer-Verlag, Berlin.
9. Sols, A., and Gancedo, C. (1972). In Biochemical Regulatory Mechanisms in Eukaryotic Cells (Kunn, E., and Grisolia, S., editors), John Wiley & Sons, pg. 85.
10. Sols, A. (1973). In Mechanisms and Control Properties of Phosphotransferases, Akademic-Verlag, Berlin, pg. 239
11. Perutz, M.F. (1978). Science 201, 1187.
12. Ricard, J., Mcunier, J. C., and Buc, J. (1974). Eur. J. Biochem. 49, 195.

13. Storer, A. C., and Cornish-Bowden, A. (1977). Biochem. J. 165, 61.
14. Viñuela, E., Salas, M.L., and Sols, A. (1963). Biochem. Biophys. Res. Commun. 12, 140.
15. Atkinson, D.E. (1966). Ann. Rev. Biochem. 35, 85.
16. Reeves, R.E., and Sols, A. (1973). Biochem. Biophys. Res. Commun. 50, 459.
17. Guilloton, M., Archambault de Vençay, J., and Rosenberg, A.J. (1970). C.R. Acad. Sci. Paris 270, 3152.
18. Barnes, L.D., McGuire, J.J., and Atkinson, D.E. (1972). Biochemistry 11, 4322.
19. Gerhart, J.C., and Pardee, A.B. (1962). J. Biol. Chem. 237, 891.
20. Perutz, M.F. (1978). Scientific American 239, nº 6, 68.
21. Passonneau, J.V., and Lowry, O.H. (1962). Biochem. Biophys. Res. Commun. 7, 10.
22. Tejwani, G.A. (1978). TIBS 3, 30.
23. Hulme, E.C., and Tipton, K.F. (1971). FEBS Letters 12, 197.
24. Hofer, H.W. (1971). H-S. Z. Physiol. Chem. 352, 997.
25. Hofer, H.W., and Krystek, E. (1975) FEBS Letters 53, 217.
26. Ramaiah, A., Hathaway, J.A., and Atkinson, D.E. (1964). J. Biol. Chem. 239, 3619.
27. Salas, M.L., Salas, J., and Sols, A. (1968). Biochem. Biophys. Res. Commun. 31, 461.
28. Pinilla, M., and Luque, J. (1977). Mol. Cell. Biochem. 15, 219.
29. Parmeggiani, A., and Bowman, R.H. (1963). Biochem. Biophys. Res. Commun. 12, 268.
30. Salas, M.L., Viñuela, E., Salas, M. and Sols, A. (1965). Biochem. Biophys. Res. Commun. 19, 371.
31. Blangy, D., Buc, H., and Monod, J. (1968). J. Mol. Biol. 31, 13.
32. Sols, A., and Salas, M.L. (1966). Methods in Enzymology 9, 436.
33. Bañuelos, M., Gancedo, C., and Gancedo, J.M. (1977). J. Biol. Chem. 252, 6394.
34. Trivedi, B., and Danforth, W.H. (1966). J. Biol. Chem. 241, 4110.
35. Tsai, M.Y., and Kemp, R.G. (1974). J. Biol. Chem. 249, 6590.
36. Colombo, G., Tate, P.W., Girotti, A.W., and Kemp, R.G. (1975). J. Biol. Chem. 250, 9404.
37. El-Bawdry, A.M., Otani, A., and Mansour, T.E. (1973). J. Biol. Chem. 248, 557.
38. Tornheim, K., and Lowenstein, J.M. (1976) J. Biol. Chem. 251, 7322.
39. Hood, K., and Holloway, M.R. (1976). FEBS Letters 68, 8.
40. Mansour, T.E. (1972). Current Topics Cell. Reg. 5, 1.
41. Bañuelos, M., and Gancedo, C. (1978). Arch. Microbiol. 117, 197.

42. Sols, A., Aragón, J.J., Lazo, P.A., and Gonzalez, J. (1978). Fed. Proc. 37. 1428.
43. Hess, B., Haeckel, R., and Brand, K. (1966). Biochem. Biophys. Res. Commun. 24, 824.
44. Gancedo, J.M., Gancedo, C., and Sols, A. (1967). Biochem. J. 102, 23c.
45. LLorente, P., Marco, R., and Sols, A. (1970). Eur. J. Biochem. 13, 45.
46. Carbonell, J., Felíu, J.E., Marco, R., and Sols, A. (1973). Eur. J. Biochem. 37, 148.
47. Sols, A., and Marco, R. (1970). Current Topics in Cell. Reg. 2, 227.
48. Ljungström, O., Hjelmqvist, G., and Engström, L. (1974). Biochem. Biophys. Acta 358, 289.
49. Riou, J.P., Claus., T.H., and Pilkis, S.J. (1978). J. Biol. Chem. 253, 656.
50. Felíu, J.E., Hué, L., and Hers, H.G. (1976). Proc. Natl. Acd. Sci. USA. 73, 2762.
51. Sillero, A., Sillero, M.A.G., and Sols, A. (1969). Eur. J. Biochem. 10, 351.
52. Otto, K., and Schepers, P. (1967). H.-S. Z. Physiol. Chem. 348, 482.
53. Brown, N.C., and Reichard, P. (1969). J. Mol. Biol. 46,39.
54. von Döbeln, V. (1977) Biochemistry 16, 4368.
55. Travers, A. (1976). Nature 263, 641.
56. Zillich, W., Mailhammer, R., Skorko, R., and Rohrer, H. (1977). Current Topics Cell. Reg. 12, 263.
57. DelaFuente, G., Lagunas, R., and Sols, A. (1970). Eur. J. Biochem. 16, 226.
58. Mayr, E. (1978). Scientific American 239, nº 3, 39.
59. Cleland, W.W. (1975). Accounts Chem. Res. 8, 145.
60. Albery, W.J., and Knowles, J.R. (1976). Biochemistry 15, 5631.
61. Rigby, P.J.W., Burleigh, B.D.Jr., and Hartley, B.S. (1974). Nature 251, 200.
62. Horowitz, N.H. (1965). In Evolving Genes and Proteins (Bryson, V. and Vogel, H.J., eds.) Academic Press, pg.15.
63. Gerhart, J.C., and Schachman, H.K. (1965). Biochemistry 4, 1052.
64. Engel, P.C. (1973). Nature 241, 118.
65. Koshland, D.E.Jr., (1976). FEBS Letters 62, E47.
66. Woese, C.R., and Fox, G.F. (1977). Proc. Natl. Acad. Sci. USA 74, 5088.
67. Waygood, E.B., and Sanwal, B.D. (1972). Biochem. Biophys. Res. Comm. 48, 402.
68. Serrano, R., Gancedo, J.M., and Gancedo, C. (1973). Eur. J. Biochem. 34, 479.
69. Heil, J.A., and Lebherz, H.G. (1978). J. Biol. Chem. 253, 6599.
70. Newsholme, E.A., Crabtree, B., Higgins, S.J., Ihornton. S. D., and Start, C. (1972). Biochem. J. 128, 89.

POLY(ADP-RIBOSE) AND ADP-RIBOSYLATION OF PROTEINS[1]

Kunihiro Ueda, Osamu Hayaishi, Masashi Kawaichi, Norio Ogata, Kouichi Ikai, Jun Oka, and Hiroto Okayama

Department of Medical Chemistry
Kyoto University Faculty of Medicine

ABSTRACT After a brief review of mono ADP-ribosylation reactions, a recent progress in poly ADP-ribosylation of nuclear proteins is discussed. New data suggest that (1) the initiation and elongation of poly(ADP-ribose) chain on a histone molecule are catalyzed by a single enzyme, poly(ADP-ribose) synthetase, (2) the ADP-ribosyl histone linkage is an ester bond between the terminal ribose and a carboxyl group of glutamate residues (Glu^2 of histone H2B, and Glu^2, Glu^{14}, and Glu^{116} of histone H1), (3) this bond is hydrolyzed by ADP-ribosyl histone hydrolase, and (4) the polymer synthesis is closely related to differentiation or neoplastic transformation of human leukocytes.

INTRODUCTION

ADP-ribosylation constitutes a unique group of posttranslational modifications of proteins (1, 2). It is unique in three ways; first, the modifying group, ADP-ribose, is derived from a respiratory coenzyme, NAD. The cleavage of the N-glycosidic linkage, which is a so-called high energy bond, provides the driving force for all types of ADP-ribosylation. Second, the ADP-ribose transferred exists as either a monomer or a polymer (Fig. 1), depending on the system and conditions. A polymer is presumed to start from a monomer attached to a protein and to be elongated sequentially. Third, this type of modification is very heterogeneous in terms of not only chain length but also natural distribution, acceptor proteins and ADP-ribosyl protein linkages. In this presentation, I briefly explain a variety of ADP-ribosylation reactions, and, then, concentrate on the poly ADP-ribosylation system in the nucleus.

[1]This work was supported in part by grants-in-aid for Cancer Research from the Ministry of Education, Science and Culture, Japan.

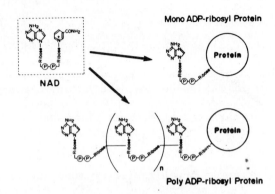

FIGURE 1. Mono and poly ADP-ribosylation.

MONO ADP-RIBOSYLATION

Table I summarizes all ADP-ribosylation reactions reported to date.

The first enzyme discovered for mono ADP-ribosylation was diphtheria toxin. This toxin transfers ADP-ribose onto eukaryotic elongation factor 2 and concomitantly inactivates protein synthesis (3). An unusual, as yet unidentified, basic amino acid has been reported to be the site of modification (4, 5). Exactly identical reaction is catalyzed by Pseudomonas toxin (6), and, recently, another modification affecting also protein synthesis was reported to be catalyzed by exoenzyme S of the same microorganism (7). Bacteriophage T4

TABLE I
NATURAL DISTRIBUTION OF VARIOUS ADP-RIBOSYLATION REACTIONS

Enzyme	Acceptor	Reporter
Mono ADP-ribosylation		
Diphtheria toxin	Elongation factor 2	Honjo et al. 1968
Pseudomonas toxin	Elongation factor 2	Iglewski & Kabat 1975
Pseudomonas exoenzyme S	Elongation factor 1 (?)	Iglewski et al. 1978
T4 phage { Viral Induced	RNA polymerase and other E. coli proteins	Zillig et al. 1976 Goff 1974
N4 phage	E. coli proteins	Pesce et al. 1976
Cholera toxin	Adenylate cyclase	Cassel & Pfeuffer 1978, Gill & Meren 1978
E. coli enterotoxin LT	Adenylate cyclase	Moss & Richardson 1978
Avian erythrocyte protein	Adenylate cyclase	Moss & Vaugham 1978
Poly ADP-ribosylation		
Nuclear enzyme	Histones	Nishizuka et al. 1968
	Nonhistone proteins	Nishizuka et al. 1968
	Mg^{++}, Ca^{++}-endonuclease	Yoshihara et al. 1974
	RNA polymerase	Müller & Zahn 1976
Mitochondrial enzyme	Mitochondrial protein	Kun et al. 1975
Cytoplasmic enzyme	Histones	Roberts et al. 1975

contains an ADP-ribosyltransferase as a virion constituent and also induces another transferase after infection (8). These enzymes produce changes termed <u>alteration</u> and <u>modification</u> in <u>E. coli</u> RNA polymerase. Arginyl residues of α-subunits of the polymerase and also of several other proteins are reported to be ADP-ribosylated (9). N4 phage has a similar enzyme (10). Furthermore, cholera toxin (11, 12) and <u>E. coli</u> enterotoxin LT (13) were recently shown to ADP-ribosylate membrane proteins including a GTP-binding component of adenylate cyclase. An arginyl residue appears to be the modification site. An avian erythrocyte protein catalyzes similar reactions (14), but its significance has not yet been elucidated.

There are two noticeable differences between mono and poly ADP-ribosylation; first, most enzymes working for mono ADP-ribosylation have a prokaryotic origin, whereas poly ADP-ribosylation is restricted in eukaryotic cells. Second, the enzymes of mono ADP-ribosylation work on proteins of different organisms, whereas those of poly ADP-ribosylation modify proteins in their own cells.

POLY ADP-RIBOSYLATION

<u>Natural Distribution of Poly(ADP-Ribose)</u>. We recently raised anti-poly(ADP-ribose) antibody in a rabbit and applied it to an immunohistochemical procedure. Fig. 2 shows the reactivity of the antiserum with poly(ADP-ribose) of various chain lengths. The polymer used for immunization had an average chain length of 24, and the immune reaction appeared most active around this size. Monomer was totally inert, and small oligomers were poorly reactive to the antibody. Table 2 shows the specificity of the antiserum, as examined by the binding competition. To the reaction mixture containing

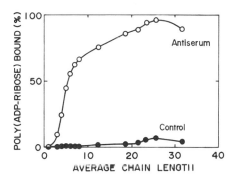

FIGURE 2. Effect of chain length on immunoreactivity.

TABLE II

NO INHIBITION OF POLY(ADP-RIBOSE) BINDING BY RELATED COMPOUNDS

Addition		Polymer bound	Inhibition
	ng	cpm	%
None	—	3253	—
Poly(ADP-ribose)	560	210	94
Poly(A)	56000	3357	0
Poly(G)	56000	3395	0
Poly(C)	56000	3399	0
Poly(U)	56000	3381	0
Yeast RNA	56000	3348	0
Calf thymus DNA	56000	3430	0
ADP-ribose	56000	3229	0
NAD	56000	3377	0
AMP	56000	3313	0
Ribose 5-P	56000	3287	0

antiserum and 5.6 ng (5000cpm) of ^{14}C-labeled poly(ADP-ribose), various unlabeled compounds were added. No compound except poly(ADP-ribosé) produced a decrease in the radioactivity.

Applying this antiserum to the so-called indirect immunofluorescence technique, we examined the existence of poly(ADP-ribose) in situ in various tissues. An example is shown in Fig. 3. In the standard condition (Fig. 3a), HeLa cells were fixed with ethanol, incubated with the anti-poly(ADP-ribose) rabbit serum, and then with fluorescein-labeled anti-rabbit IgG of swine. Diffuse fluoresecence and patchwise foci were clearly observed in nuclei. The fluorescence was interpreted to indicate natural poly(ADP-ribose) on the basis of following observations; (1) the fluorescence disappeared when the antiserum was pre-absorbed with poly(ADP-ribose), (2) it also

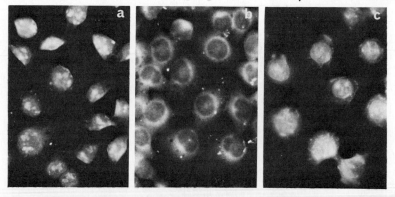

FIGURE 3. Immunostaining of HeLa cells. (a) Stained with anti-poly(ADP-ribose) serum; (b) stained with control serum; (c) preincubated with NAD.

TABLE III
REQUIREMENTS OF POLY(ADP-RIBOSE) SYNTHETASE

Addition	ADP-ribose Incorporated
	pmol
None	1.1
DNA	27
DNA + Histone H1	105
Histone H1	0.3

disappeared when the cells were pretreated with snake venom phosphodiesterase or poly(ADP-ribose) glycohydrolase, (3) no specific fluorescence was observed with control (nonimmunized) serum (Fig. 3b), and (4) the fluorescence was markedly intensified by preincubation with NAD (Fig. 3c). The latter result indicated the polymer production from exogenously added NAD in fixed cells. The intensification of fluorescence by NAD treatment was diminished by a simultaneous addition of 10 mM nicotinamide, which is a potent inhibitor of poly(ADP-ribose) synthetase.

Similar patterns were obtained with various other cells. Later in this presentation are shown some other examples.

Biosynthesis of Poly(ADP-Ribose). The biosynthesis comprises two seemingly different reactions, i.e., linkings of ribose-protein and ribose-ribose; in other words, initiation and elongation of ADP-ribose chains. Separation and reconstitution of these two reactions have been our goal for several years.

In 1974, we succeeded in purifying poly(ADP-ribose) synthetase about 5500-fold from rat liver nuclei (15). The purified enzyme required DNA almost absolutely for its activity (Table III). Histone also stimulated the activity 3- to 4-fold when added with DNA. However, the product, poly(ADP-ribose), formed in the presence of DNA and histone was hardly bound to the histone, but mostly bound to an endogenous acceptor which copurified with the enzyme. Histone served mainly as an allosteric activator and not as an acceptor in this system (16).

During subsequent studies, we took notice of the fact that 5% perchloric acid extracts poly ADP-ribosylated histone H1 as well as unmodified H1 but not the endogenous acceptor. Using this extraction as a tool for discrimination between histone H1 and the endogenous acceptor, we found that the

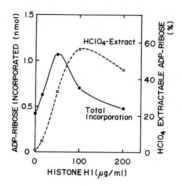

FIGURE 4. Effect of histone concentrations on poly(ADP-ribose) synthesis.

concentration of histone was critical for poly ADP-ribosylation of this protein. As shown in Fig. 4, 5% perchloric acid-extractable fraction of acid (20% trichloracetic acid)-insoluble radioactivity (namely, incorporation into the histone fraction) increased as the histone concentration was elevated up to 100 µg/ml, although the total incorporation was maximal at 50 µg/ml.

When the DNA:histone ratio was kept constant at unity and their concentrations were increased, perchloric acid-extractable material increased along with total acid-insoluble incorporation (Fig. 5). The incorporation into perchloric acid-precipitable material reached the plateau at fairly low

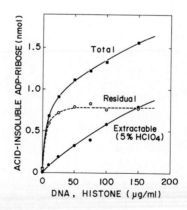

FIGURE 5. Effect of DNA and histone concentrations on poly(ADP-ribose) synthesis.

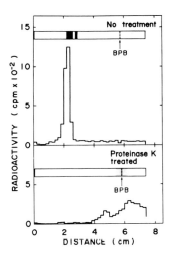

FIGURE 6. SDS-polyacrylamide gel electrophoresis of poly ADP-ribosylated histone H1.

concentrations of DNA and histone. Our previous failure to detect ADP-ribosylation of histones appeared to be ascribable to the concentrations as low as 10 μg/ml of DNA and histones used. The concentrations of substrate, NAD, had little effect on the extractability with perchloric acid.

Using sufficiently high concentrations of DNA and histone H1 and extraction with perchloric acid, we were able to demonstrate direct ADP-ribosylation of histone. Fig. 6 shows a profile of SDS-polyacrylamide gel electrophoresis of the material thus obtained. The incorporated ADP-ribose migrated closely with histone bands, and moved to the position of a tracking dye after treatment with proteinase K.

In order to verify true initiation, the experiment illustrated in Fig. 7 was carried out. Histone H1 was poly ADP-ribosylated with doubly labeled NAD (^{14}C in ribose and ^{3}H in adenine), and extracted with 5% perchloric acid. The extract was treated with alkali and then with snake venom phosphodiesterase. By this procedure, both linkages of ribosyl protein and pyrophosphate were hydrolysed, and equal amounts of ribose 5-phosphate and AMP were produced from the termini. If only initiation had taken place, the ratio of [^{14}C]ribose 5-phosphate/[^{3}H]AMP would be the same as that in the substrate, NAD, whereas if elongation had taken place from pre-existing nonradioactive ADP-ribose, unlabeled ribose 5-phosphate would be produced and lower the ratio. The experimental result indicated that at least 75% of polymer chains had newly initiated. The rest might be explained by elongation from pre-

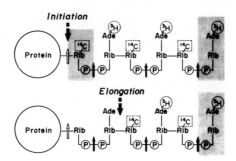

FIGURE 7. Discrimination of initiation from elongation. ⇨ , hydrolysis by alkali; ➡ , hydrolysis by snake venom phosphodiesterase.

existing molecules or by a branching structure, as recently suggested by Miwa and collaborators (17).

These results indicated that initiation of poly(ADP-ribose) chains on histones was a predominant type of reaction under certain conditions.

The subsequent stage, elongation, was studied using mono ADP-ribosyl histone as an acceptor. This type of acceptor was prepared either nonenzymatically by Schiff base formation (18), or enzymatically with chromatin and purified by borate-gel affinity chromatography (19). Fig. 8 shows the profile of SDS-polyacrylamide gel electrophoresis of [^{14}C]ADP-ribosyl histone H1 elongated with [^{3}H]NAD. The two radioactivities comigrated

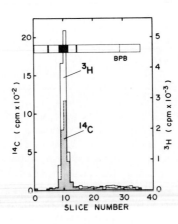

FIGURE 8. SDS-polyacrylamide gel electrophoresis of elongated ADP-ribosyl histone H1.

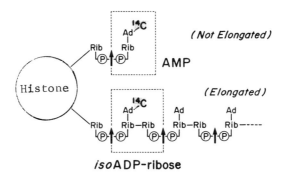

FIGURE 9. Design of elongation experiment.

in the gel, indicating ADP-ribosylation of ADP-ribosyl histone molecule added.

This ADP-ribosylation was shown to represent elongation from pre-existing ADP-ribose by a label transfer experiment (Fig. 9). [^{14}C]ADP-ribosyl histone H1, nonenzymatically prepared, was incubated with nonradioactive NAD, and the product was digested with snake venom phosphodiesterase. If elongation did not take place, ^{14}C would be recovered totally in AMP, while, if elongation took place to the free terminus, ^{14}C would be found in isoADP-ribose. Generation of [^{14}C]isoADP-ribose is thus an indication of elongation. The experimental result (Fig. 10) showed that radioactive isoADP-ribose was produced from the incubated sample. The radioactivity in iso-ADP-ribose represented about 6% of total radioactivity recovered.

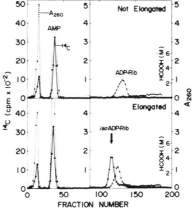

FIGURE 10. Dowex 1 column chromatography of phosphodiesterase digest of elongated and not elongated ADP-ribosyl histone H1.

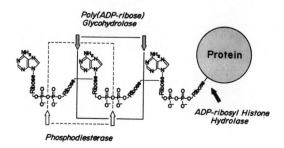

FIGURE 11. Mode of degradation of poly ADP-ribosyl protein.

These results unequivocally demonstrated that pre-attached ADP-ribose was elongated, and that the elongation proceeded by a terminal addition mechanism.

Degradation of Poly ADP-ribosyl Protein. Two types of enzymes have been known for degradation of poly(ADP-ribose) (Fig. 11). One is poly(ADP-ribose) glycohydrolase that splits the ribose-ribose linkage (20, 21), and the other is phosphodiesterase that splits the pyrophosphate bond (22). Several lines of evidence have suggested that glycohydrolase plays a principal role in in vivo degradation of poly(ADP-ribose). The third enzyme, ADP-ribosyl histone hydrolase, was recently discovered in our laboratory (23).

The hydrolase has been purified about 400-fold from rat liver cytosol, and its properties investigated (24). Fig. 12 shows the time courses of hydrolysis of various ADP-ribosyl proteins. All three substrates tested, ADP-ribosyl histone

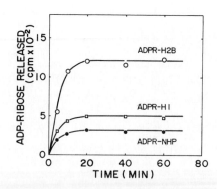

FIGURE 12. Time courses of hydrolysis of various ADP-ribosyl proteins.

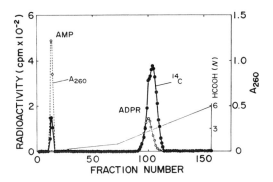

FIGURE 13. Dowex 1 column chromatography of split products.

H2B, H1 and nonhistone proteins, were hydrolyzed. However, their rates and extents differed considerably. The reason for the cessation of hydrolysis after about 20 min was not clear, since more than a half of the substrate remained unhydrolyzed and the enzyme remained active throughout this incubation period.

Analysis of split products is shown in Fig. 13. Although the main product coincided with authentic ADP-ribose in several paper chromatographic systems, it turned out to be different from ADP-ribose on a Dowex 1 column. Judging from the facts that all portions of ADP-ribose were included in this product, that its molecular weight and ionic property were very similar to those of ADP-ribose, and that nonenzymatic hydrolysis also produced this product, this seemed to be an ADP-ribose molecule which underwent a small rearrangement upon its release from protein.

TABLE IV
EFFECTS OF VARIOUS COMPOUNDS OF ADP-RIBOSYL HISTONE HYDROLASE

Addition	Activity
	%
None	100
ADP-ribose (10 mM)	3
NAD (10 mM)	20
NADP (10 mM)	109
3',5'-Cyclic AMP (10 mM)	103
AMP (10 mM)	107
Histone H2B (0.5 mg/ml)	95
DNA (0.5 mg/ml)	16
Histone H2B + DNA	67
pCMB (1.0 mM)	23

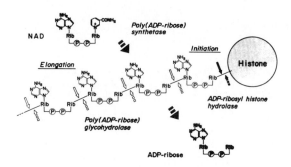

FIGURE 14. Biosynthesis and degradation of poly ADP-ribosyl histone.

The hydrolase was inhibited by ADP-ribose monomer and NAD (Table IV). Cyclic AMP, which is a potent inhibitor of poly-(ADP-ribose) glycohydrolase, and AMP, which is an inhibitor of nuclear phosphodiesterase, were totally inert. DNA was inhibitory; this effect appeared to be due to binding of substrate, since the effect was partly reversed by the addition of histone. p-Chloromercuribenzoic acid was also inhibitory.

Fig. 14 summarizes enzymatic reactions of poly(ADP-ribose) metabolism. Both biosynthesis and degradation take place at ribose-ribose or ribose-protein linkages and nowhere else, suggesting that frequent modification and degradation may be possible in vivo.

Acceptor Proteins and Acceptor Sites. Fig. 15 shows SDS-polyacrylamide gel electrophoretograms of ADP-ribose acceptor proteins of rat liver. These proteins were ADP-ribosylated in nuclei and isolated with borate-gel affinity chromatography. Major acceptors were identified as histones H2B and H1. Some other histones and nonhistone proteins such as A24 protein (25) and HMG proteins (26) were also identified as minor acceptors. We took up histone H2B for further chemical studies.

ADP-ribosyl histone H2B was purified from chromatin incubated with [^{14}C]ribose-labeled NAD. The purified preparation gave a single protein band on SDS-polyacrylamide gel electrophoresis, and contained about 1 mole of ADP-ribose per 12 histone molecules; the average length of ADP-ribose chains was 1.0.

Fig. 16 shows chemical stabilities of ADP-ribosyl histone linkage. The linkage was stable at acidic pH, but was very unstable in alkaline conditions; its half life was about 45 min at pH 9.6 and 25°, and was shorter than 5 min in 1/10 N NaOH. The ADP-ribosyl histone bond was also sensitive to

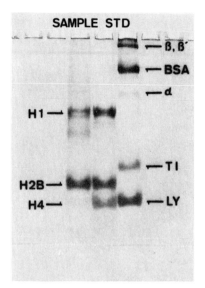

FIGURE 15. SDS-polyacrylamide gel electrophoresis of nuclear ADP-ribose acceptor proteins. Standard proteins (STD) contained either a calf thymus histone mixture (H1, H2B and H4) (left) or various proteins of known molecular weights (right).

neutral hydroxylamine. At pH 7 and 25°, 2 M hydroxylamine rapidly released ADP-ribose from histone. These results indicated that the bond was different from any of known N- or O-glycosidic bonds of glycoproteins, and suggested an ester bond.

Digestion of ADP-ribosyl histone H2B was carried out with trypsin and snake venom phosphodiesterase. The digests were

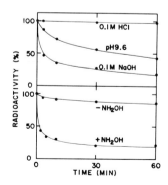

FIGURE 16. Stabilities of ADP-ribosyl histone H2B bond in various pH values (upper) or neutral hydroxylamine (lower).

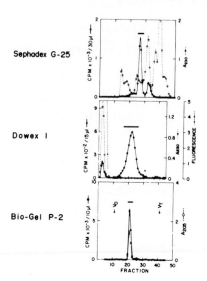

FIGURE 17. Purification of 5-phosphoribosyl tryptic peptide. Bars indicate pooled fractions.

purified with columns of Sephadex G-25, Dowex 1 and Bio-Gel P-2 (Fig. 17). The final peptide, obtained with an overall yield of 42%, contained equimolar [^{14}C]ribose 5-phosphate.

The amino acid sequence of this peptide was Pro-Glu-Pro-Ala-Lys, as determined by a dansyl-Edman technique and carboxypeptidase digestion. This sequence coincided exactly exactly with the amino-terminal pentapeptide of three known histone H2B's (calf thymus, trout testis and human spleen). Further digestion with carboxypeptidases revealed that ribose 5-phosphate was bound to somewhere in the terminal tripeptide, Pro-Glu-Pro. From this amino acid composition, it seemed most plausible that the glutamate residue was the modification site.

Stabilities of phosphoribosyl peptide bond were identical with those of original ADP-ribosyl histone H2B. Furthermore, an estimation of electric charge of ribosyl tripeptide, prepared by phosphatase treatment, revealed that it had no net charge at pH 6.5.

All these results were in accord with the view that γ-carboxyl group of the glutamate residue in the second amino-terminal position formed an ester bond with ribose. The final confirmation of this structure is in progress.

In addition to histone H2B, we found glutamate at ADP-ribosylation sites of histone H1. Preliminary results indicated that there were three ADP-ribosylation sites on histone H1, and all isolated phosphoribosyl peptides contained glutamate residues; they were tentatively identified as Glu^2, Glu^{14}

and Glu^{116}. Their chemical stabilities were identical to those of ADP-ribosyl H2B, supporting, again, the view of an ester bond.

It is of interest that all four ADP-ribosylation sites of histones H2B and H1 resided in so-called polar regions that have interaction with DNA. Poly ADP-ribosylation will presumably alter DNA-histone interactions.

Biological Significance of Poly(ADP-Ribose). A number of observations have so far been reported suggesting a close correlation of poly(ADP-ribose) synthesis to DNA replication, DNA repair, neoplastic transformation, cell proliferation or cell differentiation. None of the observations, however, have been conclusive, because of a lack of sensitive and direct methods to examine in vivo events. We applied our immunohistochemical procedure, described earlier, to this problem. Application to blood cell analysis is shown herein. In these experiments, all samples were preincubated with NAD after fixation to increase the poly(ADP-ribose) content.

Fig. 18 shows the immunostaining profile of normal human, peripheral blood. Lymphocytes and a monocyte emitted fluorescence, indicating the activity of poly(ADP-ribose) synthesis. In contrast, erythrocytes and polymorphonuclear leukocytes (granulocytes) were negative in the fluorescence. In the latter cells, the nuclei appeared darker than the cytoplasm, since the cytoplasmic granules emitted nonspecific fluorescence. The absence of poly(ADP-ribose) synthetase activity in granulocytes was confirmed by assaying the isolated nuclei from these cells. This type of cell is the first example lacking the polymer synthesis among eukaryotic cells.

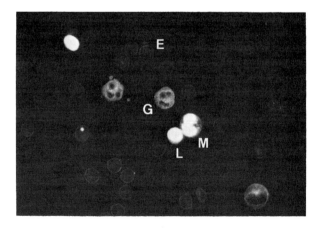

FIGURE 18. Immunostaining of human peripheral blood. E, erythrocyte; G, granulocyte; L, lymphocyte; M, monocyte.

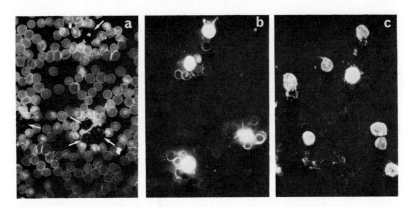

FIGURE 19. Immunostaining of peripheral blood of patients with chronic myeloid leukemia (a), acute myeloid leukemia (b), or chronic myeloid leukemia in a blastic crisis (c). Arrows indicate leukemic cells.

Fig. 19 shows the profiles of peripheral blood of patients with chronic or acute myeloid leukemia. The leukemic cells appearing in a chronic patient (Fig. 19a) was negative in fluorescence as in the case of normal granulocytes. In marked contrast, the leukemic cells appearing in an acute patient (Fig. 19b) mostly emitted fluoresecence. Whether this difference reflects different maturity of leukocytes or some fundamental difference between the two diseases is now under investigation. In this context, it was noteworthy that the leukemic cells of a patient with chronic myeloid leukemia in the phase of blastic crisis (Fig. 19c) emitted fluorescence of various intensities. It probably reflected the diversity of maturation or differentiation of proliferating cells.

Although much work has to be done before proper understanding of these findings, it is apparent that poly(ADP-ribose) synthetic capacity _in situ_ is closely related to differentiation, maturation or neoplastic transformation of leukocytes.

Investigation of various cells using both _in vitro_ and _in vivo_ approaches as developed in this study will help unveiling biological significance of this curious modification reaction in chromatin.

REFERENCES

1. Hilz, H., and Stone, P. (1976). In "Rev. Physiol. Biochem. Pharmacol." (R.H.Adrian, et al., eds.), Vol. 76, pp. 1-59. Springer-Verlag, New York.
2. Hayaishi, O., and Ueda, K. (1977). Annu. Rev. Biochem. 46, 95.
3. Honjo, T., Nishizuka, Y., Hayaishi, O., and Kato, I. (1968). J. Biol. Chem. 243, 3553.
4. Robinson, E. A., Hendriksen, O., and Maxwell, E. S. (1974). J. Biol. Chem. 249, 5088.
5. Van Ness, B. G., Howard, J. B., and Bodley, J. W. (1978). J. Biol. Chem. 253, 8687.
6. Iglewski, B. H., and Kabat, D. (1975). Proc. Nat. Acad. Sci. USA 72, 2284.
7. Iglewski, B. H., Sadoff, J., Bjorn, M. H., and Maxwell, E. (1978). Proc. Nat. Acad. Sci. USA 75, 3211.
8. Zillig, W., Mailhammer, R., Storko, R., and Rohrer, H. (1977). Curr. Top. Cell. Regul. 12, 263.
9. Goff, C. G. (1974). J. Biol. Chem. 249, 6181.
10. Pesce, A., Casoli, C., and Schito, C. G. (1976). Nature 262, 412.
11. Cassel, D., and Pfeuffer, W. (1978). Proc. Nat. Acad. Sci. USA 75, 2669.
12. Gill, D. M., and Meren, R. (1978). Proc. Nat. Acad. Sci. USA 75, 3050.
13. Moss, J., and Richardson, S. H. (1978). J. Clin. Invest. 62, 281.
14. Moss, J., and Vaughan, M. (1978). Proc. Nat. Acad. Sci. USA 75, 3621.
15. Ueda, K., Okayama, H., Fukushima, M., and Hayaishi, O. (1975). J. Biochem. 77, 1.
16. Okayama, H., Edson, C. M., Fukushima, M., Ueda, K., and Hayaishi, O. (1977). J. Biol. Chem. 252, 7000.
17. Miwa, M., Saikawa, N., Ohashi, Z., Nishimura, S., and Sugimura, T. (1979). Proc. Nat. Acad. Sci. USA in press.
18. Ueda, K., Kawaichi, M., Okayama, H., and Hayaishi, O. (1972). Biochem. Biophys. Res. Commun. 46, 516.
19. Okayama, H., Ueda, K., and Hayaishi, O. (1978). Proc. Nat. Acad. Sci. USA 75, 1111.
20. Ueda, K., Oka, J., Narumiya, S., Miyakawa, N., and Hayaishi, O. (1972). Biochem. Biophys. Res. Commun. 46, 516.
21. Miwa, M., Tanaka, M., Matsushima, T., and Sugimura, T. (1974). J. Biol. Chem. 249, 3475.
22. Futai, M., Mizuno, D., and Sugimura, T. (1968). J. Biol. Chem. 243, 6325.
23. Okayama, H., Honda, M., and Hayaishi, O. (1978). Proc. Nat. Acad. Sci. USA 75, 2254.

24. Oka, J., Okayama, H., Ueda, K., and Hayaishi, O. (1978). Seikagaku 50, 921.
25. Okayama, H., and Hayaishi, O. (1978). Biochem. Biophys. Res. Commun. 84, 755.
26. Kawaichi, M., Ueda, K., and Hayaishi, O. (1978). Seikagaku 50, 920.

SELECTIVE INACTIVATION AND DEGRADATION OF ENZYMES IN SPORULATING BACTERIA[1]

Robert L. Switzer, Michael R. Maurizi[2], Joseph Y. Wong and Kerry J. Flom

Department of Biochemistry, University of Illinois, Urbana, Illinois 61801

ABSTRACT A number of enzymes are selectively inactivated when microbial cells undergo a shift in nutritional state. Detailed studies of two illustrative cases are summarized in this article, namely, the inactivation of aspartate transcarbamylase and of glutamine phosphoribosylpyrophosphate amidotransferase in sporulating Bacillus subtilis. The two inactivation processes differ in mechanism. Aspartate transcarbamylase inactivation requires metabolic energy, while inactivation of amidotransferase involves reaction of oxygen with an iron-sulfur center. Immunochemical evidence is presented for degradation of both enzymes in vivo. Attempts to reconstruct the inactivation of these enzymes in vitro are described. While far from complete, studies of these systems contribute to an understanding of the mechanism and regulation of protein degradation in bacteria.

INTRODUCTION

Of the various mechanisms of modulation of protein function in living cells, the least well understood is enzyme degradation. Protein degradation occurs in virtually all types of cells. It appears that both the rate and selectivity of degradation are subject to regulation, but the manner in which this regulation is accomplished is obscure. We have found the bacteria to be useful experimental objects for studying these problems. Nearly all of the proteins of rapidly growing bacteria are stable, but certain proteins are selectively inactivated and/or degraded following a shift in nutritional state (1). For example, a number of the

[1]This work was supported by USPHS Grant No. AI 11121
[2]Present address: Laboratory of Biochemistry, NHLBI, Bethesda, Md. 20014

enzymes of de novo amino acid and nucleotide biosynthesis are rapidly inactivated when Bacillus subtilis cells are starved for an essential nutrient (1). Presumably, inactivation of these enzymes serves as part of the metabolic differentiation that occurs during transformation of vegetative cells into dormant endospores.

We have chosen to examine in detail the inactivation of two enzymes of nucleotide biosynthesis in stationary phase B. subtilis cells: aspartate transcarbamylase (ATCase, EC 2.1.3.2) and glutamine phosphoribosylpyrophosphate amidotransferase (amidotransferase, EC 2.4.2.14). It is hoped that a complete biochemical description of the mechanism and regulation of the inactivation of these enzymes will provide insight into the general problem of protein degradation. This paper summarizes our current progress toward these goals.

RESULTS

Characteristics of the Inactivation of Aspartate Transcarbamylase In Vivo. Following the initial report by Deutscher and Kornberg (2) of the disappearance of ATCase from sporulating B. subtilis cells, this process was studied in detail by Waindle and Switzer (3). ATCase was synthesized rapidly in cells growing exponentially on a glucose-nutrient broth medium and disappeared rapidly when the cells exhausted the glucose from the medium. During the inactivation the cells derive metabolic energy exclusively from oxidation of organic acids in the medium. Interference with the continuous generation of metabolic energy during this time prevented inactivation of ATCase, a property that is shared with protein degradation in many systems. The inactivation was initiated not only by conditions that lead to sporulation, such as glucose or ammonia starvation, but also when growth was prevented by antibiotic inhibitors of macromolecule biosynthesis or by other conditions that do not permit sporulation. The inactivation did not appear to require RNA or protein synthesis either prior to or during inactivation. The enzyme was not excreted from the cells, nor was it entrapped in a subcellular compartment or debris (3,4). ATCase synthesis has been shown to cease prior to inactivation (5), but inactivation does not simply result from degradation of a continually unstable enzyme, because ATCase is completely stable during exponential growth (3,4).

Relation of the Inactivation of Aspartate Transcarbamylase to Bulk Protein Turnover. It is well known that the rate of general protein degradation increases in starving bacteria, and, in particular, that intracellular protein is degraded at a rate of about 18% per hour in B. subtilis during sporulation (6). The relation between this general protein degradation and ATCase inactivation was examined in the experiments depicted in Fig. 1 (4). It was observed that

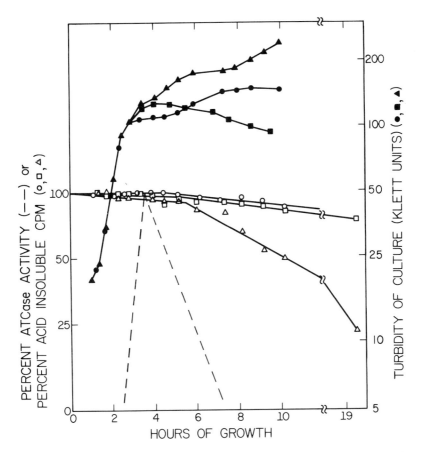

FIGURE 1. Degradation of intracellular protein and inactivation of ATCase in B. subtilis. Cells were labelled with [^{3}H]phenylalanine, washed, and transferred into fresh non-radioactive nutrient broth. Cell density (closed symbols) and trichloroacetic acid-insoluble radioactivity (open symbols) were followed in wild type (△, ▲), protease-deficient (○, ●), and wild type cells to which 5μg rifampin per ml was added at 3 h (□, ■). The activity of ATCase in a parallel control with wild type cells is shown with a dotted line. (From reference 4, reprinted with permission of the American Society of Biological Chemists, Inc.).

intracellular protein was degraded quite slowly during exponential growth and the first 2 to 3 hours of glucose starvation. Subsequently, the rate of protein degradation increased sharply. This increase was blocked by addition of rifampin and also did not occur in a mutant known to be deficient in a intracellular serine protease (7). However, the inactivation of ATCase was more than 50% complete by the time of the increase in protein degradation. Furthermore, ATCase inactivation was normal under the conditions that blocked increased protein degradation in the stationary phase. The rate of inactivation of ATCase was much faster than the overall rate of protein degradation. Since it will be shown below that inactivation and degradation of ATCase are indistinguishable, the results show that ATCase degradation is a selective process that appears to involve a degrading system which is distinct from that involved in bulk protein turnover.

Immunochemical Studies of Aspartate Transcarbamylase Inactivation In Vivo. Since studies of the disappearance of ATCase activity gave no information about the fate of the enzyme protein, Maurizi et al. used immunochemical approaches to determine whether inactivation involved, or was followed by, degradation (4,8). The primary technique used to determine the amount of ATCase protein present in crude extracts was specific immunoprecipitation from extracts of cells labeled with tritiated amino acids, followed by electrophoretic analysis of the immunoprecipitates in SDS-containing polyacrylamide gels. The specificity of precipitation is documented in Fig. 2. The only bacterial protein that was precipitated in siginificant quantities was ATCase. No radioactive protein was precipitated when the experiment was repeated with cells grown in the presence of uracil to repress ATCase. Fig. 2 also shows that immunoprecipitable material that was cross-reactive with ATCase disappeared as the enzyme was inactivated in vivo and that no precipitable smaller peptides were detected. The ability of the antibodies to precipitate denatured and chemically modified, inactive forms of ATCase was documented (4,8). Fig. 3 presents the results of a complete study such as shown in Fig. 2. It is evident that the loss of ATCase activity and cross-reactive protein were simultaneous. This conclusion was confirmed by microcomplement fixation analysis of uncentrifuged extracts and by activity neutralization assays (4,8). Two additional immunochemical procedures were shown to be more effective than those described above at detecting proteolytic fragments of ATCase. These were immunoprecipitation with antibody directed against denatured ATCase and double immunoprecipitation with goat anti-rabbit IgG antibody (4,8). With the latter

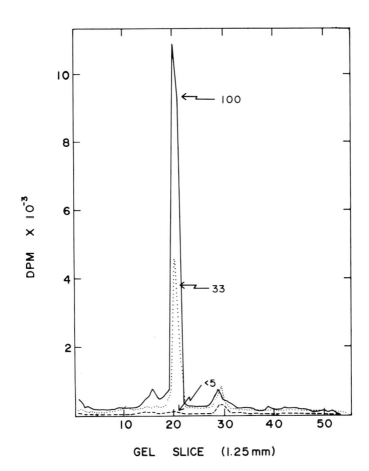

FIGURE 2. Direct immunoprecipitation of ATCase during inactivation in vivo. Wild type B. subtilis cells were grown on glucose-nutrient broth containing mixed tritiated amino acids. Samples were harvested at various times during the inactivation of ATCase, extracts were prepared and incubated with antiATCase antibody. Immunoprecipitates were washed and analyzed by electrophoresis in SDS-containing polyacrylamide gels. The profiles of radioactivity are shown from cells that were harvested when ATCase was 100%, 33%, and 5% of the maximal activity.

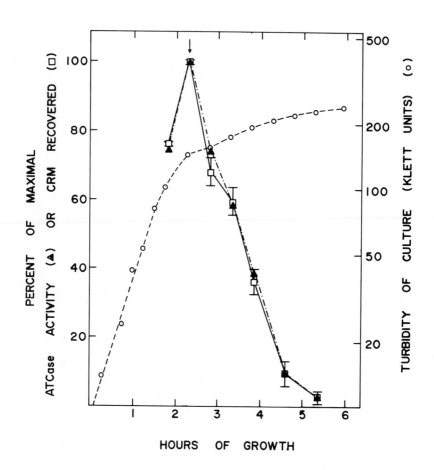

FIGURE 3. Comparison of the loss of ATCase activity and cross-reactive protein during inactivation in vivo. The experiment was performed as in Fig. 2. Each extract was supplemented with pure non-radioactive ATCase to 7 µg/ml to ensure constant efficiency of precipitation. (□), radioactivity in the immunoprecipitate, as resolved on SDS gels; (▲) ATCase catalytic activity; (---) cell density. From reference 4, reprinted with permission of the American Society of Biological Chemists, Inc.

MODULATION OF PROTEIN FUNCTION

technique some evidence for the transient formation of a cross-reactive peptide of molecular weight (25,000) smaller than native ATCase (33,000) was obtained, but identification of this peptide as a fragment of ATCase requires further documentation. As pointed out above, the use of conditions designed to decrease the rate of intracellular protein degradation generally did not affect the rate of loss of protein that was cross-reactive with ATCase.

Although the case could only be made conclusive by the unequivocal identification of a peptide fragment of ATCase that was formed during its inactivation in vivo, the results strongly suggest that ATCase inactivation involves, or is immediately followed by degradation.

Attempts to Reconstruct the Inactivation of Aspartate transcarbamylase In Vitro. ATCase is quite stable when the enzyme is incubated with crude extracts of cells harvested during the inactivation. Supplementation of the extracts with a variety of metabolites did not alter this stability (3,9). This situation has frustrated attempts to reconstruct the inactivation of ATCase in vitro. The availability of the highly purified enzyme (10) and the development of procedures for labeling native ATCase with [^{14}C]HCHO (4) have permitted further experiments directed toward this goal. Two interesting systems have been found, but it is far from certain that either reconstructs portions of the inactivation process as it occurs in vivo.

Purified ATCase was inactivated and degraded by a variety of proteases, including the major extracellular protease of bacilli, subtilisin (8). The enzyme was protected by the substrate carbamyl phosphate. We have not succeeded in demonstrating proteolysis of ATCase by crude extracts of B. subtilis cells, but the enzyme was attacked by an extracellular serine protease that accumulated in the culture fluid during the stationary phase of growth (8). This activity rapidly converted ATCase from its native subunit molecular weight of 33,000 to a value about 1,500 daltons smaller. This "nicking" activity could only be demonstrated on slab gels containing SDS and a gradient of polyacrylamide from 7.5 to 15%. Longer incubation resulted in further degradation of the "nicked" ATCase. The "nicking" action of the extracellular protease could be inhibited by crude cell extracts -- even after boiling, centrifugation and dialysis -- but not by 10 mg bovine serum albumin per ml. These results suggest the presence of a heat-stable protease inhibitor in B. subtilis extracts similar to that reported by Millet (11). The inhibitor reported by Millet, however, is specific for tho intracellular serine protease. Possibly, our failure to detect degradation of ATCase by intracellular activity results from the presence of such a protease inhibitor. It must be pointed out that the "nicked" ATCase has full catalytic activity, has never been detected in immunoprecipitates of ATCase during inactivation in vivo, and has so far been detected only after incubation with extracellular activity. Thus, it is unlikely -- although still possible -- that the "nicked" species is an intermediate in the physiological inactivation process.

Very recently, Kerry Flom discovered that ATCase can be inactivated in vitro by two systems derived from B. subtilis cells. The first system was a preparation of washed membrane vesicles, and the second was found in the soluble fraction of the cells. Both systems had an absolute requirement for O_2 and NADH (or NADPH) and were stimulated by azide. These systems brought about apparently irreversible inactivation of ATCase without significant degradation of the protein. The soluble system has been extensively purified by Flom. The best preparations contained a highly purified ($\geq$ 90%) protein, which has the spectrum of a flavoprotein and a native molecular weight of about 75,000 (subunit molecular weight of 35,000). This system was no longer stimulated by azide and was saturated by 2-4 mM NADH. The enzyme catalyzed formation of H_2O_2 from NADH and O_2. Inactivation of ATCase was inhibited by catalase, but not by superoxide dismutase or by hemoglobin. These observations point to a role for H_2O_2 in the inactivation of ATCase, but H_2O_2 did not replace NADH in the process. However, 2 mM NADH and H_2O_2 together brought about substantial inactivation in the absence of the flavoprotein. A great deal more must be learned about the chemistry of the inactivation of ATCase by NADH and H_2O_2 in the absence and presence of the flavoprotein before we can decide whether it plays a role in the inactivation and degradation of ATCase in vivo. Clearly, in an area such as this, one must expect the unexpected!

Inactivation of Glutamine Phosphoribosylpyrophosphate Amidotransferase: An Oxygen Dependent Process. The inactivation of ATCase at the end of exponential growth of B. subtilis cells suggested that the first enzyme of purine biosynthesis, which we will call simply "amidotransferase," would also be inactivated at the same time. Turnbough showed this to be the case (12). Amidotransferase was synthesized and was stable in exponentially growing cells, but disappeared at about the same time and rate as ATCase. In this case, however, the inactivation in vivo was specifically blocked by anaerobiosis. Other means of inhibiting energy metabolism did not block the inactivation. Thus, a specific requirement for O_2 was indicated (12).

Turnbough also showed that amidotransferase was susceptible to O_2 dependent inactivation in crude extracts (13). This inactivation was observed with the enzyme from all stages of growth, was not reversed by reducing agents, and was not blocked by catalase, superoxide dismutase or exclusion of light. The nature of the inactivation has been clarified by recent studies of Wong et al. with highly purified ($\geq$ 98% pure) amidotransferase (14,15). The purified enzyme retained its sensitivity to O_2, but was quite stable when protected from O_2 (Fig. 4). Thus, no O_2 dependent inactivaing protein appears to be involved. The unexpected discovery that amidotransferase is an iron-sulfur protein (14) has provided an explanantion for the O_2 sensitivity of the enzyme:

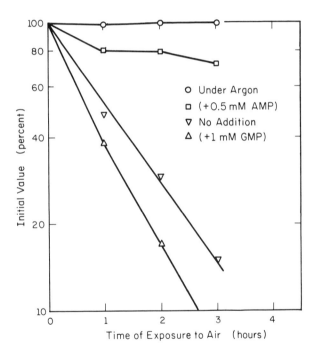

FIGURE 4. Inactivation of purified amidotransferase by reaction with O_2. A sample of 0.5 mg amidotransferase per ml of 50 mM Tris/HCl - 10 mM $MgCl_2$ - 0.1 mM EDTA, pH 7.9 was exposed to air at 37° under the conditions shown. At the times shown samples were removed and assayed for amidotransferase activity (15).

inactivation by O_2 results from oxidation of the iron-sulfur center. The data presented in Fig. 5 describe the changes in amidotransferase that accompany oxidative inactivation of the purified enzyme. The loss of activity correlated well with oxidation of enzyme-bound S^{2-} to unidentified products and partial bleaching of the yellow-brown chromophore. All of the iron in the enzyme was eventually oxidized to Fe^{3+}, but inactivation was virtually complete by the time 2/3 of the iron was oxidized. The oxidized iron remained bound to the protein. The oxidized amidotransferase had a much lower solubility than the native enzyme. It behaved on gel filtration as though it were highly aggregated, and -- unlike native amidotransferase -- it was precipitated by 10 mM $MgCl_2$. These observations indicate that the oxidized protein has undergone significant changes in tertiary and quaternary structure. We speculate that such changes may lead to enhanced susceptibility to proteolysis, but we have not yet tested this idea.

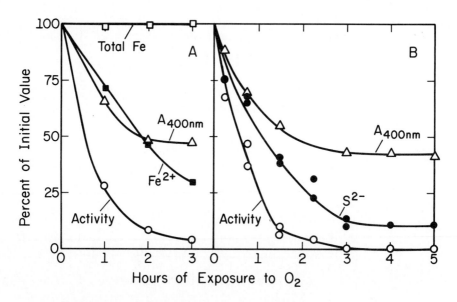

FIGURE 5. Reaction of the iron-sulfur center in amidotransferase with O_2. Samples of purified amidotransferase in the buffer described in Fig. 4 were flushed with H_2O-saturated O_2, stoppered, and incubated at 37°. Samples were withdrawn as indicated, flushed again, and incubated further. Total Fe, Fe^{2+}, S^{2-}, and activity were assayed as in reference 14. (From reference 17, reprinted with permission of the U.S. Government Printing Office.)

A complete description of the reaction of amidotransferase with O_2 must await a better understanding of the structure of the iron-sulfur center in the enzyme. The occurrence of such a center in an amidotransferase represents a novel catalytic function for iron-sulfur proteins. This and a number of other properties have led us to suggest that amidotransferase may contain a new type of iron-sulfur center. In particular, many preparations of the enzyme appear to contain 3 atoms of iron and 2 atoms of inorganic sulfide per subunit (14). Very recently, however, a preparation with higher specific activity and 3.3 atoms of iron per subunit was obtained by Steven Vollmer in our laboratory. This result and the results of Mössbauer spectroscopy (unpublished experiments conducted in collaboration with Drs. Dwivedi and P. Debrunner, Department of Physics, University of Illinois) indicate that a more conventional Fe_4S_4 center is possible.

A possible means for regulation of the reaction of amidotransferase with O_2 was provided by the discovery of Turnbough (13) that substrates and allosteric inhibitors strongly affect the rate of inactivation. Two examples are shown in Fig. 4. The allosteric inhibitor AMP strongly stabilized the enzymes, while the inhibitor GMP destabilized it. Both effects were seen at inhibitor concentrations below those needed to inhibit catalytic activity fully. These antagonistic effects of AMP and GMP may reflect their effects on the quaternary structure of amidotransferase: AMP stabilizes a dimeric form, while GMP stabilizes a tetramer. Of possibly greater significance is the observation that the substrates glutamine and PRPP together stabilized the enzyme, but neither alone was effective (13). It certainly is easy to see how such effects could mediate inactivation of amidotransferase during carbon or nitrogen starvation. More detailed studies of kinetic (16) and stabilizing effects of ligands on purified amidotransferase are in progress.

Immunochemical Evidence for the Proteolysis of Inactivated Amidotransferase In Vivo. Is the O_2 dependent inactivation of amidotransferase followed by proteolysis of the inactive protein? The question may be answered by immunochemical techniques like those already described above for study of ATCase degradation. Wong (15) has prepared antibodies directed against purified amidotransferase and has documented their ability to precipitate the enzyme specifically from extracts of cells labeled with [^{3}H]leucine. The antibodies precipitated both native and O_2-inactivated enzyme efficiently. Thus, it was possible to determine whether inactivated amidotransferase accumulates in the cells during inactivation. The results of such an experiment are shown in Fig. 6. It will be seen that amidotransferase activity and cross-reactive protein increased in parallel during exponential growth and that both declined in the stationary phase. However, catalytic activity declined faster than did the immunoprecipitable protein of native subunit molecular weight. Late in the inactivation as much as half of the precipitated protein was catalytically inactive. It is likely that this inactive protein is oxidized enzyme, although this

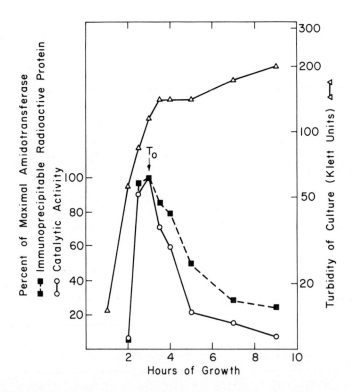

FIGURE 6. Loss of amidotransferase activity and cross-reactive protein during inactivation in vivo. B. subtilis cells were grown in the presence of [^{3}H]leucine, and samples were harvested at the times indicated. Amidotransferase activity was determined as in reference 14, and immunoprecipitates were collected and analyzed by electrophoresis on SDS-containing polyacrylamide gels. Each extract was supplemented with pure amidotransferase to 15-18 µg/ml to ensure constant efficiency of precipitation. (■) radioactivity in amidotransferase region of gels; (O) amidotransferase activity; (△) cell density.

point is not certain. More importantly, both active and inactive amidotransferase eventually disappeared. No smaller protein fragments were detected on SDS gels of immunoprecipitates. We believe that it is likely that the disappearance of inactive amidotransferase results from proteolysis, but a number of further experiments are in progress to test this conclusion.

If proteolysis of inactive amidotransferase can be demonstrated more convincingly, the way will be clear to characterizing the degradation more thoroughly. Does it have a dependence on metabolic energy, for instance? It is already known that the disappearance of protein cross-reactive with amidotransferase proceeds normally in the mutant (7) which is deficient in bulk protein turnover and intracellular serine protease. Another major goal is to reconstruct the degradation of O_2-inactivated purified amidotransferase in vitro.

DISCUSSION

A number of experimental approaches have been taken to the study of protein degradation in bacteria. The overall features of degradation and its regulation have been revealed by determining the effects of nutritional state, inhibitors of metabolism, and mutational loss of specific biochemical functions on the release of radioactive amino acids from total cellular protein (reviewed in references 18,19,20). Another valuable approach has been to examine the effects of mutational loss of specific proteases and peptidases, especially when multiple mutations are selected in a single strain (for example, see references 21 and 22). There are two possible limitations to these methods. First, they may yield more information about the degradation of already inactive proteins and degradation of peptides than about the regulation of the initial inactivating or proteolytic step of degradation. Second, if protein degradation turns out to be a mechanistically heterogeneous process, the results may be made very complicated by the averaging out of different effects on individual proteins when bulk protein degradation is followed. For these reasons, we have chosen to take the much more laborious approach of identifying specific proteins that are inactivated under well-defined physiological conditions. The fate of these proteins has been followed in in vivo experiments by measuring catalytic activity and by immunochemical procedures. The immunochemical procedures can provide clear evidence for the degradation of specific proteins in vivo, if degraded fragments are detected or controls have been performed to demonstrate that other possible reasons for failure to detect cross-reactive material have been eliminated. After characterization of such an inactivation and degradation process, the purified enzyme, especially in a radioactively labeled form, can be used to attempt reconstruction of the process in vitro. Clearly, the reconstructed process must possess qualities that reflect the physiological requirements and regulation of the process in vivo. A fully convincing demonstration that a reconstructed inactivation/degradation occurs in vivo may rely on isolation and characterization of mutants that fail to carry out the inactivation.

Clearly, the experiments presented here on the inactivation of ATCase and amidotransferase in sporulating B. subtilis cells do not as yet meet the ambitious goals outlined above. We believe, however, that they offer a promising beginning. The general physiological requirements of the two inactivation processes have been determined and shown to be significantly different. Indeed, it may well develop that many mechanisms of enzyme inactivation will be found (1). Both enzymes have been purified, and immunochemical evidence for their degradation in vivo has been obtained. Further evidence is desirable to establish degradation conclusively, especially in the case of amidotransferase. The reconstruction of ATCase inactivation and degradation in vitro has proven elusive. This is a most important objective, however, because clarification of the requirement for metabolic energy in this system might have implications for many other systems in which energy-dependent protein degradation has been reported (19,20). In the case of amidotransferase, inactivation in vivo and in vitro requires O_2, rather than metabolic energy. We believe that this inactivation process can be reconstructed in vitro. We are currently studying the chemical and physical details of this inactivation. We are especially interested in learning how it may be regulated by the metabolites that bind to amidotransferase. Experiments to study the second step of amidotransferase inactivation -- degradation -- in an in vitro system are just beginning. While it is too early to draw generalizations from our studies, certainly it is our hope that the findings will have implications for the study of protein degradation in many types of cells.

REFERENCES

1. Switzer, R. L. (1977). Annu. Rev. Microbiol. 31, 135.
2. Deutscher, M. P., and Kornberg, A. (1968). J. Biol. Chem. 243, 4653.

3. Waindle, L. M., and Switzer, R. L. (1973). J. Bacteriol. 114, 517.
4. Maurizi, M. R., Brabson, J. S., and Switzer, R. L. (1978). J. Biol. Chem. 253, 5585.
5. Maurizi, M. R., and Switzer, R. L. (1978). J. Bacteriol. 135, 943.
6. Spudich, J. A., and Kornberg, A. (1968). J. Biol. Chem. 243, 4600.
7. Hageman, J. H., and Carlton, B. C. (1973). J. Bacteriol. 114, 612.
8. Maurizi, M. R. (1978) Ph.D. thesis, University of Illinois, Urbana.
9. Waindle, L. M. (1974) Ph.D. thesis, University of Illinois, Urbana.
10. Brabson, J. S., and Switzer, R. L. (1975). J. Biol. Chem. 250, 8664.
11. Millet, J. (1977). FEBS Lett. 74, 59.
12. Turnbough, C. L., Jr., and Switzer, R. L. (1975). J. Bacteriol. 121, 108.
13. Turnbough, C. L., Jr., and Switzer, R. L. (1975). J. Bacteriol. 121, 115.
14. Wong, J. Y., Meyer, E., and Switzer, R. L. (1977). J. Biol. Chem. 252, 7424.
15. Wong, J. Y. (1978) Ph.D. thesis, University of Illinois, Urbana.
16. Meyer, E., and Switzer, R. L. (1979). J. Biol. Chem. in press.
17. Switzer, R. L., Maurizi, M. R., Wong, J. Y., Brabson, J. S., and Meyer, E. (1979) "Limited Proteolysis in Microorganisms -- Proceeding of Fogarty International Conference" (G. Cohen and H. Holzer, eds.) U.S. Government Printing Office, in press.
18. Pine, M. J. (1972). Annu. Rev. Microbiol. 26, 103.
19. Goldberg, A. L., and Dice, J. F. (1974). Annu. Rev. Biochem., 43, 835.
20. Goldberg, A. L., and St. John, A. C. (1976). Annu. Rev. Biochem. 45, 747.
21. Miller, C. G. (1975). Annu. Rev. Microbiol. 29, 485.
22. Heiman, C., and Miller, C. G. (1978). J. Bacteriol. 135, 588.

ENDOGENOUS PROTEOLYTIC MODULATION OF YEAST ENZYMES

Heidrun Matern and Helmut Holzer

Biochemisches Institut der Universität Freiburg (D-7800 Freiburg, Germany) and Institut für Toxikologie und Biochemie der Gesellschaft für Strahlen- und Umweltforschung (D-8042 Neuherberg bei München, Germany)

ABSTRACT The proteolytic system in yeast consists of at least 7 intracellular proteinases and 3 specific macromolecular proteinase inhibitors. The proteinases have been shown to play a role in the regulation of yeast enzymes. The following 4 types of proteolytic processes are discussed by which selected yeast enzymes might be regulated: 1) activation of inactive proforms of chitin synthase and carboxypeptidase Y by limited proteolysis; 2) activation of the inactive inhibitor complexes of the proteinases A and B and of carboxypeptidase Y by proteolytic degradation of the respective inhibitors; 3) inactivation of enzymes by proteolytic degradation as shown for the glucose-induced "catabolite inactivation" of cytoplasmic malate dehydrogenase and for the glucose starvation-promoted inactivation of NADP-dependent glutamate dehydrogenase; 4) irreversible glucose-effected decrease in the affinities of the galactose uptake system and the maltose uptake system for galactose and maltose, respectively.

INTRODUCTION

Intracellular yeast proteinases have been described long ago by Dernby in 1917 (1) and by Willstätter and Grassmann in 1926 (2). However, only in the last ten years substantial progress has been achieved in the characterization of the proteolytic system in yeast, consisting of the intracellular proteinases and the specific macromolecular proteinase inhibitors. At present, at least seven different intracellular proteinases are known: two endopeptidases, the proteinases A and B, two carboxypeptidases, the carboxypeptidases Y and S, and at least three aminopeptidases (for a summary, see references (3,4)). Three classes of polypeptides specifically inhibiting proteinase A, proteinase B and carboxypeptidase Y were found and described in the laboratories of Lenney (5,6), Cabib (7) and Holzer (8-13). The inhibitors are localized in the cytosol (14,15), whereas most of the proteinases are found in the vacuoles of the yeast cell (16).

In regard to the function of the intracellular yeast proteinases a general role in the regulation of protein levels has been recognized (for a recent review, see reference (4)), since it is obvious that the turnover of proteins (17) is the result of a complicated interplay of synthetic reactions and degradative processes (18). In the course of studies on the regulation of tryptophan synthase in yeast (19,20) the interest of our laboratory was attracted to a more specific role of the intracellular yeast proteinases in the regulation of yeast enzymes. In the present paper, some proteolytic processes in yeast are discussed which can be correlated with specific events occuring in vivo and may therefore be of biological significance in the regulation of yeast enzymes.

ACTIVATION OF PROENZYMES

Proteinases have been shown to be involved in the maturation of enzymes by converting inactive proenzymes (zymogens) into the biologically active products (21). The activation of chitin synthase (22,23) and carboxypeptidase Y (24,25) are examples for the maturation of enzymes from inactive proenzymes by limited proteolysis in yeast.

Chitin synthase participates in the formation of the septum between mother and daughter cells and the budscar in budding yeast. Cabib and coworkers demonstrated that the enzyme is attached as inactive precursor to the yeast plasma membrane (26). In vitro studies showed that the activation of chitin synthase is a proteolytic process (22,23), catalyzed in vitro by proteinase B (27,28). A heat-stable protein, inhibiting the activation in vitro, was identified as proteinase B-inhibitor (27,28). The finding that proteinase A activates inhibitor-bound proteinase B by proteolysis of the inhibitor (10,29) and that furthermore the inactive proteinase A-inhibitor complex can be activated by lowering the pH (8,30) suggests a cascade mechanism of activation of chitin synthase as shown in Figure 1. In vivo, the activation of chitin synthase might occur through the action of an as yet unknown proteinase since recently isolated yeast mutants lacking proteinase B (31,32) showed a normal budding behaviour (32) and a normal synthesis of chitin (31).

Hasilik and Tanner (24,25) showed that a precursor protein of carboxypeptidase Y exists of about 6 000 Daltons higher molecular weight as compared to carboxypeptidase Y.

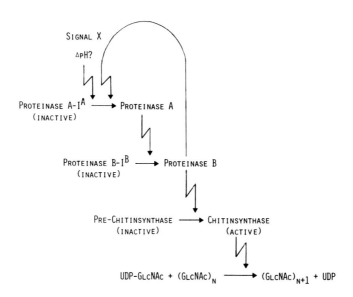

FIGURE 1. Cascade mechanism of activation of chitin synthesis. I^A, proteinase A-inhibitor; I^B, proteinase B-inhibitor.

Trypsin and proteinase B from yeast were able to convert the precursor protein to the mature enzyme by limited proteolysis (25). In contrast to carboxypeptidase Y which is localized in the vacuoles (16) the precursor protein has been found to be associated with a membranous fraction (24). It is assumed that the precursor is an inactive proenzyme which is required during the transport from the site of synthesis to the vacuole where it is proteolytically activated (25).

ACTIVATION OF INHIBITOR COMPLEXES

Proteinases not only activate inactive proenzymes (zymogens) but also inactive complexes of proteinases and their endogenous inhibitors by limited proteolysis. Such an activation mechanism (29,33) might be of biological significance in the regulation of enzyme activity as discussed above for the activation of chitin synthase (see Figure 1).

Studies on the regulation of proteinases and inhibitors during growth showed a parallel derepression of proteinase and inhibitor activities upon glucose exhaustion of the cells (29,34). Since furthermore the inhibitors are always found in a surplus over the proteinases (29), inactive proteinase-inhibitor complexes would be formed whenever the proteinases leave the vacuolar compartment.

In vitro experiments showed that activation of the inactive proteinase-inhibitor complexes can be started in crude yeast extracts by adjusting the pH to around 5, which liberates small amounts of proteinase A from its inactive complex (8,35). Proteinase A activates inhibited carboxypeptidase Y and proteinase B by hydrolysis of the respective inhibitors, and the active proteinase B itself digests the proteinase A- and carboxypeptidase Y-inhibitors, thereby leading to more free proteinase A and carboxypeptidase Y (33).

The interaction of the proteolytic components of a yeast crude extract was investigated more detailed with the purified proteins. The inactivation of the proteinase B-inhibitor $I^B 2$ by proteinase A occurs by a strictly limited proteolytic process (37). As shown in Figure 2, after 30 minutes of incubation with proteinase A $I^B 2$ has completely lost its inhibitor activity and two new peptides appear in disc gel electro-

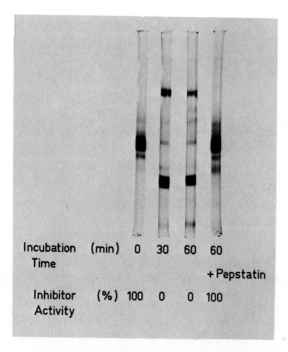

FIGURE 2. Anionic disc gel electrophoresis (pH 9.5) of $I^B 2$ incubated at 25° with proteinase A in the absence and presence of the acid proteinase inhibitor pepstatin for the indicated time intervals. After 30 minutes an additional amount of proteinase A was added to the incubation mixture. After the respective incubation times proteinase B-inhibitor activities were measured by the Azocoll method (19).

phoresis. Another incubation for 30 minutes with additional proteinase A causes no further proteolysis as indicated by the unchanged band pattern in disc gel electrophoresis. The two cleavage products exhibit molecular weights of 5 900 and 3 300 in sodium dodecyl sulfate gel electrophoresis. Further studies on the properties of these two peptides formed from I^B2 by proteinase A are in progress.

Studies on the purified carboxypeptidase Y-inhibitor complex showed that proteolytic inactivation of the inhibitor is only possible after its dissociation from carboxypeptidase Y by a peptide substrate (36). From many peptide substrates tested only a few peptides were able to dissociate the inhibitor from carboxypeptidase Y. This finding suggests that metabolic formation of certain peptides might trigger and control the activation of the inactive inhibitor complexes of proteinases.

INACTIVATION OF ENZYMES

In yeast, there is evidence that proteinases participate in the regulation of metabolism by degradation of selected enzymes. Glucose induced catabolite inactivation (38) and inactivation of enzymes upon carbon starvation (39-42) are probably examples for such a selective proteolytic degradation of yeast enzymes dependent on a change in the metabolic conditions. These inactivation mechanisms have the common characteristic to be irreversible processes. In all cases of enzymes so far examined (38,40-43) reactivation is prevented by cycloheximide or other inhibitors of protein synthesis suggesting that de novo protein synthesis is necessary for reestablishment of the active enzymes.

Addition of glucose to yeast cells grown on acetate or ethanol causes a rapid disappearance of certain enzyme activities. This phenomenon has been called "catabolite inactivation" (38). The enzymes known at present to be subject to catabolite inactivation are listed in Table 1. Good evidence that proteolysis is responsible for the mechanism of catabolite inactivation has been obtained for cytoplasmic malate dehydrogenase (49). Using immunological techniques, Neeff et al. (49) demonstrated that the amount of precipitable enzyme disappeared parallel to catalytic activity following addition of glucose to yeast cells. Furthermore, in vitro studies showed that cytoplasmic malate dehydrogenase, in contrast to its mitochondrial isoenzyme, is extremely sensitive against proteolytic inactivation by the yeast proteinases A and B (50).

An inactivation of enzymes upon glucose starvation has been reported for malate dehydrogenase from Schizosaccharomyces pombe (39), glutamine synthetase from Candida utilis

TABLE 1
GLUCOSE-EFFECTED INACTIVATION AND IRREVERSIBLE MODIFICATION OF ENZYMES

I   Catabolite Inactivation

       Cytoplasmic malate dehydrogenase (44)
       Fructose 1,6-bisphosphatase (45)
       δ-Aminolevulinate synthetase (43)
       Uroporphyrinogen synthetase (43)
       Phosphoenolpyruvate carboxykinase (46,47)
       Uridine nucleosidase (48)

II   Catabolite Modification

       Galactose uptake system (51)
       α-Glucoside (maltose) permease (52,53)

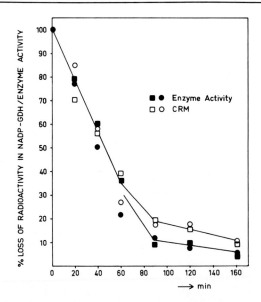

FIGURE 3. Decrease of activity of NADP-dependent glutamate dehydrogenase (NADP-GDH) and radioactively labeled NADP-GDH cross-reacting material (CRM) following transfer of Candida utilis cells to glucose starvation medium. Modified from Hemmings (41).

(40), and NADP-dependent glutamate dehydrogenase from Candida utilis (41) and from Saccharomyces cerevisiae (42). In the case of glucose starvation-promoted inactivation of NADP-dependent glutamate dehydrogenase from Candida utilis

Hemmings (41) obtained strong evidence for a proteolytic mechanism of inactivation. As depicted in Figure 3 the rapid decrease in enzyme activity upon glucose starvation was shown to be paralleled by a loss of antigenically precipitated radioactively labeled enzyme.

MODIFICATION OF ENZYMES

As discussed in the preceeding chapter, proteolytic "inactivation" of enzymes is the consequence of total degradation and therefore results in a complete loss of enzyme activity. In contrast proteolytic "modification" causes only a change in the properties of an enzyme without complete loss of enzyme activity. The glucose-effected decrease in the activities of the galactose uptake system (51) and the maltose uptake system (52,53) are examples for a regulation by modification of enzymes (Table 1). When assayed at low substrate concentrations the glucose induced decrease in activities of the two uptake systems is almost complete. Therefore the two uptake systems had originally been ranged in the group of enzymes which are subject to catabolite inactivation (38,51).

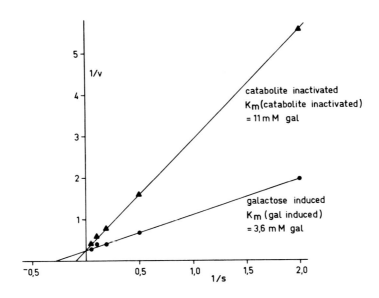

FIGURE 4. Lineweaver-Burk plot of the dependence of the activity of the galactose uptake system on galactose concentration after treatment (4 h) of galactose-grown yeast cells with (▲-▲) and without (●-●) glucose (51).

However, in contrast to cytoplasmic malate dehydrogenase (44), fructose 1,6-bisphosphatase (45), δ-aminolevulinate synthetase (43), uroporphyrinogen synthetase (43), phosphoenolpyruvate carboxykinase (46,47), and uridine nucleosidase (48) (Table 1), addition of glucose to yeast cells does not lead under all conditions to a complete loss of activity of the two uptake systems but only to a decrease in the affinity of the two systems for galactose and maltose, respectively, without effect on the maximal rates of uptake at substrate saturation (51,53). As shown in Figure 4, addition of glucose to galactose-grown cells caused an increase in the apparent K_m of the galactose uptake system from 3.6 to 11 mM galactose whereas V_{max} remained constant (51). Very similar results have been observed earlier in studies by Görts (53) with the maltose uptake system. For this phenomenon the term "catabolite modification" is proposed to reflect the similarity to "catabolite inactivation" which is also caused by glucose and/or its catabolites on the one hand and to differentiate from total inactivation on the other hand.

A proteolytic mechanism of catabolite modification is supported by the observation that reestablishment of the high affinity form of the galactose uptake system and the maltose uptake system in a glucose-free medium is inhibited by cycloheximide suggesting that reactivation of the two uptake

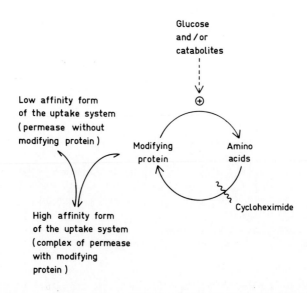

FIGURE 5. Proposed mechanism for the glucose-effected "catabolite modification" of the galactose and maltose uptake systems, respectively.

systems is dependent on de novo protein synthesis (51,53). Examples for a change in the K_m of an enzyme by limited proteolysis are well known, however, reversal of limited proteolysis by de novo protein synthesis would be hard to understand. Therefore, as shown in Figure 5, we assume that a "modifying protein" which increases the affinity of the system by protein-protein interaction participates in control of the uptake system. If glucose or its catabolites dissociate the modifying protein from the high affinity form of the uptake system and thereby render this protein susceptible to proteolysis one could understand that reestablishment of the high affinity form of the uptake system is dependent on de novo synthesis of protein, i.e. modifying protein.

CONCLUSIONS

In yeast, there is strong evidence that proteinases play a role in the regulation of metabolism by activation, inactivation or modification of selected enzymes. This evidence has been obtained in studies on specific events occuring in vivo such as activation of chitin synthase in budding yeast (22, 23), inactivation of NADP-dependent glutamate dehydrogenase at release from glucose protection (41), and glucose induced catabolite inactivation (38) and catabolite modification (51, 53) of certain enzymes.

In the last ten years a variety of proteolytic in vitro effects on yeast enzymes have been reported (for a summary, see reference (54)) which might also be of biological significance in vivo. Several yeast enzymes have been shown to be contaminated by tightly bound proteinases even in highly purified preparations as in the cases of alcohol dehydrogenase (55) and phosphoenolpyruvate carboxykinase (56). These tightly bound proteinase-enzyme complexes are perhaps not only artifacts which occur upon disruption of the vacuoles during the preparation of crude extracts but might also be of biological significance in the intracellular regulation of enzymes by proteinases, since some of these enzymes have been shown to be specifically modified by the action of proteinases, e.g. pyruvate decarboxylase (57) and phosphofructokinase (58).

Little is known at present about the mechanisms of control of proteinase action. Specific endogenous proteinase inhibiting polypeptides seem to play a role in the control of some of the proteinases. Concerning the mechanism(s) of selection of enzymes for proteolytic degradation an effector-mediated specific uptake into the vacuoles (which contain most of the proteinases) has been discussed for cytoplasmic enzymes (59). In regard to proteolytic modulation of membrane-bound enzymes such as chitin synthase, the galactose uptake system and the maltose uptake system, fusion of small vacuoles with the plasma membrane might precede the mechanisms of activation or modification. This hypothesis is supported by the finding of Wiemken (60) who reported that the large vacuoles divide into numerous small vesicles at bud initiation.

ACKNOWLEDGMENTS

This work was supported by the Deutsche Forschungsgemeinschaft (SFB 46) and the Fonds der Chemischen Industrie.

REFERENCES

1. Dernby, K. G. (1917). Biochem. Z. 81, 107.
2. Willstätter, R., and Grassmann, W. (1926). Hoppe-Seyler's Z. Physiol. Chem. 153, 250.
3. Holzer, H., Betz, H., and Ebner, E. (1975). Curr. Top. Cell. Reg. 9, 103.
4. Wolf, D. H., and Holzer, H. (1979). In "Transport and Utilization of Amino Acids, Peptides and Proteins by Microorganisms" (J. W. Payne, ed.), John Wiley & Sons Limited, Chichester-England, in press.
5. Lenney, J. F., and Dalbec, J. M. (1967). Arch. Biochem. Biophys. 120, 42.
6. Lenney, J. F. (1975). J. Bacteriol. 122, 1265.
7. Ulane, R. E., and Cabib, E. (1974). J. Biol. Chem. 249, 3418.
8. Saheki, T., Matsuda, Y., and Holzer, H. (1974). Eur. J. Biochem. 47, 325.
9. Núñez de Castro, I., and Holzer, H. (1976). Hoppe-Seyler's Z. Physiol. Chem. 357, 727.
10. Betz, H., Hinze, H., and Holzer, H. (1974). J. Biol. Chem. 249, 4515.
11. Bünning, P., and Holzer, H. (1977). J. Biol. Chem. 252, 5316.
12. Matern, H., Hoffmann, M., and Holzer, H. (1974). Proc. Nat. Acad. Sci. U.S. 71, 4874.
13. Matern, H., Barth, R., and Holzer, H. (1979). Biochim. Biophys. Acta, in press.
14. Lenney, J. F., Matile, Ph., Wiemken, A., Schellenberg, M., and Meyer, J. (1974). Biochem. Biophys. Res. Commun. 60, 1378.
15. Matern, H., Betz, H., and Holzer, H. (1974). Biochem. Biophys. Res. Commun. 60, 1051.
16. Matile, Ph., and Wiemken, A. (1967). Arch. Mikrobiol. 56, 148.

17. Schoenheimer, R. (1942). "The Dynamic State of Body Constituents". Cambridge, Harvard Univ.Press.
18. Schimke, R. T., and Doyle, D. (1970). Ann. Rev. Biochem. 39, 929.
19. Saheki, T., and Holzer, H. (1974). Eur. J. Biochem. 42, 621.
20. Katsunuma, T., and Holzer, H. (1975). In "Intracellular Protein Turnover" (R. T. Schimke and N. Katunuma, eds.), pp. 207-212. Academic Press, New York.
21. Neurath, H., and Walsh, K. A. (1976). Proc. Nat. Acad. Sci. U.S. 73, 3825.
22. Cabib, E., Ulane, R., and Bowers, B. (1974). Curr. Top. Cell. Regul. 8, 1.
23. Cabib, E. (1976). Trends Biochem. Sci. 1, 275.
24. Hasilik, A., and Tanner, W. (1976). Biochem. Biophys. Res. Commun. 72, 1430.
25. Hasilik, A., and Tanner, W. (1978). Eur. J. Biochem. 85, 599.
26. Duran, A., Bowers, B., and Cabib, E. (1975). Proc. Nat. Acad. Sci. U.S. 72, 3952.
27. Cabib, E., and Ulane, R. (1973). Biochem. Biophys. Res. Commun. 50, 186.
28. Hasilik, A., and Holzer, H. (1973). Biochem. Biophys. Res. Commun. 53, 552.
29. Saheki, T., and Holzer, H. (1975). Biochim. Biophys. Acta 384, 203.
30. Meussdoerffer, F. (1978). Doctoral Dissertation. Faculty of Biology, University of Freiburg.
31. Jones, E. W. (1979). In "Proceedings of the International Conference on Limited Proteolysis in Microorganisms" (G. N. Cohen and H. Holzer, eds.), United States Government Printing Office, Washington D.C., in press.
32. Wolf, D. H., Beck, I., and Ehmann, C. (1979). In "Proceedings of the International Conference on Limited Proteolysis in Microorganisms" (G. N. Cohen and H. Holzer, eds.), United States Government Printing Office, Washington D.C., in press.
33. Holzer, H. (1975). In "Advances in Enzyme Regulation", Vol. 13 (G. Weber, ed.), pp. 125-134. Pergamon Press, Oxford-New York.
34. Hansen, R. J., Switzer, R. L., Hinze, H., and Holzer, H. (1977). Biochim. Biophys. Acta 496, 103.
35. Holzer, H., Bünning, P., and Meussdoerffer, F. (1977). In "Acid Proteases" (J. Tang, ed.), pp. 271-289. Plenum Publishing Corporation, New York.
36. Barth, R., Wolf, D. H., and Holzer, H. (1978). Biochim. Biophys. Acta 527, 63.
37. Bünning, P., and Holzer, H., unpublished results.

38. Holzer, H. (1976). Trends Biochem. Sci. 1, 178.
39. Flury, U., Heer, B., and Fiechter, A. (1974). Arch. Mikrobiol. 97, 141.
40. Ferguson, A. R., and Sims, A. P. (1974). J. Gen. Microbiol. 80, 173.
41. Hemmings, B. A. (1978). J. Bacteriol. 133, 867.
42. Mazon, M. J. (1978). J. Bacteriol. 133, 780.
43. Labbe, P., Dechateaubodeau, G., and Labbe-Bois, R. (1972). Biochimie 54, 513.
44. Witt, I., Kronau, R., and Holzer, H. (1966). Biochim. Biophys. Acta 118, 522.
45. Gancedo, C. (1971). J. Bacteriol. 107, 401.
46. Gancedo, C., Schwerzmann, N., and Molano, J. (1974). In "Proc. Fourth International Symposium on Yeasts", Vienna 1974, Part I, p. 9, A5.
47. Haarasilta, S., and Oura, E. (1975). Eur. J. Biochem. 52, 1.
48. Magni, G., Santarelli, J., Natalini, P., Ruggiero, S., and Vita, A. (1977). Eur. J. Biochem. 75, 77.
49. Neeff, J., Hägele, E., Nauhaus, J., Heer, U., and Mecke, D. (1978). Biochem. Biophys. Res. Commun. 80, 276.
50. Jušič, M., Hinze, H., and Holzer, H. (1976). Hoppe-Seyler's Z. Physiol. Chem. 357, 735.
51. Matern, H., and Holzer, H. (1977). J. Biol. Chem. 252, 6399.
52. Robertson, J. J., and Halvorson, H. O. (1957). J. Bacteriol. 73, 186.
53. Görts, C. P. M. (1969). Biochim. Biophys. Acta 184, 299.
54. Pringle, J. R. (1975). In "Methods in Cell Biology", Vol. 12 (D. M. Prescott, ed.), pp. 149-184. Academic Press, New York.
55. Grunow, M., and Schöpp, W. (1978). FEBS Lett. 94, 375.
56. Müller, M., and Holzer, H., unpublished results.
57. Juni, E., and Heym, G. A. (1968). Arch. Biochem. Biophys. 127, 89.
58. Hofmann, E. (1976). Rev. Physiol. Biochem. Pharmacol. 75, 1.
59. Holzer, H. (1976). In "Metabolic Interconversion of Enzymes 1975" (S. Shaltiel, ed.), pp. 168-174. Springer Verlag, Berlin Heidelberg.
60. Wiemken, A. (1975). In "Methods in Cell Biology", Vol. 12 (D. M. Prescott, ed.), pp. 99-110. Academic Press, New York.

THIOREDOXIN AND ENZYME REGULATION IN PHOTOSYNTHESIS

Bob B. Buchanan

Department of Cell Physiology, University of California
Berkeley, California 94720

ABSTRACT Thioredoxin, a hydrogen carrier protein that functions in the synthesis and replication of DNA and in the transformation of sulfur metabolites, has been found to serve as a regulatory protein in linking light to the activation of enzymes during photosynthesis. In this system, thioredoxin is reduced photochemically via ferredoxin and the enzyme ferredoxin-thioredoxin reductase. The enzymes activated by this mechanism (designated the ferredoxin/thioredoxin system) include four enzymes of the reductive pentose phosphate cycle of CO_2 assimilation (fructose 1,6-bisphosphatase, sedoheptulose 1,7-bisphosphatase, NADP-glyceraldehyde 3-phosphate dehydrogenase, and phosphoribulokinase) as well as three enzymes not associated with the cycle (NADP-malate dehydrogenase, phenylalanine ammonia lyase, and PAPS sulfotransferase). Two different enzyme-specific thioredoxins (thioredoxins f and m) function in chloroplasts as part of the ferredoxin/thioredoxin system. A third thioredoxin of unknown function (thioredoxin c) resides outside of the chloroplasts, possibly in the cytoplasm.

The enzymes photochemically activated by thioredoxin appear to be deactivated in the dark by a mechanism that differs as to the enzyme. The mechanisms include deactivation by (i) a soluble oxidant such as oxidized glutathione or dehydroascorbate; (ii) an unidentified membrane-bound oxidant; and (iii) an unknown chloroplast reaction. Each of the enzymes regulated by the ferredoxin/thioredoxin system shows hysteretic properties, i.e., its rate of activation or deactivation is slow relative to the rate of catalysis.

In addition to its role in light mediated enzyme regulation, thioredoxin has been found to increase the dithiothreitol-dependent ATPase activity of heated chloroplast coupling factor (CF_1). Recent experiments have demonstrated that the δ subunit fraction isolated from purified CF_1 may replace thioredoxin in this activation. The isolated δ subunit fraction also showed thioredoxin activity. The physiological signifiance of the thioredoxin-mediated ATPase activation remains to be determined.

In sum, thioredoxin seems to act in chloroplasts as a regulatory messenger between light and the enzymes of different biosynthetic processes that utilize the products formed by light, i.e., ATP and NADPH. We currently visualize that the ferredoxin/thioredoxin system operates in conjunction with other light-actuated regulatory mechanisms of chloroplasts.

INTRODUCTION

One type of covalent enzyme modification that has been used in studies on the structure and mechanism of action of enzymes is a change in the oxidation state of a "nonprosthetic" part of the protein, such as a sulfhydryl group. Limited attention has been given to the idea that oxidation-reduction changes may be used to regulate the activation of enzymes in a manner analogous to other types of reversible covalent modification, such as phosphorylation-dephosphorylation and adenylation-deadenylation. We summarize in this report evidence that enzyme regulation due to a reversible oxidation-reduction change is a process fundamental to photosynthetic and perhaps to other types of living cells. In this article we will first describe work from our own laboratory that led to the finding of a redox-based regulatory mechanism in photosynthesis; we will then relate these findings to other mechanisms of light-dependent enzyme regulation in chloroplasts.

RESULTS

Our first indication that the activity of an enzyme may be altered by redox change came more than 10 years ago when studies with isolated spinach chloroplasts revealed that the enzyme fructose 1,6-bisphosphatase (Fru-P_2ase)--a key enzyme of the reductive pentose phosphate cycle of CO_2 assimilation--is activated by reduced soluble ferredoxin (1). [Soluble ferredoxin is an iron-sulfur protein that functions as an electron carrier for the light-dependent reduction of chloroplast metabolites such as NADP in the presence of specific ferredoxin-linked reductase enzymes.] In our experiments, ferredoxin, which was reduced photochemically with chloroplast membranes, activated the Fru-P_2ase enzyme component of the stroma or soluble phase of chloroplasts ('chloroplast extract'). Fru-P_2ase was also found, by Bassham and his colleagues (2), to be light-activated in whole cells. At about the same time, the Zieglers showed that another enzyme of the reductive pentose phosphate cycle, NADP-glyceraldehyde 3-phosphate dehydrogenase, is light-activated (3). We had at the time no premonition of findings to be made years later that the NADP-glyceraldehyde 3-phosphate dehydrogenase, as well as other light-activated enzymes that had been studied by Bassham's group, share the capacity for activation by reduced ferredoxin.

Following the demonstration of the ferredoxin-linked Fru-P_2ase activation, we began attempts to purify Fru-P_2ase and to study its mechansim of activation. This study revealed that reduced ferredoxin does not interact directly with the enzyme but that another soluble chloroplast component is required (4). We separated this component [which, because of its properties, was designated the protein factor] from both ferredoxin and Fru-P_2ase (Table 1). Somewhat later, we showed that sedoheptulose 1,7-bisphosphatase (Sed-P_2ase), another enzyme of the reductive pentose phosphate cycle of chloroplasts, resembles Fru-P_2ase in its capacity for activation by ferredoxin and the protein factor (5).

<u>Identification of the Chloroplast Protein Factor</u>. Once the protein factor was isolated, we began an investigation aimed at its purification and identification. These studies led to the finding that the protein factor consists not of one but of two components that we ultimately isolated and gave the names "assimilation regulatory protein a" (ARP_a) and "assimilation regulatory protein b" (ARP_b) (6) (Table 2).

To gain insight into the nature of the newly separated regulatory proteins, we exploited our earlier finding (4) that the sulfhydryl reagent dithiothreitol (DTT) can replace reduced ferredoxin in the activation of Fru-P_2ase, provided the then-called protein factor is present. Our experiments with the resolved protein factor components showed that ARP_b is the active component and that ARP_a has no effect on activation in the presence of DTT (6). This finding not only provided new

TABLE 1
REQUIREMENTS FOR ACTIVATION OF HOMOGENEOUS CHLOROPLAST FRU-P_2ASE BY FERREDOXIN

Treatment	P_i Released (μmol)
Complete	4.0
Minus Ferredoxin	0.1
" Fru-P_2ase	0.1
" $MgCl_2$	0.1
" Protein factor	0.1
" Fructose bisphosphate	0.0
Complete, Ferredoxin not reduced (dark)	0.1

TABLE 2
REQUIREMENT FOR ARP_a AND ARP_b
FOR ACTIVATION OF CHLOROPLAST $FRU-P_2ASE$
BY REDUCED FERREDOXIN

Treatment	P_i Released (nmol/min)
Control	0
+ ARP_a	14
+ ARP_b	0
+ ARP_a and ARP_b	73

information on ARP_b but also gave us a convenient assay for ARP_b that is independent of ferredoxin and ARP_a. We have taken advantage of this assay in much of our later work.

The new ARP_b assay was immediately applied in our investigation of the occurrence of ARP_b in different types of organisms. This study revealed that ARP_b is not confined to photosynthetic cells but occurs in other types of cells and is widespread, if not ubiquitous, in nature (7). We found ARP_b not only in chloroplasts, algae, and photosynthetic bacteria, but also in roots, seeds, etiolated plant shoots, fermentative bacteria, aerobic bacteria, and even in animal cells. In studying mammalian ARP_b we were able to obtain a highly purified ARP_b preparation from rabbit liver.

Just after we observed the wide natural distribution of ARP_b, an article on the protein thioredoxin caught our attention. This article initiated a search that led ultimately to the identification of ARP_b as chloroplast thioredoxin (Table 3)

TABLE 3
IDENTIFICATION OF ARP_b AS CHLOROPLAST THIOREDOXIN

	Treatment	P_i Released (nmol/min)	
		Chloroplast ARP_b	*E. coli* Thioredoxin
Light,	complete	12	10
"	$-ARP_b$ or thioredoxin	2	2
"	-ferredoxin	4	2
"	$-ARP_a$	1	1
"	-fructose 1,6-bisphosphatase	0	1
Dark,	complete	2	1

and of ARP_a as an enzyme that catalyzes the reduction of thioredoxin by reduced ferredoxin, i.e., ferredoxin-thioredoxin reductase (Table 4) (8). The ARP_b protein isolated from liver was identified as authentic liver thioredoxin (9).

Thioredoxin is a low-molecular-weight, hydrogen-carrier protein that was first isolated (under different names) in studies on the reduction of sulfoxide and sulfate compounds by enzyme systems isolated from yeast (10,11). Thioredoxin was later independently isolated, named, and extensively characterized in the pioneering work of Reichard and his colleagues on DNA synthesis (12). Experiments by those investigators revealed that thioredoxin functions as a hydrogen carrier in the reduction of ribonucleoside diphosphates to their deoxyribose derivatives in DNA synthesis by enzyme systems isolated first from bacteria and later from animal tissues (13). Work by Reichard's group also revealed that the low-molecular-weight protein isolated earlier from yeast (10,11) is the same as thioredoxin (14). At the time of our entry into the field, thioredoxin had not been reported to occur in plants.

Just prior to or parallel with out studies on the identification of chloroplast thioredoxin, new findings were made on the thioredoxin from *Escherichia coli*. Mark and Richardson showed thioredoxin to be a subunit of phage-induced DNA polymerase in that organism (15), and Pigiet reported the isolation of a phosphorylated form of *E. coli* thioredoxin (16).

Enzyme Activation by the Ferredoxin/thioredoxin System.
Thioredoxin and ferredoxin-thioredoxin reductase have emerged as components of a ferredoxin-linked regulatory mechanism (designated the ferredoxin/thioredoxin system) by which light regulates selected enzymes during photosynthesis. In the light, electrons from chlorophyll are transferred to ferredoxin and then via the enzyme ferredoxin-thioredoxin reductase to

TABLE 4
IDENTIFICATION OF CHLOROPLAST ARP_a
AS FERREDOXIN-THIOREDOXIN REDUCTASE

Component added to oxidize photoreduced ferredoxin	Ferredoxin oxidized (nmol/min)
None	0.0
ARP_a	0.6
Chloroplast thioredoxin	0.9
ARP_a +chloroplast thioredoxin	12.0

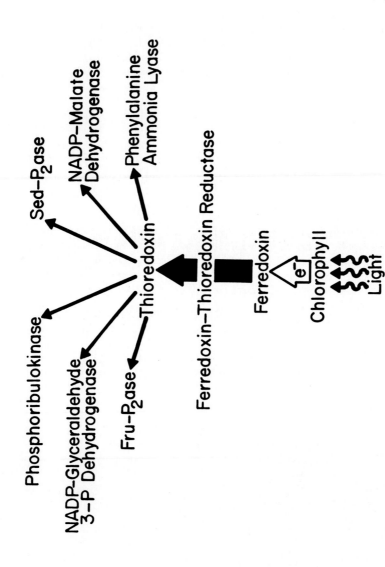

FIGURE 1. Role of thioredoxin in the light-dependent activation of chloroplast enzymes.

thioredoxin (8). Reduced thioredoxin, in turn, reduces and thereby activates a number of regulatory enzymes of chloroplasts (Figure 1). There is now evidence that thioredoxin activates four enzymes of the reductive pentose phosphate cycle [Fru-P_2ase (8), Sed-P_2ase (17), NADP-glyceraldehyde 3-phosphate dehydrogenase (18), phosphoribulokinase (19)], an enzyme of CO_2 assimilation that is not a part of the carbon cycle [NADP-malate dehydrogenase (20)], and an enzyme of secondary plant metabolism [phenylalanine ammonia lyase (21)]. The regulatory role of thioredoxin has been extended to the blue-green algae, where it was shown that an enzyme of sulfate reduction [2'-phosphoadenosine 5'-phosphosulfate (PAPS) sulfotransferase] is activated by reduced thioredoxin (22). As discussed below, reduced thioredoxin also activates the ATPase activity associated with chloroplast coupling factor (CF_1) (23).

In the light-dependent activation of enzymes by the ferredoxin/thioredoxin system, the light signal is converted via chlorophyll to a reductant signal (reduced ferredoxin), which is relayed via ferredoxin-thioredoxin reductase to thioredoxin. Thioredoxin thus may be visualized as a regulatory messenger between light and the enzymes of processes that utilize the products formed by light, i.e., ATP and NADPH. In this capacity, thioredoxin "alerts" key enzymes of diverse biosynthetic processes that the light is turned on and that biosynthesis is to proceed.

Deactivation of Thioredoxin-Activated Enzymes. An important aspect of the ferredoxin/thioredoxin system that has not yet been fully elucidated is the way by which the activated enzymes are converted to a less active (deactivated) state. The unique feature of the enzymes of this, and of other, light-mediated regulatory mechanisms of chloroplasts is that deactivation must occur in the dark. Current evidence suggests that, unlike activation, the mechanism for the dark deactivation of thioredoxin-linked enzymes can differ as to the particular enzyme involved. Based on their deactivation properties, thioredoxin-linked enzymes are of three types.

Enzymes of the first type require for deactivation a soluble oxidant, such as the oxidized form of glutathione (GSSG) (Fig. 2) or ascorbate (dehydroascorbate), both of which may be formed by chloroplasts in the dark (Fig. 3) (8,24). This group of enzymes includes Fru-P_2ase (8), phosphoribulokinase (19), and phenylalanine ammonia lyase (21). The second type of enzymes, of which NADP-malate dehydrogenase is the sole representative, appear to be deactivated by an unidentified membrane-bound oxidant in the absence of soluble oxidants (Table 5) (20). Evidence indicates this oxidant is available for deactivation only when the membranes are maintained in the dark. Enzymes of the third type, typified by NADP-glyceraldehyde 3-phosphate dehydrogenase, have no known mechanism of deactivation (18).

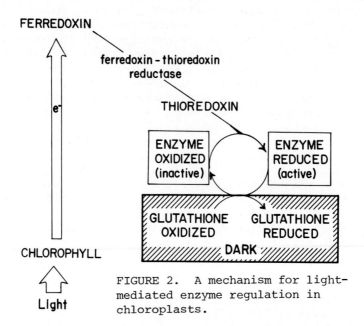

FIGURE 2. A mechanism for light-mediated enzyme regulation in chloroplasts.

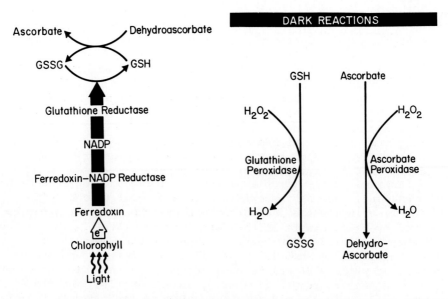

FIGURE 3. Possible mechanisms for formation of soluble oxidants functional in the deactivation of thioredoxin-activated enzymes.

TABLE 5
EVIDENCE FOR A CHLOROPLAST MEMBRANE-BOUND COMPONENT FUNCTIONAL IN THE DARK DEACTIVATION OF NADP-MALATE DEHYDROGENASE ACTIVATED BY DITHIOTHREITOL-REDUCED THIOREDOXIN

Deactivation Factor Added	Enzyme Activity (NADPH oxidized, nmol/min)
None	34
Oxidized glutathione	16
Chloroplast membranes (dark)	16
Chloroplast membranes (light)	33

Despite differences in deactivation properties, the thioredoxin linked regulatory enzymes of chloroplasts share one common feature: In each case, the rate of deactivation (in which the enzyme is converted from an active to an inactive form) and the rate of activation (in which the enzyme is converted from an inactive to an active form) are slow relative to the rate of catalysis. Frieden has designated enzymes of this type as hysteretic enzymes (25). Ribulose 1,5-bisphosphate carboxylase, the enzyme catalyzing the sole carboxylation reaction of the reductive pentose phosphate cycle, also shows hysteretic behavior (26). There is no indication that the activity of this enzyme is modulated by thioredoxin.

Multiple Forms of Thioredoxin in Leaves. The importance of thioredoxin in enzyme regulation in plants is emphasized by the recent finding that algae and leaves contain multiple forms of thioredoxin (22,27-30). Two different thioredoxins have been found in spinach chloroplasts (thioredoxins f and m) and a third thioredoxin (thioredoxin c) was found outside chloroplasts, possibly in the cytoplasm (Table 6) (27,29). The three thioredoxins are distinguished by their stability to heat (Table 7) and their molecular weight (Table 8) as well as by their enzyme specificity described below. A protein that appears to be the equivalent of thioredoxin m in maize has also been isolated in the laboratory of Hatch (31). It remains to be seen which of the thioredoxin(s) functions $in\ vivo$ in the algal ribonucleotide reductase system recently described by Follmann and his colleagues (32). It is interesting to note that a thioredoxin that may have such a function was recently isolated from wheat embryos (33).

Thioredoxins f and m show specificity in their ability to activate enzymes of chloroplasts. Thioredoxin m is more effective in the activation of NADP-malate dehydrogenase, and thioredoxin f is more effective in the activation of the other enzymes tested, i.e., Fru-P_2ase, phosphoribulokinase, NADP-

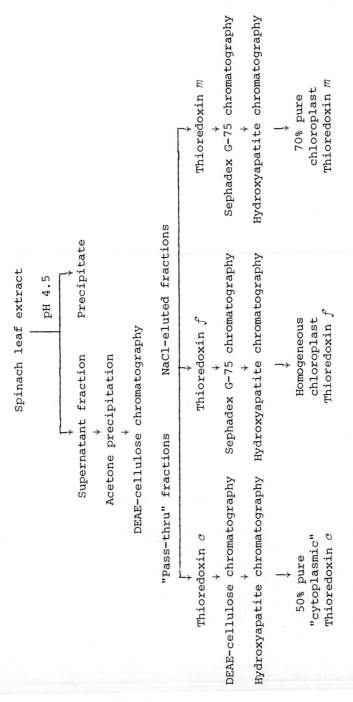

TABLE 6. Purification of three leaf thioredoxins.

MODULATION OF PROTEIN FUNCTION 103

TABLE 7
EFFECT OF HEAT ON LEAF THIOREDOXINS

	Percent remaining after 3 min at 80°C
Thioredoxin f	45
Thioredoxin m	98
Thioredoxin c	93

TABLE 8
MOLECULAR WEIGHT OF LEAF THIOREDOXINS
AS DETERMINED BY EXCLUSION CHROMATOGRAPHY
ON SEPHADEX G-75

	Molecular Weight
Thioredoxin f	16,000
Thioredoxin m	9,000
Thioredoxin c	15,000

glyceraldehyde 3-phosphate dehydrogenase, and phenylalanine ammonia lyase (Table 9) (21,27-29). Thioredoxin c poses an unsolved problem because neither its function nor its mechanism of reduction *in vivo* is known. It is noteworthy that an enzyme that reduces thioredoxin with NADPH as donor (NADP-thioredoxin reductase) was recently found in wheat embryo extracts (33). As this enzyme has not been found in photosynthetic cells it seems possible that the mechanism of thioredoxin reduction may differ in photosynthetic and nonphotosynthetic tissues: nonphotosynthetic systems would utilize NADPH and photosynthetic systems would utilize reduced ferredoxin.

Thioredoxin and CF_1. As indicated above, there is evidence that thioredoxin has the capability of interacting not only with soluble enzymes but also with membranous proteins following their solubilization. Thioredoxin was shown to stimulate the dithiothreitol-dependent ATPase activity that is associated with chloroplast coupling factor (CF_1). Although each of the three leaf thioredoxins was active in ATPase activation, chloroplast thioredoxin m was the most effective (Figure 4).

In an extension of this work, we addressed ourselves to the question of whether CF_1 itself contains a thioredoxin-like component that acts in the basal dithiothreitol-dependent ATPase activity that is observed in the absence of added thioredoxin (34). We observed that when preparations of CF_1 were

TABLE 9
EFFECTIVNESS OF DITHIOTHREITOL-REDUCED LEAF THIOREDOXINS
IN THE ACTIVATION OF REGULATORY ENZYMES FROM CHLOROPLASTS

Treatment	Relative Activity					
	Fru-P_2ase	Sed-P_2ase	NADP-Malate Dehydrogenase	Phosphoribulokinase	NADP-Glyceraldehyde-3-P Dehydrogenase	Phenylalanine Ammonialyase
Thioredoxin f	100	100	51	100	100	100
Thioredoxin m	16	18	100	43	40	25
Thioredoxin c	34	18	57	55	55	63
-Thioredoxin	3	18	28	36	42	22

fractionated by defined procedures, thioredoxin activity was recovered in the fraction enriched in the δ subunit, the 20,000-dalton component that is envisioned to be essential for the attachment of CF_1 to the chloroplast membrane (35,36). The isolated δ subunit could partially replace authentic thioredoxin in the dithiothreitol-linked activation of both chloroplast Fru-P_2ase (Figure 5) and CF_1 ATPase (Figure 6). The isolated ϵ subunit, by contrast, inhibited both of these reactions (cf. ref. 37), whereas other CF_1 subunits had no effect. These experiments raise the interesting question of whether thioredoxin [or its phosphorylated derivative (16)] functions in energy transduction in chloroplasts or other membrane systems.

DISCUSSION

The ferredoxin/thioredoxin system appears to constitute a general mechanism of enzyme regulation that operates in conjunction with other light-actuated regulatory systems in chloroplasts (Figure 7). One of the mechanisms by which light governs enzyme activity independently of thioredoxin is via specific enzyme effectors whose synthesis is increased in the light [e.g., NADPH, ATP (38,39)]. It was proposed that, by analogy to energy charge, a reductant charge (ratio of NADPH to NADP) may be operative in the light/dark control of the activity of certain chloroplast enzymes (40,41).

Light-induced ion shifts also appear to be important factors in insuring that regulatory enzymes of chloroplasts are fully active only in the light (42). In particular, the light-induced shift in stromal concentrations of H^+ (pH $7_{dark} \rightarrow$ pH 8_{light}) and Mg^{++} (increase of 1 to 3 mM in the light) seem

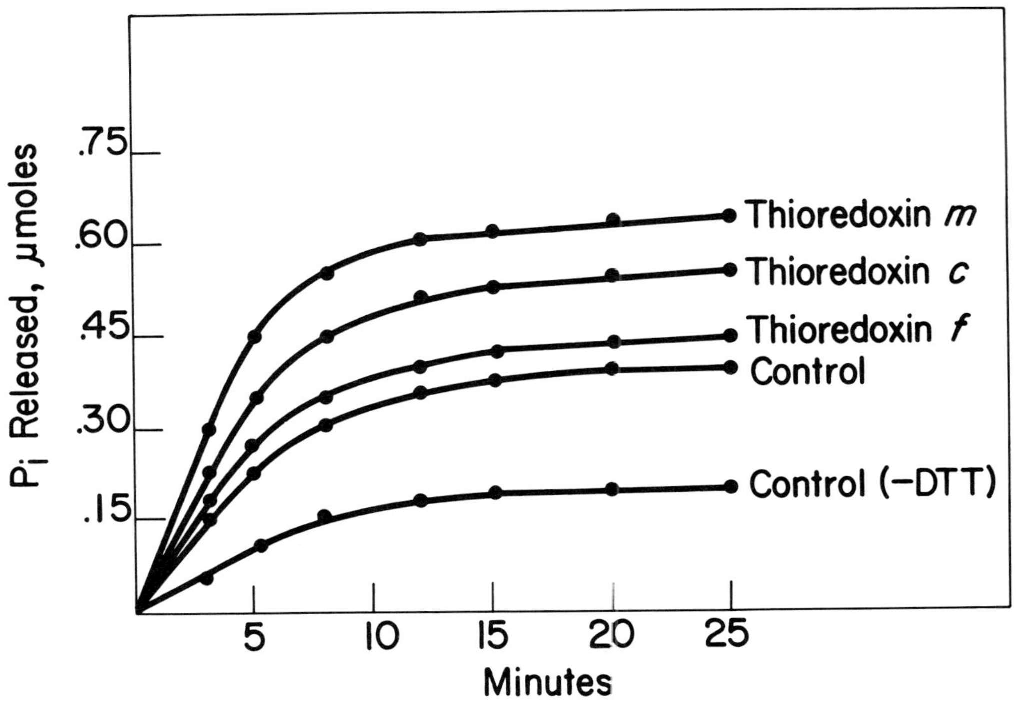

FIGURE 4. Effect of leaf thioredoxins on ATPase activity of heated CF$_1$.

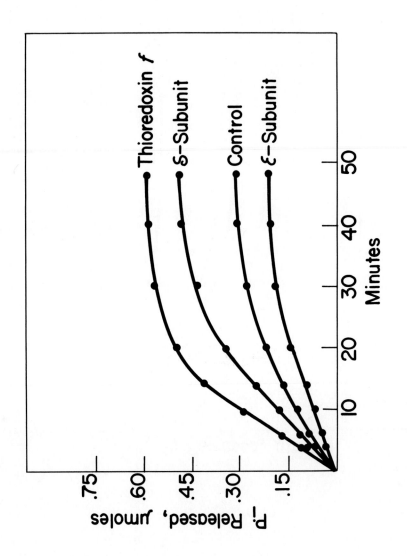

FIGURE 5. Effect of δ- and ϵ- subunits of CF_1 on thioredoxin-linked Fru-P_2

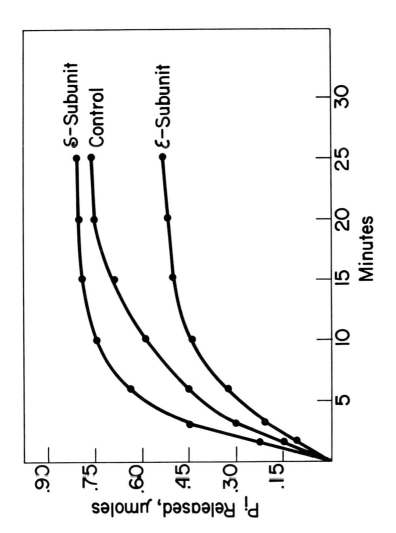

FIGURE 6. Effect of time on subunit-induced changes in ATPase activity of heated CF_1

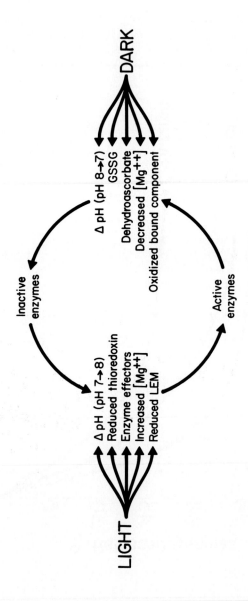

FIGURE 7. Light-dependent mechanisms of enzyme regulation in chloroplasts.

to be important in providing the environment necessary for the optimal activity of certain enzymes. Two of the thioredoxin-linked enzymes (Fru-P_2ase and Sed-P_2ase) seem to be especially sensitive to light-induced changes in H^+ and Mg^{++} concentration as well as to changes in redox state (42).

Finally, there is evidence that membrane-bound reductants may function in enzyme regulation in chloroplasts. It is envisaged that these reductants (designated light-effect mediators, or LEMs) are disulfide components that interact with the electron transport chain of chloroplasts (43). The LEM components would occur mainly in the oxidized (disulfide) state in the dark and in the reduced (sulfhydryl) state in the light. The relation of the LEM system to the ferredoxin/thioredoxin system remains to be established. It might be pointed out, nevertheless, that the above-mentioned membrane-bound component functional in the deactivation of thioredoxin-activated NADP-malate dehydrogenase appears to be analogous to LEM, with the exception that it acts in the dark rather than in the light.

A question raised by invoking a regulatory role for physiological redox agents is whether certain enzymes might behave in a manner opposite to that described above and undergo an oxidant-induced activation. Evidence that this is the case is provided by our recent finding that nonspecific acid phosphatases from spinach leaves and potato tubers were activated by oxidized glutathione or dehydroasorbate (44). Activation was accompanied by a change in the pH optimum so that the enzymes became active in the neutral as well as in the acid region. At neutral pH, the activation induced by oxidized glutathione was reversed by reduced glutathione. The results suggest that oxidized glutathione and dehydroascorbate serve not only in the previously demonstrated deactivation of thioredoxin-linked enzymes but also in the activation of other enzymes from plants (cf. ref. 45 and 46).

In view of the central regulatory role that thioredoxin seems to play in chloroplasts, the question arises as to whether thioredoxin might have an as yet unknown regulatory function in heterotrophic cells. The presence of ferredoxin is not prerequisite to such a function because, as noted above, cells of both animal and bacterial origin are enzymically fitted to reduce thioredoxin independently of ferredoxin, via an NADP-linked thioredoxin reductase (13). It seem possible that thioredoxin-dependent enzyme regulation in heterotrophic cells could be executed, for example, either by an NADP-linked mechanism, for which the thioredoxin-linked reduction of insulin provides a model (47) or by the formation of a stable thioredoxin enzyme complex, as found for the *E. coli* DNA polymerase described above. Whether these or other thioredoxin-linked mechanisms act in the regulation of enzymes of heterotrophic cells is one of the exciting problems for the future.

The work from this laboratory described herein was aided by grants from the National Science Foundation (76-82232, 76-80731, 78-15287).

REFERENCES

1. Buchanan, B.B., Kalberer, P. P., and Arnon, D. I. (1967) Biochem. Biophys. Res. Commun. 29, 74-79
2. Petersen, T. A., Kirk, M., and Bassham, J. A. (1966) Physiol. Plant. 19, 219-231
3. Ziegler, H., and Ziegler, I. (1966) Planta (Berlin) 69, 111-123
4. Buchanan, B.B., Schürmann, P., and Kalberer, P. P. (1971) J. Biol. Chem. 246, 5952-5959
5. Schürmann, P., and Buchanan, B. B. (1975) Biochim. Biophys. Acta 376, 189-192
6. Schürmann, P., Wolosiuk, R. A., Breazeale, V. D., and Buchanan, B. B. (1976) Nature 263, 257-258
7. Buchanan, B. B., and Wolosiuk, R. A. (1976) Nature 264, 669-670
8. Wolosiuk, R. A., and Buchanan, B. B. (1977) Nature 266, 565-567
9. Holmgren, A., Buchanan, B. B., and Wolosiuk, R. A. (1977) FEBS Lett 82, 351-354
10. Black, S., Harte, E. M., Hudson, B., and Wartofsky, L. (1960) J. Biol. Chem. 235, 2910-2916
11. Wilson, L. G., Asahi, T., and Bandurski, R. S. (1961) J. Biol. Chem. 236, 1822-1829
12. Laurent, T. C., Moore, E. C., and Reichard, P. (1964) J. Biol. Chem. 239, 3436-3444
13. Reichard, P. (1968) Eur. J. Biochem. 3, 259-266
14. Porque, P. G., Baldesten, A., and Reichard, J. P. (1970) J. Biol. Chem. 245, 2363-2370
15. Mark, D. F., and Richardson, C. C. (1976) Proc. Nat. Acad. Sci. USA 73, 780-784
16. Pigiet, V., and Conley, R. R. (1978) J. Biol. Chem. 253, 1910-1920
17. Breazeale, V. D., Buchanan, B. B., and Wolosiuk, R. A. (1978) Zeit. Naturforsch. 33c, 521-528
18. Wolosiuk, R. A., and Buchanan, B. B. (1978) Plant Physiol. 61, 669-671
19. Wolosiuk, R. A., and Buchanan, B. B. (1978) Arch. Biochem. Biophys. 189, 97-101
20. Wolosiuk, R. A., Buchanan, B. B., and Crawford, N. A. (1977) FEBS Lett. 81, 253-258
21. Wolosiuk, R. A., Nishizawa, A. N., and Buchanan, B. B. (1978) Plant Physiol. 61, 97S
22. Wagner, W., Follmann, H., and Schmidt, A. (1978) Zeit Naturforsch. 33c, 517-520

23. McKinney, D. W., Buchanan, B. B., and Wolosiuk, R. A. (1978) Phytochemistry 17, 794-795
24. Groden, D., and Beck, E. (1977) Abstracts, 4th Int Congr Photosynthesis, Reading, U. K., p. 139
25. Frieden, C. (1971) Annu. Rev. Biochem. 40, 653-696
26. Lorimer, G. H., Badger, M. R., and Andrews, T. J. (1976) Biochemistry 15, 529-536
27. Buchanan, B. B., Wolosiuk, R. A., Crawford, N. A., and Yee, B. C. (1978) Plant Physiol. 61, 38S
28. Jacquot, J. P., Vidal, J., Gadal, P., and Schürmann, P. (1978) FEBS Lett. 96, 243-246
29. Wolosiuk, R. A., Crawford, N. A., Yee, B. C., and Buchanan, B. B. (1979) J. Biol. Chem., in press
30. Jacquot, J. P., Vidal, J., and Gadal, P. (1976) FEBS Lett. 71, 223-227
31. Kagawa, T., and Hatch, M. D. (1977) Arch. Biochem. Biophys. 184, 290-297
32. Wagner, W., and Follmann, H. (1977) Biochem. Biophys. Res. Commun. 77, 1044-1051
33. Suske, G., Wagner, W., and Follmann, H. (1979) Zeit. Naturforsch., in press
34. McKinney, D. W., Buchanan, B. B., and Wolosiuk, R. A. (1979) Biochem. Biophys. Res. Commun., in press
35. Nelson, N. (1976) Biochim. Biophys. Acta 456, 314-338
36. Binder, A., Jagendorf, A., and Ngo, E. (1978) J. Biol. Chem. 253, 3094-3100
37. Nelson, N., Deters, H., Nelson, D. W., and Racker, E. (1973) J. Biol. Chem. 248, 2049-2055
38. Müller, B., Ziegler, I., and Ziegler, H. (1969) Eur. J. Biochem. 9, 101-106
39. Pupillo, P., and Giuliani-Piccari, G. G. (1975) Eur. J. Biochem. 51, 475-482
40. Lendzian, K., and Bassham, J. A. (1975) Biochim. Biophys. Acta 396, 260-275
41. Wildner, G. F. (1975) Z. Naturforsch. 30c, 756-760
42. Heldt, H. W., Chon, C. J., Lilley, R. McC., and Portis, A. (1978) Proc. 4th Int Congr Photosynthesis, The Biochem. Soc., pp. 469-478. London
43. Anderson, L. E., and Avron, M. (1976) Plant Physiol. 57, 209-213
44. Buchanan, B. B., Crawford, N. A., and Wolosiuk, R. A. (1979) Plant Sci. Lett., in press
45. Moreno, C. G., Aparacio, P. J., Palacian, E., and Losada, M. (1972) FEBS Lett. 26, 11-14
46. Haddox, M. K., Stephenson, J.H., Moser, M. E., and Goldberg, N. D. (1978) J. Biol. Chem. 253, 3143-3152
47. Holmgren, A. (1977) J. Biol. Chem. 252, 4600-4606

REGULATION OF THE PHOTOSYNTHETIC CARBON CYCLE,
PHOSPHORYLATION AND ELECTRON TRANSPORT IN
ILLUMINATED INTACT CHLOROPLASTS

Ulrich Heber, Ulrike Enser, Engelbert Weis,
Ursula Ziem, and Christoph Giersch

Institute of Botany, University of Düsseldorf,
4000 Düsseldorf, Germany

ABSTRACT pH-Control of the Carbon Cycle. On illumination, chloroplasts pump protons from the stroma not only into the thylakoid compartment, but also across the chloroplast envelope into the extra-chloroplast space. Increased proton back leakage from a neutral medium into the stroma as mediated by some salts of weak acids inhibited photosynthesis, and inhibition was strongly pH-dependent. It was caused by acidification of the stroma which inactivated fructose and sedoheptulose bisphosphatases. The stroma pH controlled not only activity, but also light-activation of fructose bisphosphatase, thus amplifying pH-control of photosynthesis. Light-activation of glyceraldehydephosphate dehydrogenase (NADP) and of malic dehydrogenase (NADP) was also sensitive to pH, while activation of phosphoribulokinase did not respond to salts that caused stroma acidification only. However, with potassium nitrite which decreases the stroma pH and also oxidizes ferredoxin activation became pH-dependent. Fructose bisphosphatase, phosphoribulokinase and malic dehydrogenase remained in the activated state when the stroma pH was decreased in the light, while light-activation of glyceraldehydephosphate dehydrogenase was reversed at a decreased stroma pH.
Control of Phosphorylation. Isolated intact chloroplasts contained significant ATP in the dark. Hydrolysis of endogenous ATP by the thylakoid enzyme that catalyzes synthesis in the light was extremely slow. Widely spaced microsecond flashes activated the enzyme and decreased ATP levels in darkened chloroplasts. At increased flash frequencies, ATP hydrolysis was replaced by synthesis. ATP/ADP ratios increased from 0.5 to 1 in the dark to only 2 to 4 in the light. Energy charge was about 0.4 to 0.6 in the dark and maximally 0.85 in the light. Phosphorylation potentials increased from 50-100 (M^{-1}) in the dark to

200-350 (M^{-1}) in the light. Ammonium chloride which increased the intrathylakoid pH and the rate of electron flow to nitrite or oxaloacetate failed to decrease the phosphorylation potential indicating that in intact chloroplasts electron transport is not controlled by the phosphorylation potential. In contrast to intact chloroplasts, broken chloroplasts phosphorylated added ADP up to a phosphorylation potential of 80 000 (M^{-1}). $\Delta G'_{ATP}$ was about -41 kJ/mole in darkened and -46 kJ/mole in illuminated intact chloroplasts, but -60 kJ/mole in illuminated broken chloroplasts.

Regulation of Electron Flow. The transfer of electrons from the electron transport chain of intact chloroplasts to different acceptors is under redox control. Under anaerobic conditions, cyclic electron flow was supported by far-red light and inhibited by red light. Far-red supported cyclic electron flow was inhibited and inhibition of electron transport by red light was relieved by nitrite, oxaloacetate or 3-phosphoglycerate. Bicarbonate which requires more ATP for reduction than 3-phosphoglycerate was not a good electron acceptor under anaerobic conditions. Low concentrations of oxygen were often needed to start photosynthesis. Oxygen appears to be necessary for poising the electron transport chain so that the stoichiometric requirement of carbon reduction for ATP and NADPH can be met by a proper cooperation of linear and cyclic electron transport pathways. This enables chloroplasts of different energy coupling quality to photoreduce CO_2.

INTRODUCTION

Photosynthesis is, by definition, synthesis driven by light. Light is absorbed in chloroplasts and its energy is used to split water and produce a strong reductant, NADPH, and ATP. Both are necessary to reduce CO_2 to the sugar level. The question may be asked why complex control very different from that exerted by the availability of light is necessary to regulate photosynthesis: Carbon fixation, carbon reduction and the regeneration of the carbon acceptor molecule take place in the chloroplasts, which are separated from the cytosol, the site of glycolysis, by the two membranes of the chloroplast envelope. The envelope is impermeable to many intermediates of photosynthesis but contains two translocators which have a high transfer capacity for phosphate, dihydroxyacetone-phosphate and 3-phosphoglycerate or for dicarboxylates such as oxaloacetate, malate and aspartate (1,2). These carriers

which operate on an anion counterexchange basis link the
metabolism of chloroplasts and cytosol. Efficient substrate
shuttles connect the chloroplast adenylate and pyridine
nucleotide systems with those of the cytosol. Even in the
dark chloroplasts therefore contain considerable ATP and
NADPH (3). If synthesis of the carbon acceptor molecule could
take place in the dark, CO_2 would be fixed at the expense of
energy derived from sugar degradation. To prevent waste of
energy, key enzymes of the photosynthetic carbon cycle are
inactivated on darkening and are activated on illumination
(4,5).

In the light, carbon reduction by the so-called C_3 and
C_4 plants consumes ATP and NADPH at different ratios. ATP production by the electron transport chain must obviously be
regulated so as to meet the stoichiometric requirements of
consumption. ADP can be phosphorylated in the light by thylakoid membranes with great efficiency, and high ATP/ADP ratios
are attainable at physiological phosphate concentrations.
They correspond to an energy charge of the adenylate system
which is very close to the maximal value of 1. In the cell,
the adenylate system is known to have regulatory function in
addition to its function in phosphate transfer (6). If the
full capacity of the electron transport chain for phosphorylation could be used in vivo, high ATP/ADP ratios would result
when ATP consumption is slow, and small changes in the rate of
ADP production would produce drastic changes in the ATP/ADP
ratio and in energy charge. In fact, light/dark changes in the
phosphorylation potential of chloroplasts are not large, and
energy charge is maintained at values which rarely exceed 0.85
even in the light. It thus appears that regulation is required
at at least three levels, that of carbon metabolism, that of
electron distribution between different electron acceptors to
meet different and variable requirements for ATP and reducing
equivalents, and that of energy charge. Some aspects of these
levels of regulation will briefly be considered for intact
chloroplasts.

I. pH-REGULATION OF THE PHOTOSYNTHETIC CARBON CYCLE

1. Light-dependent Alkalization of the Chloroplast Stroma.
On illumination of isolated envelope-free chloroplasts the
medium exhibits an alkaline shift as protons are pumped from
the medium into the intrathylakoid space (7). In contrast,
intact chloroplasts illuminated in the absence of substrates
such as bicarbonate acidify the suspending medium (8). They
pump protons both across the thylakoid membranes into the
intrathylakoid space and across the envelope into the medium
or, in green cells, into the cytosol. In consequence, the

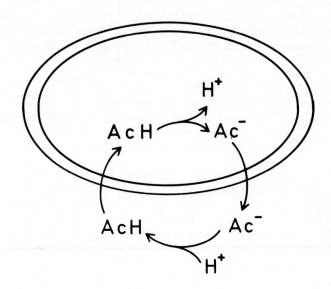

FIGURE 1. Indirect proton transfer across the chloroplast envelope by an acid/anion shuttle.

stroma pH increases on illumination by as much as one pH-unit (9). Proton transfer across the thylakoids is compensated electrically by countertransfer mainly of Mg^{++} (10) and across the envelope by countertransfer of K^+ (11).

2. <u>Effect of a Decrease in the Transenvelope Proton Gradient on Photosynthesis.</u> Werdan et al. (12) have shown that low concentrations of carbonyl cyanide-p-trifluoromethoxyphenylhydrazone which increases the proton permeability of biomembranes inhibit photosynthetic CO_2 reduction before phosphoglycerate reduction is affected. It was suggested that the decrease in the transenvelope proton gradient was primarily responsible for inhibition of photosynthesis. Photosynthesis is also decreased by several salts of weak acids such as potassium or sodium nitrite (13), glyoxylate, glycerate, formiate, glycolate, propionate and acetate or even bicarbonate (12). Sodium or potassium pyruvate are poor inhibitors. Salt inhibition is drastically pH-dependent and is strong at pH 7 and weak or absent at pH 8 of the medium in which the chloroplasts are suspended. At pH 6.9, 50 % inhibition of photosynthesis was observed at the following salt concentrations: glyoxylate 40 uM, nitrite 180 uM, formiate 5 mM, acetate 20 mM and pyruvate 90 mM. Inhibition was irreversible at a particular pH. However, when photosynthesis was inhibited at a low pH by

MODULATION OF PROTEIN FUNCTION   117

addition of salt, inhibition was relieved and photosynthetic oxygen evolution restored by simple titration to a higher pH. Inhibition can be explained for most, though not all inhibitory salts by their capability to facilitate indirect proton transfer via a shuttle of protonated and charged anions as depicted in Fig. 1. The import of the protonated anion alone is insufficient to produce irreversible inhibition of photosynthesis by acidification of the chloroplast stroma because protons derived from the dissociation of imported acid can be pumped out in exchange for K^+ (11). This would make acidification of the stroma a transient process. Back leakage into the stroma of charged anions is necessary to complete the cycle and make inhibition permanent. On both sides of the chloroplast envelope the dissociation equilibrium of the weak acid must be maintained

$$\left(\frac{(Ac^-)(H^+)}{(AcH)}\right)_i = \left(\frac{(Ac^-)(H^+)}{(AcH)}\right)_o$$

(i, inside; o, outside). If AcH is permeant, its concentration is at flux equilibrium the same on both sides of the envelope. Therefore

$$\frac{(Ac^-)_o}{(Ac^-)_i} = \frac{(H^+)_i}{(H^+)_o}$$

If the anion is permeant, anion leakage will occur. It results in more acid import which decreases the proton gradient until finally the Donnan distribution of ions is reached. In the light, active proton pumping across the envelope and proton import by the shuttle lead to a steady state of the transenvelope proton gradient which is determined by the rates of active proton pumping and of proton back leakage through the shuttle. The efficiency of the shuttle is, in turn, determined by the permeabilities of the chloroplast envelope for the acid and the anion and the p_K of the acid. All inhibitory salts have indeed been found to have permeant anions which at neutral pH are in equilibrium with significant concentrations of permeant acids. Where tested, they have been found to decrease the stroma pH. When high salt concentrations were required for inhibition, the anion permeability of the chloroplast envelope was found to be low. Examples are the bicarbonate and acetate anions (14). Nitrites are highly effective inhibitors. The nitrite anion penetrates the envelope rapidly (14). Only with glyoxylate permeability properties and effectiveness as an inhibitor did not match. At the low concentration of 50 μM, glyoxylate inhibited photosynthesis at pH 7, but did not produce much stroma acidification, and the glyoxylate anion did not penetrate fast. On addition of inhibitory concentrations of nitrite, formiate or glyoxylate levels of fructose and sedoheptulose bisphosphates in the stroma of photosynthesizing chloroplasts increased and sugar monophosphates declined. Another significant effect accompanying inhibition was the decrease in the stromal 3-phosphoglycerate/dihydroxyactone phosphate ratio(13). The latter shows that inhibition of photosynthesis cannot be explained by uncoupling. Indeed, experiments with broken chloroplasts have established that, if a proton shuttle is induced by the salts across thylakoid membranes, it is not effective enough to inhibit phosphorylation significantly. Rather, inhibition of photosynthesis is caused by inhibition of fructose and sedoheptulose bisphosphatases. This is indicated by the increase in the concentration of their substrates and the decrease in the concentration of their products. Both enzymes are known to be regulated. Fructose bisphosphatase is sensitive to pH changes and has practically no activity at pH 7. The maximum of activity is at pH 8.5 (15).

3. Effect of pH on Enzyme Activation.
a. Fructose bisphosphatase. Fructose bisphosphatase is activated in the light by a thioredoxin which is reduced by ferredoxin (4). Fig. 2 shows the time course of light-activation of fructose bisphosphatase in intact chloroplasts kept at pH 7.2 with or without some inhibitory salts present in the medium. The enzyme was always assayed at pH 8 immediately after hypotonic chloroplast rupture, i.e. close to the pH optimum of enzyme activity. In the presence of either nitrite or glyoxylate, almost no enzyme activation was observed. With acetate which is a rather poor inhibitor of photosynthesis the maximum level of activation was lower than in control chloroplasts. When the experiment of Fig. 2 was repeated at pH 8.2, acetate and low concentrations of nitrite and glyoxylate (3 mM) did not inhibit enzyme activation. At high concentrations, nitrite was inhibitory, but much less so than at low pH. Inhibition of enzyme activation as a function of salt concentration was very similar to inhibition of photosynthesis as a function of salt concentration (not shown).

When glyoxylate, nitrite or acetate were added to the assay medium in concentrations comparable to those that were sufficient to inhibit enzyme activation in chloroplasts, no significant effect on enzyme activity was observed. High concentrations were inhibitory.

Fig. 3 shows enzyme activation in chloroplasts suspended in media of different pH in the absence and in the presence of 5 mM potassium nitrite. Less than 1 mM nitrite was sufficient to cause 50 % inhibition of photosynthesis at pH 7.2 when added during illumination. When 5 mM nitrite were added to the chloroplasts in the light after 5 minutes photosynthesis, the activation state of fructose bisphosphatase was scarcely altered as compared with the control. Inactivation of photosynthesis after addition of nitrite in the light which has been

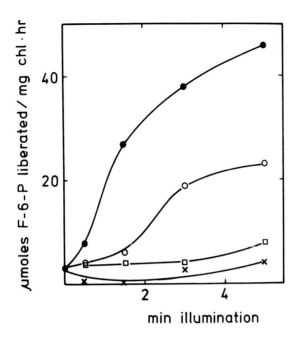

FIGURE 2. Activation of fructose bisphosphatase in intact chloroplasts illuminated with and without sodium glyoxylate, sodium acetate or potassium nitrite. After the indicated times, chloroplasts were ruptured hypotonically and immediately assayed at pH 8 for fructose bisphosphatase activity. Conditions: Chloroplast isolation and media as in (13). The chloroplast suspension contained 2 mM bicarbonate and 2500 units catalase/ml, pH 7.2. Illumination with saturating white light, enzyme assay in a reaction medium containing at pH 8 20 mM triethanolamine, 10 mM $MgCl_2$, 2.4 mM ethylene diamine tetraacetate, 1 mM fructose-1,6-bisphosphate, o.2 mM NADP, 40 µg glucose-6-phosphate isomerase/ml and 2 µg glucose-6-phosphate dehydrogenase/ml. Absorption changes were recorded at 340 nm. The experiment was performed at 20°C.

● ———— control
○ ———— 30 mM sodium acetate in chloroplast suspension
□ ———— 3 mM sodium glyoxylate in chloroplast suspension
× ———— 3 mM potassium nitrite in chloroplast suspension

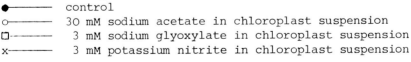

traced back to inactivation of fructose and sedoheptulose bisphosphatases (see above) must therefore be caused by pH-inactivation of the light-activated enzyme. However, when 5 mM nitrite was present before illumination, light activation was absent at low pH and was still decreased at high pH. Depending

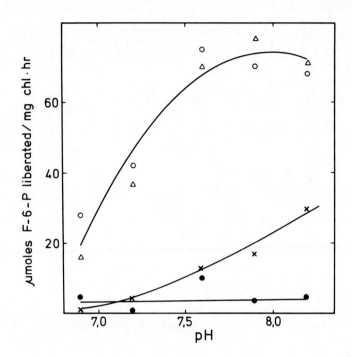

FIGURE 3. Activation of fructose bisphosphatase in intact chloroplasts as a function of the pH of the suspending medium. The chloroplasts were kept in the dark or illuminated with and without 5 mM potassium nitrite. They were subsequently ruptured hypotonically and immediately assayed at pH 8 for fructose bisphosphatase activity. For reaction media see (13) and legend to Fig. 2.

o——— chloroplast were illuminated for 5 minutes in the absence of nitrite
Δ——— chloroplasts were illuminated for 5 minutes in the absence of nitrite and then another 8 minutes with nitrite
x——— nitrite was added to the chloroplasts in the dark and the chloroplasts were then illuminated for 5 minutes
●——— controls kept for 5 minutes in the dark in the absence of nitrite

on whether the salts are added in the dark or in the light, they interfere with the activity of activated fructose bisphosphatase or with both enzyme activation and activity.

b. NADP-dependent glyceraldehydephosphate dehydrogenase.
Even though the change in the pattern of stroma metabolites

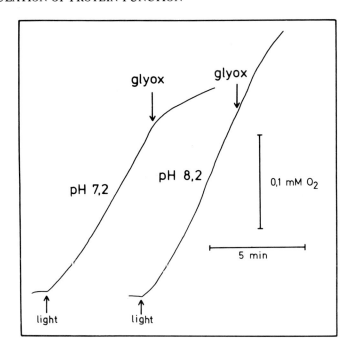

FIGURE 4. Effect of 6 mM potassium glyoxylate on phosphoglycerate-dependent oxygen evolution by intact spinach-chloroplasts. Intact chloroplasts were suspended in isotonic buffer (33), which contained 4 mM 3-phosphoglycerate and catalase (2000 units/ml) and was adjusted to pH 7.2 or 8.2, and were illuminated with saturating red light. 6 mM potassium glyoxylate was added in the light.

after salt inhibition of photosynthesis implicated fructose and sedoheptulose bisphosphatases and not glyceraldehyde-phosphate dehydrogenase as bottlenecks of photosynthesis, phosphoglycerate-dependent oxygen evolution was inhibited by glyoxylate in a pH-dependent mode (Fig. 4). NADP-dependent glyceraldehydephosphate dehydrogenase which is thought to catalyze reduction of phosphoglycerate is known to be light-activated (16). It has a broad pH optimum of activity. At pH 6.9 it was found to be only 20 % less active than at pH 8. It is therefore difficult to attribute the salt inhibition of phosphoglycerate reduction in chloroplasts to a direct pH effect on the enzyme. Rather the activation state of the anzyme is affected. Fig. 5 shows enzyme activity in chloroplasts suspended in a medium of pH 7.2 as a function of illimination time. The medium either contained 20 mM potassium formiate at the beginning of the experiment, or formiate was

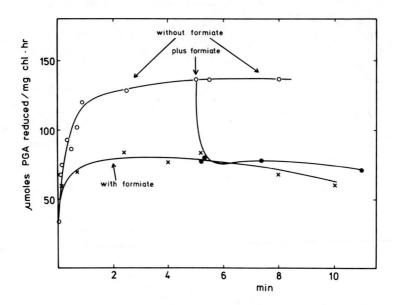

FIGURE 5. Activation of NADP-dependent glyceraldehyde-phosphate dehydrogenase in intact spinach chloroplasts illuminated with and without 20 mM potassium formiate. The chloroplast suspension contained 2 mM sodium bicarbonate and 2500 units/ml catalase, pH 7.2. After the indicated times, the chloroplasts were ruptured hypotonically and immediately assayed at pH 8. The assay medium contained 100 mM tricine, 10 mM $MgCl_2$, 5 mM ATP, 2 mM 3-phosphoglycerate, 0.3 units/ml phosphoglycerate kinase and 0.2 mM NADPH. For other conditions see legend to Fig. 2.
o———— control
x———— formiate was added in the dark 5 min before illumination
•———— formiate was added in the light, and illumination was continued

added in the light. After different illumination times, chloroplasts were hypotonically ruptured and assayed at pH 8. In the dark, activity of the enzyme was low. Illumination caused rapid activation. After about 2 minutes in the light, the enzyme had maximum activity, which was lower in the presence of formiate than in the control. When formiate was added in the light, the enzyme was rapidly inactivated. With the weak inhibitor formiate, this effect was not always as pronounced as shown in Fig. 5, but it was very clear with glyoxylate and nitrite (not shown). It contrasts to the behaviour of fructose bisphosphatase, which remained active in the light for several

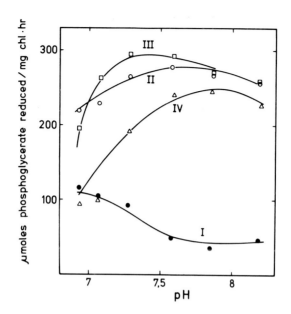

FIGURE 6. Activation of NADP-dependent glyceraldehyde-phosphate dehydrogenase in intact chloroplasts as a function of the pH of the suspending medium. The chloroplasts were kept in the dark or illuminated with saturating red light in the absence or presence of 20 mM potassium formiate. They were subsequently ruptured and immediately assayed at pH 8 for glyceraldehydephosphate dehydrogenase activity. For assay conditions see legend to Fig. 5.
  I chloroplasts kept for 3 minutes in the dark without formiate
  II after 3 minutes in the dark chloroplasts were illuminated for 5 minutes in the absence of formiate
  III after 5 minutes illumination, 20 mM formiate was added to the chloroplasts in the light and illumination was continued for further 8 minutes.
  IV After 3 minutes dark incubation with 20 mM formiate, chloroplasts were illuminated for 5 minutes.

---

minutes after inhibitory salts were added. Fig. 6 shows the pH-dependence of light-activation with and without formiate. In this particular experiment, the activation state of the enzyme was not significantly changed when formiate was added during illumination, after light-activation was complete. When it was present before illumination, it prevented activation at low, but not at high pH of the suspending medium. In chloroplasts kept in the dark, enzyme activity declined with in-

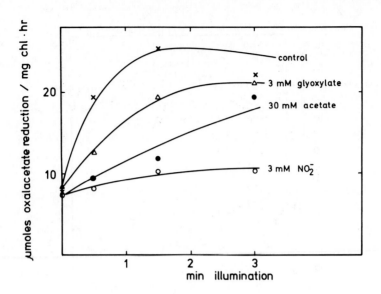

FIGURE 7. Activation of NADP-dependent malic dehydrogenase in intact chloroplasts illuminated with and without sodium glyoxylate, sodium acetate or potassium nitrite. The pH of the suspending medium was 7.2. After the indicated times, chloroplasts were osmotically ruptured in the light and immediately assayed at pH 7.8 for malic dehydrogenase activity. The assay medium contained 20 mM triethanolamine, 1 mM ethylene diamine tetraacetate, 4.5 mM $MgCl_2$, 0.5 mM oxaloacetate and 0.2 mM NADPH. For other conditions see legend to Fig. 2.

---

creasing pH of the suspending medium.

c. NADP-dependent malic dehydrogenase. On illumination, intact chloroplasts reduce added oxaloacetate to malate and oxygen is evolved (8). Oxygen evolution is maximally stimulated by 1 mM NH_4Cl. When 6 mM glyoxylate was added before illumination, oxygen evolution was decreased compared with control chloroplasts. Oxaloacetate reduction is thought to be catalyzed by NADP-dependent malic dehydrogenase which is known to be light-activated (17). This enzyme has a rather broad pH optimum of activity (18). Light-activation is pH-dependent as indicated by its sensitivity to the presence of inhibitory salts. Fig. 7 shows activity of malic dehydrogenase in intact chloroplasts as a function of illumination time. The chloroplasts were incubated at pH 7.2 with or without potassium nitrite, sodium acetate or sodium glyoxylate, then ruptured in the light and assayed at pH 7.8. All three salts inhibited enzyme activation.

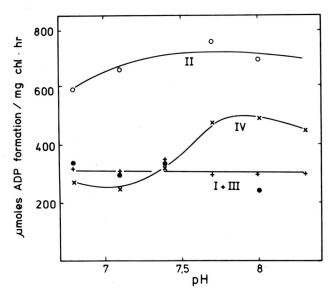

FIGURE 8. Activation of phosphoribulokinase in intact chloroplasts as a function of the pH of the suspending medium. The chloroplasts were kept in the dark or illuminated with saturating red light for 3 minutes with or without potassium nitrite present. They were subsequently ruptured hypotonically and immediately assayed at pH 7.8 for phosphoribulokinase activity. The assay medium contained 50 mM Hepes buffer, 5 mM phosphoenolpyruvate, 1 mM ATP, 1 mM ribose-5-phosphate, 0.4 mM NADH, 20 µg pyruvate kinase/ml, 20 µg lactate dehydrogenase/ml and 5 mM $MgCl_2$. The activity of ribose-5-phosphate isomerase was not rate-limiting. Chloroplast isolation and media as in (13).

●——— (I) dark controls
○——— (II) chloroplasts were kept in the dark for 3 minutes and then illuminated for 3 minutes
+——— (III) chloroplasts were incubated for 3 minutes in the dark with 2 mM nitrite and then illuminated for 3 minutes
×——— (IV) chloroplasts were incubated for 3 minutes in the dark with 0.5 mM nitrite and then illuminated for 3 minutes.

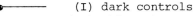

Glyoxylate was less effective than with fructose bisphosphatase (Fig. 2). At higher pH-values of the suspending medium, salt inhibition of enzyme activation was reduced or absent. When salts such as glyoxylate or acetate were added to chloroplasts in the light, no inactivation of malic dehydrogenase was ob-

served. This contrasts to the behaviour of glyceraldehyde-
phosphate dehydrogenase (Fig. 5).

d. Phosphoribulokinase. The activity of phosphoribulokinase,
which is known to be light-regulated (19), increased on illumination within little more than a minute from 100 to about
400 μmoles ribulose bisphosphate formation per mg chlorophyll
per hour. Dark inactivation was slow and had a half time of
more than one minute as compared with less than 5 seconds in
the case of NADP-dependent malic dehydrogenase. The activity
of the light-activated anzyme was pH-dependent and increased
from pH 7 to pH 8 by a factor of about 3. If the light-dependent pH shift in the chloroplast stroma and light activation
of the enzyme are both considered, the activity of phosphoribulokinase was higher by a factor of about 10 in illuminated
than in darkened chloroplasts. In contrast to the other enzymes,
light activation of phosphoribulokinase in a chloroplast suspension of pH 7.2 was not much influenced by 30 mM sodium
acetate. However, nitrite was very effective in preventing the
light-activation of the enzyme, and this effect was at appropriate nitrite concentrations pH-dependent. Fig. 8 shows activation of the enzyme as a function of the pH of the chloroplast
suspension. The assay was performed after chloroplast rupture
at pH 7.8. In the presence of 2 mM nitrite, illumination
failed to activate the enzyme over a broad pH range. When the
nitrite concentration was reduced to 0.5 mM, activation was
completely inhibited at low and much less so at high pH of the
chloroplast suspension. Nitrite oxidizes reduced ferredoxin
(20). The redox state of ferredoxin appears to be a factor in
the pH regulation of enzyme activation. Indeed, when oxaloacetate was added as electron acceptor to intact chloroplasts,
phosphoribulokinase activation at pH 7.1 became sensitive to
acetate, which was not or only scarcely inhibitory in the absence of oxaloacetate (not shown). Inhibition of phosphoribulokinase activation in the presence of oxaloacetate by acetate
was observed only at low, not at high pH. Thus, it is necessary
to consider both the stroma pH and the redox state of ferredoxin as factors which interplay to control the state of enzyme
activation. Acetate influences the stroma pH only, while nitrite
affects both the pH and the redox state of ferredoxin.

4. Amplification of pH-Regulation. As has been shown
above, photosynthesis of intact chloroplasts suspended in a
medium of low pH was inhibited when salts of weak acids such as
nitrite, glyoxylate, formiate or acetate were added in the light.
Inhibition was mainly caused by a decrease in the pH of the
chloroplast stroma which reduced the activity of light-
activated fructose and sedohetulose bisphosphatases. When the
salts were added in the dark before illumination, light-activa-

tion of the enzymes was also inhibited. Both effects together amplify pH-regulation. Under pH-restriction, only part of the total enzyme is activated, and the activated enzyme is still subject to activity regulation by pH. Indeed, at low pH photosynthesis was considerably more sensitive to acetate when the salt was added before illumination than when it was added during steady state photosynthesis. Moreover, the lag phase of photosynthesis was considerably increased by acetate indicating that the lag is not only caused by autocatalytic accumulation of photosynthetic intermediates but also by enzyme activation.

II. PHOSPHORYLATION POTENTIAL, ENERGY CHARGE AND REGULATION OF ATP SYNTHESIS

The phosphorylation potential $(ATP)/(ADP)(P_i)$ is a measure of the capacity of a system for phosphorlyating ADP, and energy charge

$$EC = 1/2 \frac{(ADP)+2(ATP)}{(AMP)+(ADP)+(ATP)}$$

indicates to which extent the adenylate system is charged with phosphate (6). Extreme values of EC are 0 (AMP the only adenylate) and 1 (ATP the only adenylate). Chloroplasts contain significant ATP in the dark not only in vivo when their adenylate system is linked to that of the cytosol, but also after isolation (21). Synthesis of ATP in the dark is possible only through triosephosphate oxidation. Triosephosphate is derived in isolated darkened chloroplasts from starch. Phosphoglycerate was formed by oxidation of endogenous triosephosphate at the slow rate of 0.1 umoles/mg chl·hr (22). This was sufficient to maintain ATP/ADP ratios between 0.5 and 1, phosphorylation potentials usually close to 50 (M^{-1}) and occasionally up to 100 (M^{-1}) and an energy charge generally between 0.4 and 0.6. The occurence of ATP in darkened intact chloroplasts implies that the thylakoid enzyme which synthesizes ATP in the light (24) is either inactive in the dark or can catalyze only the forward reaction of ATP synthesis, not the back reaction of hydrolysis. Inactivity in the dark rather than irreversibility is implied by the observation that microsecond flashes given in sufficiently long intervals (0.2 Hz) hydrolyze internal ATP (21). At higher flash frequencies (5 Hz or more), ATP was synthesized and ATP/ADP ratios increased to maximal values of 2 to 4, but still the first flashes resulted in hydrolysis which only lateron was replaced by synthesis (Fig. 9). The results indicate that the ATP synthesizing enzyme is activated by microsecond light flashes before the thylakoid system is sufficiently energized to permit synthesis. From the difference in the reaction rate of the back reaction in the dark (less than 0.1 umoles ATP hydrolyzed/mg chl·hr) and the maximum rate of the forward reaction in the light

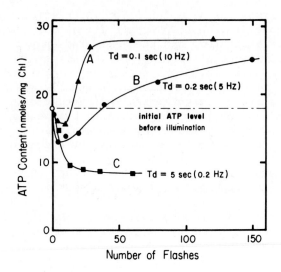

FIGURE 9. ATP level in intact spinach chloroplasts as a function of flash illumination. Saturating microsecond flashes were given to darkened chloroplasts every 0.1 sec (A), 0.2 sec (B) or 5 sec (C). Td, dark time. From (21), with permission.

(about 2000 μmoles ATP synthesized/mg chl·hr) it appears that the enzyme catalyzing ATP synthesis is regulated to an extent that by far surpasses regulation of other enzymes engaged in photosynthesis.

When intact chloroplasts were illuminated in the presence of electron acceptors such as oxaloacetate or nitrite which do not require ATP for reduction, ATP levels increased to a maximum within a few seconds. Saturation was observed at low light intensities (20 Wm^{-2}). Even in the presence of high phosphate concentrations which inhibited photosynthetic ATP turnover ATP/ADP ratios did not rise above 3 or 4 (Table 1) and ADP levels did not decrease much below 0.4 mM in the light. The phosphorylation potential increased by a factor of only about 3 on illumination, and energy charge attained values close to 0.85. Measured adenylate concentrations corresponded reasonably close to free and thermodynamically active concentrations. This was indicated by the observation that in most measurements adenylates were in adenylate kinase equilibrium. If significant binding had occured, deviations from equilibrium would be expected.

When NH_4Cl which is considered to be an uncoupling agent was added to illuminated chloroplasts, the intrathylakoid pH increased as indicated by 9-aminoacridine fluorescence measurements (Fig. 10 A) or by the distribution of ^{14}C-methylamine

TABLE I
ATP/ADP RATIOS, PHOSPHORYLATION POTENTIALS AND ENERGY
CHARGE IN INTACT CHLOROPLASTS KEPT IN THE DARK OR
ILLUMINATED WITH SATURATING LIGHT. ELECTRON ACCEPTOR
WAS 2 mM POTASSIUM NITRITE

	ATP/ADP	(ATP)/(ADP)(P_i)	energy charge
Dark	1	83 (M^{-1})	0.51
Light	3.9	325 (M^{-1})	0.84

between chloroplasts and medium, and electron transport to nitrite was stimulated (Fig. 10 B). However, ATP/ADP ratios and phosphorylation potentials did not decrease on addition of NH_4Cl up to a concentration of 5 mM (Table II) and CO_2

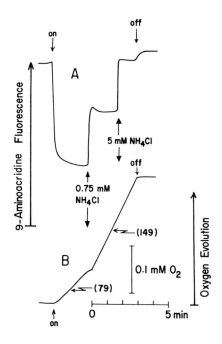

FIGURE 10. Formation of a proton gradient as indicated by the quenching of 9-aminoacridine fluorescence (A) and nitrite-dependent oxygen evolution (B) by intact spinach chloroplasts. Numbers in brackets show oxygen evolution in μmoles/mg chl·hr. Concentration of KNO_2 1 mM; illumination with saturating red light. For methods see (25).

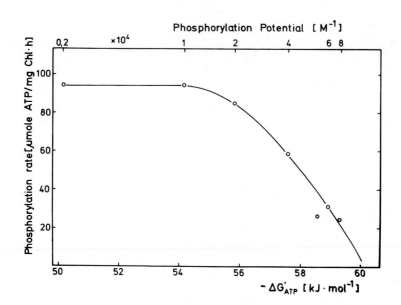

FIGURE 11. Phosphorylation of ADP in the light by broken spinach chloroplasts as a function of the phosphorylation potential. Intact chloroplasts were shocked in 5 mM $MgCl_2$ and transfered rapidly to a medium containing 330 mM sorbitol, 10 mM KCl, 5 mM $MgCl_2$, 0.5 mM phosphate, 1 mM NaN_3 and 50 /uM methylviologen. 0.23 mM ADP and 0.25-8 mM ATP were added. Nucleotide ratios were corrected for contaminations which had been determined by liquid chromatography. The pH of the reaction medium was adjusted to 8 and the steady rate of light-dependent alkalization was recorded.

reduction of intact chloroplasts was also not inhibited by NH_4Cl (25). The lacking response of the phosphorylation potential to NH_4Cl which stimulates electron transport to nitrite or oxaloacetate and increases the intrathylakoid pH shows that in intact chloroplasts the phosphorylation potential does not restrict electron flow. Rather, the intrathylakoid pH appears to be the controlling factor (23, 26). Furthermore, maximum phosphorylation potentials of not much more than 300 (M^{-1}) and ADP concentrations of about 0.4 mM indicate that phosphorylation is not limited by the availability of ADP.

However, when intact chloroplasts were ruptured and were then illuminated in the presence of ferricyanide or methylviologen and ADP, ATP and phosphate, phosphorylation was observed up to very high ATP/ADP ratios. Fig. 11 shows that the rate of phosphorylation was not controlled up to a phosphorylation potential of 10 000 (M^{-1}). Extrapolation indicated that

TABLE II

PHOSPHORYLATION POTENTIALS IN INTACT CHLOROPLASTS ILLUMINATED IN THE PRESENCE OF 1 mM OXALOACETATE AS AFFECTED BY NH_4Cl OR CARBONYL CYANIDE-m-CHLOROPHENYL-HYDRAZONE (CCCP). PHOTOSYNTHETIC ADENYLATE TURNOVER WAS INHIBITED BY 10 mM PHOSPHATE

	Control	1 mM NH_4Cl	5 mM NH_4Cl	5 μM CCCP
Ph.Pot.(M^{-1})	202	198	200	45

the high energy state of the thylakoids was in equilibrium with ATP, ADP and phosphate at the high phosphorylation potential of 80 000 (M^{-1}). Kraayenhof (27) has reported that the maximum phosphorylation potential of chloroplasts which he assumed were intact was 30 000 (M^{-1}). However, his chloroplasts were similar to our broken chloroplasts as shown by their response to ADP and ATP which penetrate intact chloroplasts only very slowly (1). At physiological phosphate concentrations (5 to 15 mM) and in the presence of sufficiently high activities of adenylate kinase, phosphorylation potentials of 30 000 or 80 000 (M^{-1}) correspond to an energy charge of more than 0.99. Thus maximal phosphorylation potentials of intact and of broken chloroplasts differed by two orders of magnitude, and energy charge in intact chloroplasts was always far below its maximal value even when photosynthetic ATP consumption was inhibited. $\Delta G'_{ATP}$ was about -41 kJ/mole in darkened and -46 kJ/mole in illuminated intact chloroplasts, but -60 kJ/mole in illuminated broken chloroplasts, if the $\Delta G'_o$ of ATP is considered to be -31 kJ/mole (28). It is unknown which factors restrict phosphorylation in intact chloroplasts so that ATP/ADP ratios and energy charge remain in the light within the metabolic control range characterized by Atkinson (6).

III. CONTROL OF ELECTRON FLOW

In C_3 plants such as spinach, CO_2 reduction requires ATP and NADPH at a ratio of at least 3 to 2. In C_4 plants such as Maize the overall ratio is believed to be 5 to 2. Up to now, the stoichiometry of coupling of ATP synthesis to electron transport is a matter of dispute, and either 1, 1.33 or 2 molecules of ATP have been assumed to be synthesized when 2 elctrons are transfered from water to an acceptor such as NADP (29-31). None of the proposed production ratios agrees with the requirements of consumption which are dictated by stoichiometry, and there is the question of how production is geared to consumption. Fig. 12 shows a scheme of vectorial photosynthetic

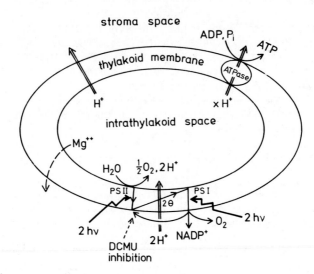

FIGURE 12. Scheme showing the sequential arrangement of 2 photosystems (PS II and PS I) and vectorial electron and proton transport across the thylakoid membrane. Proton influx during gradient formation is thought to be electrically compensated by Mg^{++} efflux (10). The proton gradient is assumed to be involved in phosphorylation (32). There is also a slow proton loss due to leakage. From (35), with permission.

electron transport which is coupled to proton translocation into thylakoid vesicles. According to Mitchell, the proton motive force represented by the proton gradient across the thylakoid membrane and a membrane potential generated by light drive ATP synthesis (32). In the scheme, three possibilities of electron transfer are envisaged. NADP has a high electron affinity and is preferentially reduced when available. Alternatively, electrons can reduce oxygen or react with the electron transport chain to form a cyclic proton pumping system. Although oxygen is reduced during photosynthesis of intact chloroplasts (33) and leaves (34), it is, probably because of kinetic restrictions, not a good electron acceptor in spinach chloroplasts, and maximum rates of oxygen reduction are only a few μmoles/mg chl·hr (35). In the absence of oxygen and oxidized $NADP^+$, linear electron transport from water cannot occur. Cyclic electron flow was indicated in intact spinach chloroplasts under anaerobic conditions by the formation of a proton gradient only under far-red illumination (Fig. 13). Under short wavelength red light which excites photosystem II no significant proton gradient was formed and ATP levels remained low

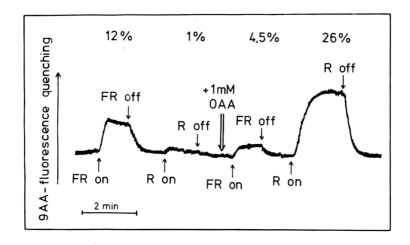

FIGURE 13. Quenching of 9-aminoacridine (9AA) fluorescence by intact chloroplasts illuminated with 100 Wm^{-2} red (R) or far-red (FR) light before and after addition of 1 mM oxaloacetate (OAA). The chloroplast suspension was made anaerobic by adding 10 mM glucose, glucose oxidase and an excess of catalase. Percent values indicate how much of the total 9-aminoacridine fluorescence was quenched. From (35), with permission.

even in the light. Inhibition of electron transport by red light is explained by over-reduction of electron carriers which in the reduced form block the cyclic path. When oxaloacetate which oxidizes NADPH was added, cyclic electron flow supported by far-red light was decreased because electrons were drained from the cyclic pathway. This caused oxidative inhibition . Under red illumination, inhibition of electron flow was released on addition of oxaloacetate, and electron flow from water to oxidized NADP supported a large proton gradient. 3-phosphoglycerate, whose reduction requires ATP and NADPH at a ratio of 1, was usually capable of replacing oxaloacetate as electron acceptor (not shown). When it was added, the proton gradient of light-inhibited chloroplasts increased indicating that inhibition of electron flow was relieved and that phosphoglycerate was reduced under anaerobiosis. Interestingly, bicarbonate was not a good electron acceptor under anaerobic conditions. Its reduction requires more ATP than that of phosphoglycerate. When bicarbonate was added in the absence of oxygen, only a short transient increase in the proton gradient was usually observed. However, when oxygen was also admitted to the system, bicarbonate reduction became possible (Fig. 14).

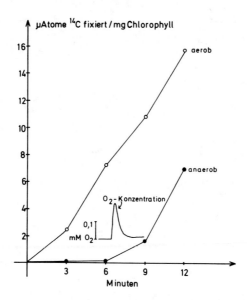

FIGURE 14. Fixation of ^{14}C from $H^{14}CO_3^-$ by intact spinach chloroplasts under illumination with red light (half bandwidth from 633 to 673 nm, intensity 110 Wm^{-2}) under aerobic and anaerobic conditions. Anaerobiosis was produced by adding 10 mM glucose, glucose oxidase and an excess of catalase. After 6 min illumination, H_2O_2 was injected into the anaerobic sample. It was decomposed by catalase and produced the transient oxygenation shown in the insert. In the aerobic sample, glucose oxidase was omitted from the reaction mixture. For methods, see (33).

Once photosynthesis was initiated, significant levels of oxygen were no longer needed. The range of oxygen concentrations that could start photosynthesis was very broad. Fig. 15 shows that 15 µM oxygen, which is equivalent to about 1 % oxygen in the gas phase, were sufficient to cause reduction of added bicarbonate as indicated by the maintenance of an increased proton gradient after the added oxygen had largely been removed by an enzymic trapping system. A low stationary level of oxygen was maintained in the light. It reflects the balance between photosynthetic oxygen evolution and oxygen consumption by the enzymic trap. A second illumination separated from the first by a brief period of darkness failed to restart photosynthesis, and the proton gradient remained small. The results show that oxygen has an important regulatory function in photosynthesis. In the absence of oxygen, the electron transport

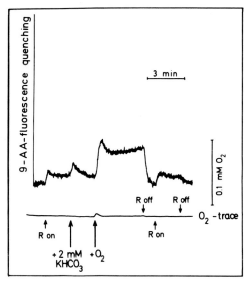

FIGURE 15. Quenching of 9-aminoacridine fluorescence by intact chloroplasts illuminated with red light. Lower trace shows oxygen concentration in the sample which before illumination was anaerobic. For conditions, see Fig. 13.

chain remains over-reduced under illumination that is capable of exciting photosystem II unless accumulated NADPH and reduced ferredoxin are oxidized. Oxaloacetate, nitrite and phosphoglycerate are capable of such oxidation, but bicarbonate is not. Bioenergetically, phosphoglycerate and bicarbonate differ only in respect to the ATP requirement of reduction. It appears that over-reduced chloroplasts can produce sufficient ATP to phosphorylate added phosphoglycerate, and electron transport induced by reduction of 1,3-diphosphoglycerate produces more ATP so that phosphorylation and reduction can continue. When bicarbonate is added the increased need for ATP to start the carbon cycle cannot be met and NADPH cannot be oxidized. Very low concentrations of oxygen relieve over-reduction, even though oxygen is an inefficient electron acceptor which is reduced by photosystem I. Once over-reduction is relieved, the system shows autoregulatory properties. Under ATP deficiency, NADPH levels increase during photosynthetic CO_2 reduction and electrons are diverted to oxygen and into the cyclic pathway. This increases ATP production which, in turn, permits photosynthetic NADPH oxidation. Autoregulation is not possible in the absence of oxygen, as over-reduction would inhibit electron flow. The mechanism of oxygen action is unclear in view of the very low rate of oxygen reduction at low oxygen concentrations which are still effective in regulating electron distribution. The physiological role of cyclic electron flow which has for some time been in doubt (3) is to supply ATP for photosynthesis that cannot be provided by linear electron flow (29, 35-37). The cooperation of linear and cyclic electron transport pathways enables chloroplasts of very different coupling quality to photoreduce CO_2 or even acceptors such as glycerate which requires more ATP for reduction than CO_2. It gives chloroplasts metabolic flexibility.

ACKNOWLEDGMENTS

Part of this work was performed during a sabbatical of U.H. at the Institute of Physical and Chemical Research, Wako-shi, Saitama, Japan. U.H. wishes to acknowledge his debt of gratitude to Prof. K. Shibata and his coworkers for hospitality and cooperation, and to the Deutsche Forschungsgemeinschaft and the Japan Society for the Promotion of Science for support. We are also very grateful to Mrs. U. Behrend and Mrs. R. Reidegeld for expert technical assistance and help in preparing the manuscript.

REFERENCES

1. Heber, U. (1974). Ann. Rev. Plant Physiol. 25, 293.
2. Heldt, H.W. (1976). In "The Intact Chloroplast" (J. Barber, ed.), pp. 215-234. Elsevier, Amsterdam.
3. Krause, G.H., and Heber, U. (1976). In "The Intact Chloroplast" (J. Barber, ed.), pp. 171-214. Elsevier, Amsterdam.
4. Wolosiuk, R.A., and Buchanan, B.B. (1977). Nature 266, 565.
5. Anderson, L.E., Nehrlich, S.C., and Champigny, M.-L. (1978) Plant Physiol. 61, 601.
6. Atkinson, D.E. (1968). Biochemistry 7, 4030
7. Neumann, J., and Jagendorf, A.T. (1964). Arch. Biochem. Biophys. 107, 109.
8. Heber, U., and Krause, G.H. (1971). In "Photosynthesis and Photorespiration" (M.D. Hatch, C.B. Osmond, and R.O. Slatyer, eds.) pp. 139-152. Wiley Interscience, New York.
9. Heldt, H.W., Werdan, K., Milovancev, M., and Geller, G. (1973). Biochim. Biophys. Acta 314, 224.
10. Krause, G.H. (1977). Biochim. Biophys. Acta 460, 500.
11. Gimmler, H., Schäfer, G., and Heber, U. (1974). In: Proc. of the 3rd Int. Congress on Photosynthesis (M. Avron, ed.), pp. 1381-1392. Elsevier, Amsterdam.
12. Werdan, K., Heldt, H.W., and Milovancev, M. (1975). Biochim. Biophys. Acta 396, 276.
13. Purczeld, P., Chon, C.J., Portis, A.R., Heldt, H.W., and Heber, U. (1978). Biochim. Biophys. Acta 501, 488.

14. Heber, U., and Purczeld, P. (1978). In: Proc. of the 4th Int. Congress on Photosynthesis 1977. (D.O. Hell, J. Coombs, and T.W. Goodwin, eds.), pp. 107-118. The Biochemical Society, London.
15. Garnier, R.V., and Latzko, E. (1972). In: Proc. of the 2nd Int. Congress on Photosynthesis Research (G. Forti, M. Avron, and A. Melandri, eds.), pp. 1839-1845. Dr. W. Junk, The Hague.
16. Wolosiuk, R.A., and Buchanan, B.B. (1978). Plant Physiol. 61, 669.
17. Wolosiuk, R.A., Buchanan, B.B., and Crawford, N.A. (1978). FEBS-Letters 81, 253.
18. Johnson, H.S., and Hatch, M.D. (1970). Biochem. J. 119, 273.
19. Wolosiuk, R.A., and Buchanan, B.B. (1978). Arch. Biochem. Biophys. 189, 97.
20. Leech, R.M., and Murphy, D.J. (1976). In "The Intact Chloroplast" (J. Barber, ed.), pp. 365-401. Elsevier, Amsterdam.
21. Inoue, Y., Kobayashi, Y., Shibata, K., and Heber, U. (1978). Biochim. Biophys. Acta 504, 142
22. Heldt, H.W., Chon, C.J., Maronde, D., Herold, A., Stankovic, Z.S., Walker, D.A., Kraminer, A., Kirk, M.R., and Heber, U. (1977). Plant Physiol. 59, 1146.
23. Kobayashi, Y., Inoue, Y., Shibata, K., and Heber, U. (1979). Manuscript submitted to Planta.
24. Jagendorf, A.T. (1977). In: Photosynthesis I. Photosynthetic Electron Transport and Photophosphorylation, Encyclopedia of Plant Physiol. New Series, Vol. 5 (A. Trebst, and M. Avron, eds.) pp. 307-337, Springer, Heidelberg.
25. Tillberg, J.-E., Giersch, Ch., and Heber, U. (1977). Biochim. Biophys. Acta 461, 31
26. Siggel, U. (1975). In: Proc. of the 3rd Int. Congress on Photosynthesis (M. Avron, ed.), pp. 645-654, Elsevier, Amsterdam.
27. Kraayenhof, R. (1969). Biochim. Biophys. Acta 180, 213.
28. Melandri, A. (1977). In: Encyclopedia of Plant Physiol., New Series, Vol. 5 (A. Trebst, and M. Avron, eds.), pp. 358-368, Springer, Heidelberg.
29. Arnon, D.I., and Chain, R.K. (1977). In "Photosynthetic Organelles" (S. Miyachi, S. Katoh, Y. Fujita, and K. Shibata, eds.), pp. 129-147. Japanese Society of Plant Physiologists and Center for Academic Publications Japan, Tokyo.
30. Hall, D.O. (1976). In "The Intact Chloroplast" (J. Barber, ed.), pp. 135-170. Elsevier, Amsterdam.
31. Heber, U. (1976). J. Bioenerg. Biomembr. 8, 157.
32. Mitchell, P. (1966). Biol. Rev. 41, 445.
33. Egneus, H., Heber, U., Matthiesen, U., and Kirk, M. (1975). Biochim. Biophys. Acta 408, 252.

34. Heber, U. (1969). Biochim. Biophys. Acta 180, 302.
35. Heber, U., Egneus, H., Hanck, U., Jensen, M., and Köster, S. (1978). Planta 143, 41.
36. Slovacek, R.E., Mills, J.D., and Hind, G. (1978). FEBS-Letters 87, 73-76.
37. Mills, J.D., Slovacek, R.E., and Hind, G. (1978). Biochim. Biophys. Acta 504, 298.

REGULATION OF PHOTOSYNTHETIC CARBON METABOLISM
AND PARTITIONING OF PHOTOSYNTHATE[1]

James A. Bassham

Laboratory of Chemical Biodynamics, University of
California, Berkeley, California 94720

ABSTRACT The rate-limiting steps in the Reductive
Pentose Phosphate Cycle (RPP Cycle) were identified by
measurement of metabolite concentrations during steady-
state photosynthesis and calculation of free energy
changes associated with all the steps in the cycle. The
rate-limiting, regulated steps are: the carboxylation of
ribulose 1,5-bisphosphate (mediated by RuBPCase), the
conversion of ribulose-5-phosphate to ribulose 1,5-bis-
phosphate with ATP (PR kinase) and the conversion of
fructose 1,6-bisphosphate and sedoheptulose 1,7-bisphos-
phate to fructose-6-phosphate and sedoheptulose-7-phos-
phate, respectively (mediated by FBPase and SBPase).
These enzymes plus PR kinase and to some extent triose
phosphate dehydrogenase are inactivated in intact chloroplasts in the dark. The light activation of these
enzymes in chloroplasts involves specific changes in the
metabolic environment which occur when the light is on.
These include increased levels of reduced cofactors,
especially reduced ferredoxin and NADPH, an increased
level of magnesium ion, and an increase in pH. RuBPcase-
oxygenase is activated in the presence of Mg^{++} and CO_2,
is further activated by 6-phosphogluconate or NADPH, and
is inactivated by free RuBP (or an impurity formed
spontaneously from RuBP). In the light, reduced carbon
is exported from the chloroplast in the form of triose
phosphate, but hexose monophosphates may be converted to
starch within the chloroplast as a storage product.
Since the ratio of these two types of carbon drain from
the cycle can vary, a fine tuning of the activity of FBP-
ase and SBPase compared to RuBPCase is required. With
isolated chloroplasts, triose phosphate export increases
with external inorganic phosphate (P_i) concentration, and
this may be an important regulatory mechanism in vivo as

[1] This work was supported by the Division of Biomedical
and Environmental Research of the U.S. Department of Energy
under contract No. W-7405-ENG-48.

well. In developing cells exported triose phosphate is mainly converted via phosphoenolpyruvic acid (PEPA) to amino acids, fatty acids, and other substances required for cell growth. In more mature cells, the triose phosphate is largely converted to sucrose which is translocated to other parts of the plant. Rate-limiting steps for these branching pathways include pyruvate kinase and PEPA carboxylase as well as sucrose phosphate synthetase.

INTRODUCTION

The Reductive Pentose Phosphate Cycle (RPP Cycle)(1,2) is ubiquitous to all photoautotrophic green plants (3). Higher plants which contain either the 4-carbon acid cycle (C_4 cycle) or the crassulacean acid metabolism (CAM) also employ the RPP Cycle as the only pathway for the net reduction of CO_2 to sugars.

The C_4 cycle (4,5,6) apparently utilizes ATP from photochemical reactions to fix CO_2 in chloroplasts of the outer leaf cells and bring that carbon into the chloroplasts of the inner or bundle sheath cells where the RPP cycle is functioning. Through this process, C_4 plants maintain a higher level of CO_2 at the local environment of the carboxylation enzyme of the RPP cycle, ribulose 1,5-bisphosphate carboxylase (RuBP-Case). The result is to diminish the oxygenase activity of this enzyme (7,8,9) which can convert the carboxylation substrate, ribulose 1,5-bisphosphate (RuBP) to glycolate, a substrate of the apparently wasteful process of photorespiration (10,11). In addition, the C_4 pathway can recycle CO_2 that is produced within the leaves by photorespiration (10).

The function of CAM is to collect CO_2 at night for plants in semi-arid or arid regions in which the stomata of the plants must be closed during the hot daylight hours in order to conserve water. The carbon, stored at night as C_4 acids, is released as CO_2 to be refixed by the RPP cycle during the daytime.

The RPP cycle in eucaryotic cells occurs only inside the chloroplasts (12), the subcellular organelles which are also the site of the primary photochemical reactions of photosynthesis in eucaryotic cells. The process of photosynthesis is complete within the chloroplasts, from the capture and conversion of light energy and the oxidation of water to molecular oxygen, to the uptake and reduction of CO_2 to sugar phosphates. The primary reactions of photosynthesis occur in the lamellae, or thylakoid membranes which contain the pigments, electron carriers, and other constituents involved in light absorption and conversion to chemical energy. The chemical energy derived from the primary reactions is used to drive electron transport and photophosphorylation (13). The overall result of electron transport in the membranes is the oxidation of water, giving O_2, and the reduction of the non-heme iron protein ferredoxin, to its reduced form (14). Photophosphorylation converts ATP and inorganic phosphate to ATP (12). The soluble cofactors, reduced ferredoxin and ATP, are the source of reducing equivalents and energy for the conversion of CO_2 to sugar phosphates in the chloroplasts.

The enzymes catalyzing steps of the RPP cycle are water soluble and are located in the stroma region of the chloroplast (15). Only 3 steps in the cycle out of a total of 13 actually use up cofactors (1,2) (ATP and NADPH) which must be regenerated at the expense of cofactors formed by the light reaction in the thylakoids. ATP is used directly, while NADPH is formed by reduction of $NADP^+$ at the expense of oxidation of two equivalents of ferredoxin.

There are no photochemical steps in the RPP cycle. There are, however, indirect effects of the light reactions on steps of the RPP cycle other than those requiring ATP or NADPH (16, 17,18,19). These indirect effects on the catalytic activities of enzymes of the cycle are apparently mediated via changes in levels of cofactors and perhaps other electron carriers, and by changes in concentrations of certain ions such as Mg^{++} and H^+ in the stroma regions of the chloroplast where the soluble enzymes of the RPP cycle are located.

Following the mapping of the RPP cycle, not much effort was made to investigate its metabolic regulation for several years. Possibly it was assumed that the cycle would operate or not depending on the availability of light energy and other favorable physiological conditions for photosynthesis. In time it became apparent that the metabolism of the chloroplast is not limited to the RPP cycle, nor is reduced carbon withdrawn from the cycle at a single point to make a single product. Within the chloroplast, hexose monophosphates may be converted with the ATP from the light reaction to make starch which is stored in the chloroplast. Triose phosphates are known to be exported from the chloroplasts to the cytoplasm where they serve as a major source of carbon and reducing power for subsequent biosynthetic reactions. Intermediates of the RPP cycle, such as the triose phosphates and pentose phosphates, may be used for biosynthetic reactions within the chloroplasts. In the dark, an oxidative metabolism occurs whereby stored starch may be degraded via either glycolysis or an oxidative pentose phosphate cycle (OPP Cycle) to give various sugar phosphates and PGA, some of which may be exported from the chloroplasts. Metabolic regulation of the cycle was clearly needed at rate-limiting steps to allow for shifts in metabolism in the chloroplasts. Such shifts would accompany transitions between light and dark (16,17) and changes in the needs for utilization of RPP cycle intermediates for biosynthesis in accordance with the physiological requirements of the plant (20).

Our interest in metabolic regulation of the RPP cycle came from two kinds of experimental findings. One was the apparently anomalous transient changes in pool sizes of intermediate compounds in the chloroplasts observed when physiological conditions were suddenly changed. The other reason was the apparent inadequacy in the activities of some of the isolated enzymes catalyzing steps in the cycle (21,22). The key carboxylation enzyme (RuBPCase) appeared for years to have too high a K_m CO_2 value to accommodate the physiological levels of CO_2 normal for plants. The changes in enzyme activities with physiological conditions and the apparent inadequacies of activities under suboptimal conditions have now been mostly explained as resulting from the needs of metabolic regulation (18,23,24). Initially, information about regulation of the RPP cycle has come from studies of the pool sizes of intermediate metabolites (16,17,19,25,26). Other information was obtained by measurements of O_2 evolution, phosphorylation, etc. in leaves, whole cells, isolated chloroplasts, and reconstituted chloroplasts (e.g., 27-33). Much important information has come from studies of the properties of the individual enzymes, both in crude extracts and in isolated pure form. (Review, see Kelley, et al., ref. 34). Much of the recent information about the regulation of rate-limiting enzymes of the RPP cycle is covered in another paper at this meeting by Buchanan. I will focus on the general pattern of *in vivo* regulation of the RPP cycle and of biosynthetic pathways leading from this cycle in the green cell.

The complete RPP cycle, in which each reaction occurs at least once, is depicted in Fig. 1. The number of arrows represent the number of products and reactants participating in one complete cycle. In such a complete cycle, three molecules of RuBP are carboxylated to give six molecules of PGA, which after reduction, yield six molecules of GAl3P. Five of these GAl3P molecules (15 carbon atoms) are required to regenerate the three RuBP molecules. The sixth GAl3P molecule, equivalent in carbon to the three CO_2 molecules fixed, can either be converted by a reverse of glycolysis to glucose phosphate for starch synthesis, or can be exported from the chloroplast to the cytoplasm for extra chloroplastic reactions (35-39). Other biosynthetic uses in the chloroplasts are also possible; for example, conversion of GAl3P to glycerol phosphate and eventually fats.

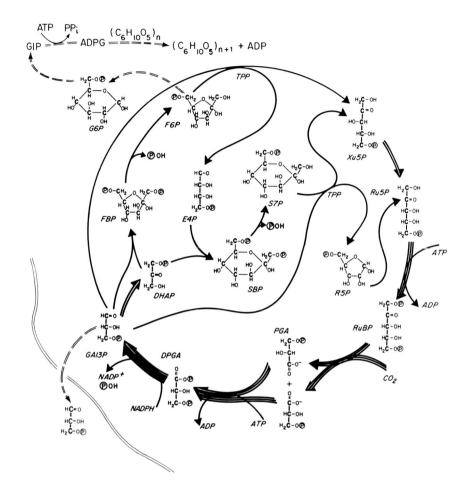

FIGURE 1. The Reductive Pentose Phosphate Cycle. The heavy lines are for reactions of the RPP cycle, the light double lines indicate removal of intermediate compounds of the cycle for biosynthesis. The number of heavy lines in each arrow equals the number of times that each reaction occurs for one complete turn of the cycle, in which 3 molecules of CO_2 are converted to 1 molecule of GAl3P. Abbreviations: RuBP, ribulose 1,5-bisphosphate; PGA, 3-phosphoglyceraldehyde; DHAP, dihydroxyacetone phosphate; FBP, fructose 1,6-bisphosphate; F6P, fructose 6-phosphate; G6P, glucose 6-phosphate; E4P, erythrose 4-phosphate; SBP, sedoheptulose 1,7-bisphosphate; S7P, sedoheptulose 7-phosphate; Xu5P, xylulose 5-phosphate, R5P, ribose 5-phosphate; Ru5P, ribulose 5-phosphate; TPP, thiamine pyrophosphate. ⓟ represents the $-PO_3^{2-}$ and $-PO_3H^-$ radicals present at physiological pH.

In the initial step of the RPP cycle, RuBPCase catalyzes the addition of CO_2 to the No. 2 carbon atom of RuBP (40-43). It is believed that an unstable enzyme-bound six carbon intermediate results, and that this intermediate is hydrolytically split with a concurrent transfer of a pair of electrons from C-3 of the RuBP to C-2.

The enzyme, RuBPCase can also function as an oxygenase. In this case O_2 binds at the catalytic binding site normally intended for CO_2 and subsequently reacts with the C-2 carbon atom of RuBP. The products of this RuBP oxygenase reaction are phosphoglycolate derived from carbon atoms 1 and 2 and PGA derived from carbon atoms 3, 4 and 5 (7-9). This oxygenase reaction is not a normal part of the RPP cycle, but is thought to be the starting point of the process of photorespiration, discussed later.

The product of the carboxylation reaction, PGA, is first converted to acylphosphate, a reaction using ATP and mediated by PGA kinase. The acylphosphate is then reduced with NADPH in the presence of triose phosphate dehydrogenase yielding GAl3P. The overall reaction proceeds in the light under highly reversible conditions. During photosynthesis this conversion of PGA to GAl3P is probably not subject to metabolic regulation. When the light is off however, conversion of GAl3P to PGA would occur with substantial negative free energy change, and it appears that this oxidative sequence may well be regulated by mechanisms similar to those involved in other light-dark controlled steps.

A series of isomerizations and rearrangements are required for the conversion of five triose phosphate molecules to three pentose phosphate molecules. None of these reactions utilize light generating cofactors (ATP and NADPH), and most of the steps are highly reversible. The two steps which liberate inorganic phosphate (those mediated by FBPase and SBPase) are rate-limiting and are subject to metabolic regulation.

The final step in the RPP cycle is the conversion of Ru5P to RuBP with ATP and phosphoribulose kinase. This is also a rate-limiting reaction and there is evidence that it is metabolically regulated.

Within the chloroplast, the principal direct product of the cycle is starch (44). The pathway to starch begins with the conversion of GAl3P to F6P by reactions that are also involved in the RPP cycle. F6P is then converted to glucose-6-phosphate (G6P) with hexose phosphate isomerase. Little or no free glucose is formed in photosynthesizing chloroplasts. G6P is converted to GlP with phosphoglucomutase in a reversible reaction such that _in vivo_ the ratio of G6P/GlP is about 20. GlP is then converted with ATP to ADP glucose and inorganic pyrophosphate (PP_i). As shown first by Ghosh and Preiss

(45) this reaction is an important regulatory point. Once formed, ADP glucose can transfer glucose to lengthen the amylose chain of the starch molecule.

In the light the principal export of carbon from the chloroplast occurs in the form of triose phosphates, GA13P and DHAP. A specific phosphate translocator apparently exists whereby a transport out of the chloroplast of triose phosphate is balanced by a movement into the chloroplast of inorganic phosphate (38,29,46). It appears that PGA also can move through the chloroplast envelope via the phosphate translocator mechanism.

RESULTS AND DISCUSSION

Determination of the rate-limiting and hence regulated steps of the RPP cycle and subsequent biosynthetic pathways in green cells has been greatly facilitated by the techniques of measuring levels of intermediates by steady-state tracer analysis (25). The unicellular algae, Chlorella pyrenoidosa, were allowed to photosynthesize under steady state conditions with air levels of CO_2 in which the CO_2 was replaced by $^{14}CO_2$. Samples were taken from time to time and analyzed by two-dimensional paper chromatography and radioautography. After some minutes of photosynthesis, the level of radioactive labeling became constant, indicating that the specific activity of each carbon atom position in the intermediary metabolites was equal to that of the $^{14}CO_2$ taken up by the cells. The ^{14}C content may then be used as a measure of the actual concentration of the compound in the actively turning over pools in the cells. These concentrations were used together with standard free energy change data to calculate the steady state free energy changes accompanying each reaction of the cycle under the chosen physiological conditions. This information provides a direct measure of the reversibility of the reactions as they are occurring in vivo (25). It can easily be shown that the relation between ΔG^s and the reversibility of the reaction is given by $\Delta G^s = -RT \ln (f/b)$, where f is the forward reaction rate and b the back reaction. In order for such measurements to be meaningful, accurate procedures for the measurements of steady state conditions and continuous measurement of CO_2, specific radioactivity, rapid sampling and killing, and quantitative analysis of radioactivity in each compound as a function of the amount of tissue sampled were developed (47).

The physiological free energy changes thus obtained for photosynthesizing Chlorella are summarized in Table 1 (25). The reactions shown to be rate-limiting in the light (during photosynthesis) were those mediated by RuBPCase, FBPase,

TABLE 1
FREE ENERGY CHANGES OF THE RPP CYCLE

The standard physiological Gibbs free energy changes ($\Delta G'$) were calculated for units activities, except $[H^+]=10^{-7}$. The physiological free energy changes at steady-state are for a 1% w/v suspension of <u>Chlorella pyrenoidosa</u> photosynthesizing with 0.04% $^{14}CO_2$ in air and with other conditions as described by Bassham and Krause (25). The stroma concentrations were assumed to be four times the total cellular concentrations, and are used as approximations for activities.

Reaction	$\Delta G'$	ΔG^S
$CO_2 + RuBP^{4-} + H_2O \rightarrow 2\ PGA^{3-} + 2\ H^+$	-8.4	-9.8
$H^+ + PGA^{3-} + ATP^{4-} + NADPH \rightarrow ADP^{3-} + GA13P^{2-} + NADP^+ + P_i^{2-}$	+4.3	-1.6
$GA13P^{2-} \rightarrow DHAP^{2-}$	-1.8	-0.2
$GA13P^{2-} + DHAP^{2-} \rightarrow FBP^{4-}$	-5.2	-0.4
$FBP^{4-} + H_2O \rightarrow F6P^{2-} + P_i^{2-}$	-3.4	-6.5
$F6P^{2-} + GA13P^{2-} \rightarrow E4P^{2-} + Xu5P^{2-}$	+1.5	-0.9
$E4P^{2-} + DHAP^{2-} \rightarrow SBP^{4-}$	-5.6	-0.2
$SBP^{4-} + H_2O \rightarrow S7P^{2-} + P_i^{2-}$	-3.4	-7.1
$S7P^{2-} + GA13P^{2-} \rightarrow R5P^{2-} + Xu5P^{2-}$	+0.1	-1.4
$R5P^{2-} \rightarrow Ru5P^{2-}$	+0.5	-0.1
$Xu5P^{2-} \rightarrow Ru5P^{2-}$	+0.2	-0.1
$Ru5P^{2-} + ATP^{4-} \rightarrow RuBP^{4-} + ADP^{3-} + H^+$	-5.2	-3.8
$F6P^{2-} \rightarrow G6P^{2-}$	-0.5	-0.3

SBPase, and phosphoribulose kinase. As it happens, these are four of the five reactions for which there is evidence for light-dark regulation. The three reactions with the highest negative free energy changes were catalyzed by RuBPCase, FBPase, and SBPase, the enzymes which in their isolated states had been reported to have insufficient catalytic activity to accommodate the requirements of the RPP cycle (21,22).

The relationships between the positions of the rate-limiting regulated steps of the RPP cycle and the points at which reduced intermediates are removed from the cycle for subsequent biosynthesis are shown in Figure 2. As mentioned

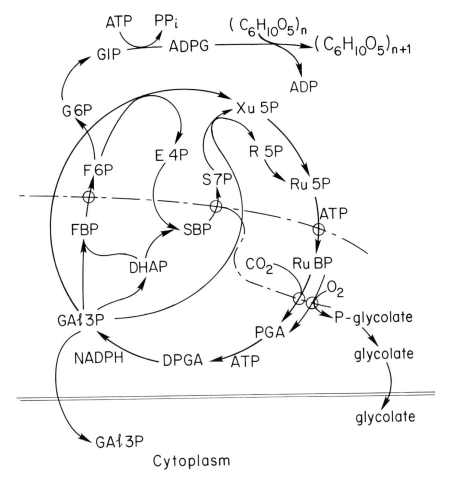

FIGURE 2. Principal Regulation Sites of RPP Cycle. Circles on arrows indicate regulation sites at reactions mediated by RuBPCase/oxygenase, FBPase, SBPase, and phosphoribulokinase.

earlier, the two major points at which carbon is withdrawn from the cycle (besides the photorespiratory loss of phosphoglycolate) are the export of triose phosphate from the chloroplast and the conversion of fructose-6-phosphate via G6P and G1P to starch. A line through the regulatory sites at phosphoribulose kinase and RuBPCase on one side of the cycle and FBPase and SBPase on the other divides the cycle into two stages. Following the RuBPCase and proceeding the steps mediated by SBPase and FBPase are PGA and triose phosphates which can be exported from the chloroplasts for biosynthesis. In the other half of the cycle are found the hexose phosphates which are used for starch biosynthesis. Clearly, the concentrations of metabolites in the two halves of the RPP cycle may be kept in balance through small adjustments in the activities of FBPase and SBPase as compared with the activity of the most rate limiting reaction, the RuBPCase. This fine control of these rate-limiting reactions in the light may be viewed as superimposed on the on-off type regulation of these same enzymes which appears to occur in the transition between light and dark metabolism.

Probably several mechanisms are involved in the fine control of activity of FBPase and SBPase in the light. One possible mechanism can be found in the sigmoidal dependence of FBPase activity on FBP concentration described by Preiss et al. (48). Many kinetic tracer studies in our laboratory (e.g., 16,19) show that concentrations of triose phosphates, FBP and SBP rise and fall together under a variety of physiological conditions, indicating rapid reversibility of the reactions catalyzed by aldolase. A drop in the level of GA13P therefore would result in a corresponding drop in the levels of FBP and SBP. In the case of FBPase and perhaps SBPase as well, this drop in concentration of substrate would result in diminished enzymic activity thereby decreasing the rate of utilization of the triose phosphate and hexose-heptose diphosphate pools, thus allowing these pools to build up again. Probably other mechanisms are required as well for maintaining the levels of triose phosphate and the diphosphates under changing physiological conditions of photosynthesis.

Major on-off types of regulation of the FBPase, SBPase, and RuBPCase were revealed by kinetic tracer studies during the light-dark and dark-light transitions (16,17,19). When the light was turned off following steady state photosynthesis, the concentration of RuBP declined but not to below detectable levels in some experiments. After about 100 seconds there was no further decline, indicating that the carboxylase had become essentially inactive since the large negative free energy change accompanying that reaction would ensure complete utilization of the RuBP substrate if the enzyme remained active.

Further studies of dark metabolism following photosynthesis under steady state conditions were carried out using both ^{32}P-labeled phosphate and $^{14}CO_2$ (16,17). There is an immediate appearance of labeled 6-phosphogluconate in the dark and an equally rapid disappearance in the light. This intermediate is unique to the OPP cycle. Heber et al. (49) found that the unique enzymes of the OPP cycle, glucose phosphate dehydrogenase and 6-phosphogluconate dehydrogenase, are both present in cytoplasm and chloroplast of spinach and Elodea, with larger amounts located in the cytoplasm. Other studies of photosynthesis in isolated spinach chloroplasts in the presence of vitamin K_5, which is thought to divert electrons from the light reaction and thus mimic aspects of dark metabolism, demonstrated the formation of 6-phosphogluconate in chloroplasts (19).

In the experiments with both ^{14}C and ^{32}P, labeling of various intermediates approached saturation in the light. When the light was turned off, the curves for the labeling of sugar phosphates by the two isotopes diverged, with ^{14}C labeling diminishing compared to ^{32}P-labeling with time in the dark (16,17,26). While intermediate pools were saturated with both labels in the light, and ^{32}P labeling continues to be saturated in the dark, it was clear that dark respiration utilized endogenous compounds that were not fully ^{14}C-labeled during the period of photosynthesis with $^{14}CO_2$. Most likely, the source of this partially labeled substrate is the stored starch in the chloroplasts.

When the light was first turned off, the levels of both FBP and SBP dropped precipitously during the first minute of darkness, and then returned to higher levels in the dark than in the light over a 5 min period, indicating the formation of these diphosphates from starch via either the hexose monophosphate shunt, phosphofructose kinase, or both. When the light was again turned on, there was a very rapid build-up in the levels of FBP and SBP (as well as DHAP) for about 30 sec, with the levels reaching higher than steady state levels. Then there was an equally rapid drop in these levels for another 30 sec, followed by damped oscillations leading to a steady state light level equal to that achieved in the previous light period.

The interpretation of these interesting kinetics is that when light was turned on, there was a rapid reduction of PGA to triose phosphates which are rapidly converted to FBP and SBP. The "overshoot" in the levels is attributed to the bisphosphatase having become inactive in the dark, and requiring about 30 sec in the light to become reactivated. During this period the levels of F6P and S7P also dropped--further indication that diphosphatases were still inactive.

After 30 seconds, when the bisphosphatases became fully active, levels of these bisphosphates fell sharply as these compounds were converted to sugar monophosphates and eventually to RuBP. With the removal of the bisphosphate blocks and other blocks, the RPP cycle reached full velocity, and soon the rate of reduction of PGA to triose phosphates and consequent formation of SBP and FBP were sufficient to bring the levels of these compounds to their steady state light values.

The inactivation of FBPase and SBPase activities in the dark are required to prevent the operation of futile cycles in the dark when respiratory metabolism occurs in the chloroplast. As is well known in other cells, phosphofructokinase and FBPase activity, when both present in the same cellular compartment, must be so regulated that they are not active at the same time, lest a futile cycle operate to hydrolyze ATP.

FBPase and SBPase, like other enzymes of the RPP cycle and the OPP cycle in chloroplasts, appear to be regulated by more than one factor that changes in concentration between light and dark. In common with some other regulated chloroplast enzymes (e.g., RuBPCase) FBPase and SBPase respond to changes in pH and Mg^{++}. Increased Mg^{++} lowers the pH optimum of these enzymes (48,50). Since both Mg^{++}(51-55) and pH (56) increase in the light in chloroplasts, the combined change has a substantial effect on enzyme activity.

A second major regulation of FBPase and SBPase depends on another important change between light and dark: the level of reduced ferredoxin. This regulation by small protein factors which are reduced by the ferredoxin has been worked out largely in the laboratory of Buchanan (57-61). This work is described by Buchanan elsewhere in this Symposium.

Like the activities of FBPase and SBPase, RuBPCase activity depends in part on pH, Mg^{++}, and reduced cofactors, but the mechanisms are different. RuBPCase activity regulation is complicated by the necessity for the plants to avoid, at least to some extent, the wasteful conversion of RuBP by oxygenase activity to phosphoglycolate and PGA. As mentioned earlier, O_2 binds competively at the CO_2 binding site. This O_2 binding is thus favored by a high O_2 and low CO_2 pressures. Low CO_2 pressure in the light can cause an increased level of RuBP, providing optimal conditions for the oxygenase reaction. It is therefore advantageous for the enzyme to be inactivated with respect to O_2 binding by a combination of high concentration of RuBP and low concentration of CO_2. Apparently, the binding of O_2 can only be decreased by a change in conformation of the enzyme which results in increased binding constants for both CO_2 and O_2.

Isolated RuBPCase is activated by the incubation with CO_2 or bicarbonate plus high levels of Mg^{++} (e.g. 10 mM), before the enzyme is exposed to RuBP (62-65). Preincubation of the enzyme with physiological levels of RuBP in the absence of either bicarbonate or Mg^{++} results in conversion of the enzyme to its inactive form with high K_m values for CO_2. The enzyme does not recover its activity for many minutes upon subsequent exposure to physiological levels of bicarbonate and Mg^{++} (63,64). Full activation of the isolated purified enzyme requires that the preincubation with CO_2 and Mg^{++} also be carried out in the presence of either 0.5 mM NADPH or 0.05 mM 6-phosphogluconate (64,65).

During the past year, it was reported by Paech et al. (66) that even freshly prepared solutions of RuBP rapidly develop small quantities of an inhibitor which is responsible for the inactivation of RuBPCase in kinetic studies employing added RuBP as substrate. It is claimed that it is this inhibitor rather than the RuBP itself which is responsible for many or all of the recorded inhibitions of the RuBPCase by the substrate RuBP. While it seems likely that this is indeed a complication, with the inhibitor being responsible for some of the inhibition seen upon the addition of RuBP to the enzyme, it also seems doubtful that the inhibitor can account for all of the observed effects. Certainly, it has made the kinetic study of this enzyme highly difficult.

Although it appeared for many years that the K_m CO_2 for RuBP carboxylase was too high to support the reductive pentose phosphate cycle, several laboratories have found evidence in recent years that the $K_m CO_2$ is sufficiently low. In particular, Bahr and Jensen (67) found that a low K_m CO_2 form of the enzyme obtained from freshly lysed spinach chloroplasts could be stabilized with dithioerythritol, ATP, $MgCl_2$, and R5P. Lilley and Walker (28) have shown that the activity and K_m CO_2 for the enzyme for spinach chloroplasts are more than adequate to support photosynthesis.

A common problem for some of the reported studies on the biochemical constants of RuBPCase outside the intact chloroplast is the changing value of K_m CO_2 and fixation rate during the time of the enzyme assays. Even with the purified enzyme preincubated with Mg^{++} and bicarbonate in the presence of NADPH or 6-phosphogluconate as effectors, the greatest rate of reaction was always during only the first 3-5 minutes followed by a decline to a much slower rate. In some of the studies reported, linear fixation was obtained for less than 2 min. Determinations of K_m CO_2 were reported using only the rate during the first minute after the addition of RuBP to preincubated enzyme. The rapid falloff in rate may be due to the inactivation of the enzyme by the quantity of the substrate RuBP which has to be added to support an extended kinetic experiment, or it could be due in part to the reported impurity which forms rapidly in solutions in RuBP and functions as a very strong inhibitor. Nevertheless, it was desirable to demonstrate outside the chloroplast that the low K_m form of the enzyme could be maintained.

To accomplish this, we made use of reconstituted spinach chloroplasts (68). Previously isolated spinach chloroplasts capable of high rates of photosynthesis, were lysed and the soluble contents separated from the thylakoids by high speed centrifugation. Soluble components were then added back to lamellae at a ratio of 7:4 (compared with the intact chloroplast) which was found to be suitable for photosynthesis in the reconstituted system. Flasks containing this reconstituted system were connected with the steady state apparatus which permits us to maintain selected constant levels of $^{14}CO_2$ for the duration of an experiment. The flasks were illuminated with sufficient intensity to produce light-saturated photosynthesis. Samples were withdrawn at timed intervals by needles through the serum stoppers. These samples were killed and the acid-stable ^{14}C fixation was determined. In each case some samples were analyzed by two-dimensional paper chromatography and the concentration of RuBP was carefully determined to ensure that RuBP (which was continuously regenerated) was not limiting. Experiments were carried out with various levels of CO_2 ranging from 0.013% to 0.128% CO_2 in air. Linear $^{14}CO_2$ fixation was obtained in each case during the period from 5 min to 20 min of photosynthesis.

The reciprocals of rates from these curves vs. reciprocals of substrate concentrations were then plotted and yielded a K_m CO_2 value of 0.023%, which is equivalent to 8.5 µM of CO_2 dissolved at pH 8.0 (69). Thus we were able to demonstrate that the enzyme can operate outside the chloroplast with a low K_m equivalent to that required under physiological conditions and for an extended period of time.

Attempts to carry out kinetic experiments with just the soluble enzymes in the dark and with added RuBP demonstrated that extended fixation could be carried out at respectable rates by preincubation with Mg^{++} and CO_2, and that substantial stimulation of the enzyme activity was obtained in the presence of NADPH (69). However, linear fixation could not be obtained nor could the low K_m form be maintained beyond the first minute. Presumably this is due to the inactivation by RuBP or its inhibitory contaminant.

Since this inhibitory contaminant is said to form quickly even under physiological conditions, it is perhaps a moot question as to whether the conversion of RuBPCase to a less active high K_m CO_2 form in the presence of RuBP is due to RuBP itself or to its contaminant. Presumably, even in the chloroplast, if RuBP were to accumulate, the inhibition would occur. Whatever the mechanism of inhibition, it can provide for increased K_m for both CO_2 and O_2 thus minimizing the deleterious effect of photorespiratory reactions at times when the supply of CO_2 is very low, as in the case of closed stomates on a dry, hot day. As a safety mechanism, this could greatly reduce the burning of endogenous starch in the chloroplast through the process of photorespiration.

Briefly, it should be mentioned that there are two other steps in the RuBP cycle which appear to be subject to light-dark regulation; namely, the phosphoribulose kinase step and steps involved in the reduction of PGA to triose phosphate. In both cases there may be involved mechanisms similar to those being described by Buchanan for the regulation of FBP-ase and SBPase; namely, control by small protein factors which are in turn reduced by ferredoxin in the light. The function of inactivation of phosphoribulokinase presumably is to prevent the unnecessary accumulation of RuBP and wasteful utilization of ATP in the dark. The function of regulation of the triose phosphate dehydrogenase may be to limit the rate of the reaction in the dark. This would allow triose phosphate levels to be maintained high enough to permit alternative biosynthetic pathways such as the formation of glycerol phosphate to occur in the chloroplast in the dark.

As mentioned earlier, there is a specific phosphate translocator whereby the transport of triose phosphate out of the chloroplasts is balanced by a movement into the chloroplasts of inorganic phosphate (29,38,46). The level of P_i in the cytoplasm thus appears to play an important role in the regulation of allocation of photosynthate from the RPP cycle to subsequent biosynthesis. Increased P_i in the cytoplasm can be expected to induce an increased rate of transport of triose phosphate out of the chloroplast. This in turn will lower the pool sizes of triose phosphates, FBP and SBP, resulting in a decreased activity of FBPase and SBPase. At the same time, the increased level of P_i inside the chloroplast reduces the rate of formation of starch from fructose-6-phosphate, since as first shown by Ghosh and Preiss (45), the rate-limiting step in this pathway mediated by ADP glucose pyrophosphorylase is strongly inhibited by high levels of P_i. This is a useful mechanism in that it reduced the drain of carbon into starch synthesis at a time when there is an increased drain of carbon as triose phosphate export. Conversely, we may suppose that when the level of P_i in the cytoplasm falls to a very low level, triose phosphate export decreases along with the level of P_i inside the chloroplast and the ADPG pyrophosphorylase can then be reactivated and allowed to convert a greater amount of reduced carbon from the RPP cycle into starch. These biosynthetic alternatives and some other stages of biosynthesis and its regulation are shown in Figure 3.

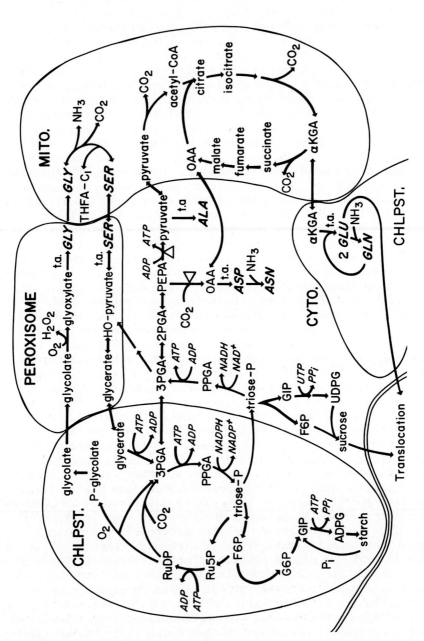

FIGURE 3. Metabolic Pathways among Subcellular Organelles in Green Cells.

The regulation of starch synthesis in chloroplasts depends also on other factors. Ghosh and Preiss found that the activity of the enzyme is stimulated by PGA. Also there is a rapid drop of ATP in the dark inside the chloroplast which is believed to be responsible for the immediate drop in level of ADP glucose when the light is turned off suddenly (23).

How might the level of P_i be controlled in the cytoplasm? Synthesis of sucrose in the cytoplasm and synthesis of other macromolecules results in the formation of large amounts of inorganic pyrophosphate (PP_i). It is often assumed that PP_i is rapidly hydrolyzed to P_i, thus providing additional energy for the biosynthetic reactions. In kinetic experiments with Chlorella pyrenoidosa several years ago we found that there is a measurable level of PP_i during photosynthesis and that that level drops rapidly in the dark. Experiments with isolated photosynthesizing chloroplasts have generally failed to reveal the presence of any PP_i. Thus we might conjecture that there is in fact a pool of PP_i in the cytoplasm and that the activity of inorganic pyrophosphatase regulates the conversion of PP_i back to P_i. One way in which inorgainc pyrophosphatase activity is regulated is through the concentration of Mg^{++}. In experiments with photosynthesizing isolated spinach chloroplasts we commonly employ a medium containing PP_i from which P_i is released slowly through the action of pyrophosphatase. Several years ago, we found (70) that a small increase in Mg^{++} in the medium of the isolated chloroplasts resulted in a very dramatic change in the rates of photosynthesis which was related to the export of triose phosphates resulting from the increased liberation of P_i when activated by Mg^{++} (27). It will be interesting in the future to investigate relationships between hormonal control of photosynthesis and levels of P_i, Mg^{++}, and PP_i in the cytoplasm.

In very young leaf cells, triose phosphate coming from the chloroplasts must be utilized to a considerable extent in the synthesis of new materials such as proteins and fats for the building of new cellular material. As leaf cells mature, their function is to synthesize and export sucrose and certain key amino acids. A cycling of biosynthetic function in unicellular algae was demonstrated using synchronously growing Chlorella cells several years ago (20). In the case of Chlorella, it was found that addition of 1 mM NH_4^+ in the medium during photosynthesis by cells which had been starved for nitrogen resulted in an immediate cessation of sucrose synthesis and a rapid increase in the synthesis of various amino acids. Accompanying this dramatic change was an increased rate of conversion of phosphoenolpyruvate to pyruvate as revealed by a drop in the steady-state level of phosphoenolpyruvate and an increase in the steady-state level of pyruvate (26).

Recently it has been possible to show similar effects of NH_4^+ on suspensions of photosynthesizing isolated cells from leaves of <u>Papaver somniferum</u> (71). In addition, it was demonstrated that there was a rapid increase in the carboxylation of phosphoenolpyruvic acid to produce oxaloacetic acid. Thus both of the rate-limiting steps involved in the anaplerotic synthesis of amino acid from sugar phosphate were immediately and substantially increased upon the addition of 1 mM NH_4^+ to the medium in which the cells were suspended.

Unpublished studies with pyruvate kinase isolated from leaves show quite clearly that these effects are not produced directly on the enzyme by NH_4^+. The studies with <u>Chlorella</u> did demonstrate a change in the steady-state level of ATP and ADP upon addition of NH_4^+ to the medium and perhaps it is via this mechanism that the pyruvate kinase is regulated. Some different but as yet unexplained regulatory mechanism must be required for the phosphoenolpyruvate carboxylase reaction. At this point all that can be said is that we are beginning to map out the kinds of changes in regulation of rate-limiting steps that must occur when biosynthetic metabolism of green cells is adjusted in accordance with the physiological needs of the plant.

REFERENCES

1. Bassham, J. A., Benson, A. A., Kay, L. D., Harris, A. A., Wilson, A. T., Calvin, M. (1954) J. Am. Chem. Soc. 76, 1760.
2. Bassham, J. A., and Calvin, M. (1957) In "The Path of Carbon in Photosynthesis" pp. 1-107. Prentice-Hall, Englewood Cliffs, New Jersey.
3. Norris, L., Norris, R. E., Calvin, M. (1955) J. Exp. Bot. 6, 64.
4. Kortschak, H. P., Hartt, C. E., Burr, G. O. (1965) Plant Physiol. 40, 209.
5. Hatch, M. D., Slack, C. R. (1966) Biochem. J. 101, 103.
6. Hatch, M. D., Slack, C. R. (1970) Annu. Rev. Plant Physiol. 21, 141.
7. Andrews, T. J., Lorimer, G. H., Tolbert, N. E. (1973) Biochem. 12, 11.
8. Bowes, G., Ogren, W. L., Hageman, R. H. (1971) Biochem. Biophys. Res. Commun. 45, 716.
9. Lorimer, G. H., Andrews, T. J., Tolbert, N. E. (1973) Biochem. 12, 18.
10. Tolbert, N. E. (1971) Annu. Rev. Plant. Physiol. 22, 45.

11. Tolbert, N. E. (1963) NSF-NRC Publication 1145, 648.
12. Arnon, D. I., Allen, M. B., Whatley, F. R. (1954) Nature 174, 394.
13. Arnon, D. I. (1958) In "The Photochemical Apparatus: Its Structure and Function," Brookhaven Symposium in Biology, No. 11, pp. 181-235. Upton, New York.
14. Whatley, F. R., Tagawa, K., Arnon, D. I. (1963) Proc. Nat. Acad. Sci. US 49, 266.
15. Allen, M. B., Whatley, F. R., Rosenberg, J. R., Capindale, J. B., Arnon, D. I. (1957) In "Research in Photosynthesis" (H. Gaffron, A. H. Brown, C. C. French, R. Livingston, E. I. Rabinowitch, B. C. Strehler, N. E. Tolbert, eds.) pp. 288-295. Interscience Publ. Inc., New York.
16. Pedersen, T. A., Kirk, M., Bassham, J. A. (1966) Biochim. Biophys. Acta 112, 189.
17. Bassham, J. A., and Kirk, M. (1968) In "Comparative Biochemistry and Biophysics of Photosynthesis" (K. Shibata, A. Takamiya, A. T. Jagendorf, R. C. Fuller, eds.) pp. 364-378. Univ. of Tokyo Press.
18. Bassham, J. A. (1971) Science 172, 526.
19. Krause, G. H. and Bassham, J. A. (1969) Biochim. Biophys. Acta 172, 553.
20. Kanazawa, T., Kanazawa, K., Kirk, M. R., Bassham, J. A. (1970) Plant Cell Physiol. 11, 149.
21. Stiller, M. (1962) Annu. Rev. Plant Physiol. 13, 151.
22. Peterkofsky, A., and Racker, E. (1961) Plant Physiol. 36, 409.
23. Bassham, J. A. (1971) In "Proc. 2nd Int. Congr. Photosynthesis Res. Stresa (G. Forti, M. Avron, A. Melandri, eds.) pp. 1723-1735. W. Junk N.V. Publ., The Hague.
24. Bassham, J. A. (1973) In "Symp. Soc. Exp. Biol. XXVII, Rate Control of Biological Processes", pp. 461-483. Cambridge Univ. Press, Cambridge.
25. Bassham, J. A. and Krause, G. H. (1969) Biochim. Biophys. Acta 189, 207.
26. Kanazawa, T., Kanazawa, K., Kirk, M. R., Bassham, J. A. (1972) Biochim. Biophys. Acta 256, 656.
27. Lilley, McC. R., Schwenn, J. D., Walker, D. A. (1973) Biochim. Biophys. Acta 325, 596.
28. Lilley, McC. R., Walker, D. (1975) Plant Physiol. 55, 1087.
29. Walker, D. A. (1976) In "The Intact Chloroplast" (J. Barber, ed.) pp. 236-278. Elsevier/North Holland Biomedical Press, The Netherlands.
30. Walker, D. A., Cockburn, W., Baldry, C. W. (1976) Nature 216, 597.
31. Slabas, R. R., and Walker, D. A. (1976) Biochim. Biophys. Acta 430, 154.

32. Schacter, B., Eley, J. H., Gibbs, M. (1971) Plant Physiol. 48, 707.
33. Stokes, D. M., and Walker, D. A. (1971) In "Photosynthesis and Respiration" (M. D. Hatch, C. B. Osmond, R. O. Slayter, eds.) pp. 226-231. John Wiley & Sons, Inc., New York.
34. Kelley, G. J., Latzko, E., Gibbs, M. (1976) Am. Annu. Rev. Plant Physiol. 27, 181.
35. Bassham, J. A., Kirk, M., Jensen, R. G. (1968) Biochim. Biophys. Acta 153, 211.
36. Stocking, C. R., and Larson, S. (1969) Biochem. Biophys. Res. Commun. 3, 278.
37. Heber, V. W., and Santarius, K. A. (1970) Z. Naturforsch. 256, 718.
38. Werden, K., and Heldt, H. W. (1971) In "Proc. 2nd Int. Congr. Photosynthesis Res. Stresa (G. Forti, M. Avron, A. Melandri, eds.) 2, pp. 1337-1344.
39. Bamberger, E. S., Ehrlich, B. A., Gibbs, M. (1975) Plant Physiol. 55, 1023.
40. Weissbach, A., and Horecker, B. L. (1955) Fed. Proc. 14, 302.
41. Weissbach, A., Horecker, B. L., Hurwitz, J. (1956) J. Biol. Chem. 218, 795.
42. Quale, J. R., Fuller, R. C., Benson, A. A., Calvin, M. (1954) J. Am. Chem. Soc. 76, 3610.
43. Müllhoffer, G. and Rose, I. A. (1965) J. Biol. Chem. 240, 1341.
44. Ghosh, H. P., and Preiss, J. (1965) J. Biol. Chem. 240, 960.
45. Ghosh, H. P., and Preiss, J. (1966) J. Biol. Chem. 241, 4491.
46. Heldt, H. W., and Rapley, L. (1970) FEBS Lett. 10, 143.
47. Bassham, J. A., and Kirk, M. R. (1960) Biochim. Biophys. Acta 43, 447.
48. Preiss, J., Biggs, M., Greenberg, E. (1967) J. Biol. Chem. 242, 2292.
49. Heber, U., Hallier, U. W., Hudson, M. A. (1967) Z. Naturforsch. 22, 1200.
50. Garnier, R. V., and Latzko, E. (1972) In "Proc. 2nd Int. Congr. Photosynthesis Res. (G. Forti, M. Avron, A. Melandri, eds.) pp. 1839-1845. W. Junk, The Hague.
51. Lin, D. C., and Nobel, P. S. (1971) Arch. Biochem. Biophys. 145, 622.
52. Krause, G. H. (1974) Biochim. Biophys. Acta 333, 301
53. Hind, G., Nakatani, H. Y., Izawa, S. (1974) Proc. Natl. Acad. Sci. USA 71, 1484.
54. Barber, J. (1976) Trends Biochem. Sci. 1, 33.
55. Barber, J., Mills, J., Nicholson, J. (1974) FEBS Lett 49, 106.

56. Heldt, H. W., Werdan, K., Milovancev, M., Geller, G. (1973) Biochim. Biophys. Acta 314, 224.
57. Buchanan, B. B., Kalberer, P. O., Arnon, D. I. (1967) Biochem. Biophys. Res. Commun. 29, 74.
58. Buchanan, B. B., and Schurmann, P. (1973) J. Biol. Chem. 248, 4956.
59. Buchanan, B. B., Schurmann, P., Kalberer, P. P. (1971) J. Biol. Chem. 246, 5952.
60. Buchanan, B. B., Schurmann, P., Wolosiuk, R. A. (1976) Biochem. Biophys. Res. Commun. 69, 970.
61. Buchanan, B. B., and Wolosiuk, R. A. (1976) Nature 264, 669.
62. Pon, N. G., Rabin, B. R., Calvin, M. (1963) Biochem. Z. 338, 7.
63. Chu, D. K., and Bassham, J. A. (1973) Plant Physiol. 52, 373.
64. Chu, D. K., and Bassham, J. A. (1974) Plant Physiol. 54, 556.
65. Chu, D. K., and Bassham, J. A. (1975) Plant Physiol. 55, 720.
66. Paech, C., McCurry, S. D., Pierce, J., Tolbert, N. E. (1978) In "Photosynthetic Carbon Assimilation" (H. W. Siegelman, G. Hind, eds.) p. 422. Brookhaven National Laboratory, Plenum Press, New York.
67. Vaklinova, S. and Popova, L. (1971) In "Proc. 2nd Intl. Congr. on Photosynthesis, Stresa (G. Forti, M. Avron, A. Melandri, eds.) pp. 1879-1874. W. Junk N. V., The Hague.
68. Bassham, J. A., Levine, G., Forger, J. (1974) Plant Sci. Lett. 2, 15.
69. Bassham, J. A., Krohne, S., Lendzian, K. (1978) In "Photosynthetic Carbon Assimilation" (H. W. Siegelman and G. Hind, eds.) pp. 77-93. Plenum Press, New York
70. Bassham, J. S., El-Badry, A. M., Kirk, M. R., Ottenhyem, H. C. J., Springer-Lederer, H. (1970) Biochim. Biophys. Acta 223, 261.
71. Paul, J. S., Cornwell, K. L., Bassham, J. A. (1978) Planta 142, 49.

COMPARATIVE REGULATION OF α1,4-GLUCAN SYNTHESIS IN PHOTOSYNTHETIC AND NON-PHOTOSYNTHETIC SYSTEMS[1]

Jack Preiss, William K. Kappel and Elaine Greenberg

Department of Biochemistry and Biophysics
University of California
Davis, California 95616

ABSTRACT ADPglucose (ADPG) pyrophosphorylase catalyzes the biosynthesis of ADPglucose from ATP and α-glucose-1-P and is the first unique enzyme in the bacterial glycogen and plant starch biosynthetic pathways. It is activated by glycolytic intermediates in most if not all bacterial and plant extracts and may be inhibited by 5'-AMP, ADP and/or P_i (see Preiss, J., in Adv. in Enzymol., Vol. 46, pp. 317-381 (1978)). The major activators are pyruvate (usually observed for the photosynthetic bacterial enzyme), 3-P-glycerate (the major activator for the higher plant, algal and blue green bacterial ADPG pyrophosphorylases), fructose-6-P and fructose-P_2. The last two activators are usually observed for ADPG pyrophosphorylases from bacteria that catabolize sugars via the Entner Doudoroff and glycolytic pathways, respectively. Thus, the activator specificity of the bacterial and plant ADP-glucose pyrophosphorylases roughly correlates with the ability of the organism to metabolize various carbon sources for energy and growth. The activator specificity may also be used in certain classes of organisms as a taxonomic marker. Causal data with plants and bacteria correlating fluctuation of inhibitor and activator concentrations *in vivo* with rates of α-glucan synthesis strongly suggest that the activator and inhibitor effects observed *in vitro* are physiologically operative. Furthermore isolation of a class of mutants of *E. coli* and *S. typhimurium* altered in their ability to accumulate α-glucan can be correlated with altered affinities of the inhibitor AMP and the activator, fructose-P_2 for the organism's ADPglucose pyrophosphorylase. Covalent attachment of [^{3}H]pyridoxal-P (PLP) to the activator binding site of the *E. coli* enzyme has enabled us to determine the amino acid sequence around the activator site and determine it to be a lysyl residue 37 amino acid residues

---

[1]This work was supported by NIH Grant AI05520.

from the amino-terminus of the enzyme. The sequence is:
NH$_2$-Val-Ser-Leu-Glu-Lys-Asn-Asp-His-Leu-Met-Leu-Ala-Arg-Gln-Leu-Pro-Leu-Lys-Ser-Val-Ala-Leu-Ile-Leu-Ala-Gly-Gly-Arg-Gly-Thr-Arg-Leu-Lys-Asp-Leu-Thr-Asn-Lys-(PLP)-Arg-Ala-Lys-Pro-Ala-Val-His-Phe-Gly-Gly-Lys....

INTRODUCTION

The biosynthesis of α1,4-glucosidic linkages of both bacterial glycogen and plant starch occur by the following reactions (1).

$$\text{ATP} + \alpha\text{-glucose-1-P} \rightleftharpoons \text{ADPglucose} + \text{PP}_i \tag{1}$$

$$\text{ADPglucose} + \alpha 1,4\text{-glucan} \longrightarrow \text{ADP} + \alpha 1,4\text{-glucosyl-glucan} \tag{2}$$

Synthesis of these reserve polysaccharides differs in two respects from mammalian glycogen synthesis. One, the effective glucosyl donor in bacterial and plant systems is ADPglucose rather than UDPglucose. Two, allosteric regulation occurs in the bacterial and plant systems at the enzyme level where ADPglucose synthesis occurs (reaction 1). This is in contrast to the animal and yeast systems, where allosteric regulation occurs at the mammalian glycogen synthase level. Moreover, the major regulatory mechanism for modulating glycogen synthase activity is exerted via phosphorylation or dephosphorylation of the synthase itself (2). At present there is no evidence for regulation of the bacterial or plant α1,4-glucan synthesizing systems via phosphorylation or dephosphorylation of the enzymes involved.

We will attempt to describe the regulatory properties associated with the bacterial and plant ADPglucose pyrophosphorylases and the evidence indicating that the regulatory phenomena are operative *in vivo*. The most recent studies on the isolation and characterization of an allosteric site of the enzyme will be described.

Regulatory Effects on ADPglucose Pyrophosphorylase. A notable feature of ADPglucose pyrophosphorylase, the enzyme catalyzing the first unique reaction in the synthesis of the α1,4 glucosidic linkage in bacteria and plants, is its regulatory properties. With very few exceptions, the enzyme from many bacterial and plant sources is activated by glycolytic intermediates and inhibited by either AMP, ADP or P$_i$. Thus, the enzyme appears to be modulated by the energy charge state of the cell (3,4) and requires the presence of glycolytic intermediates for full activity. The physiological rationale

MODULATION OF PROTEIN FUNCTION   163

for glycolytic intermediates as activators for ADPglucose pyrophosphorylase apparently is due to their being indicators of carbon excess in the cell. Glycogen accumulation for most bacteria occurs when they cease to grow either due to a limitation of nitrogen or some other nutrient besides carbon (5-7). A considerable portion of the excess carbon in the media is then converted into glycogen. In plant leaves starch synthesis seems to occur only in the light, when both CO_2 is quickly assimilated into glycolytic intermediates and the energy charge of the leaf cell is high.

A most remarkable finding is that a consistent pattern is usually observed between the source of the ADPglucose pyrophosphorylase, the glycolytic intermediates that are most effective as activators, and the carbon assimilatory pathways prevalent in the microorganism or tissue.

Table 1 lists the major metabolite effectors observed for ADPglucose pyrophosphorylases isolated from numerous organisms or leaf tissues. On the basis of activator specificity, the ADPglucose pyrophosphorylases can be classified into seven groups. The enzyme isolated from the bacteria belonging to the genus *Rhodospirillum* is solely activated by pyruvate (1). These organisms cannot metabolize glucose but can grow either as heterotrophs in the light or dark on various tricarboxylic acid intermediates and associated metabolites, or as autotrophs on CO_2 and H_2. A group of organisms that catabolize glucose via the Entner-Doudoroff pathway contains an ADPglucose pyrophosphorylase activated both by fructose-6-P and pyruvate (8-10). The photosynthetic organisms belong to the families Rhodospirillaceae, Chromatiaceae and Chlorobiaceae, while the non-photosynthetic organisms that have an ADPglucose pyrophosphorylase with this activator specificity are *Agrobacterium tumefaciens* and *Arthrobacter viscosus*. ADPglucose pyrophosphorylases isolated from three Rhodopseudomonads (*Rps. spheroides*, *Rps. globiformis* and *Rps. gelatinosa*) are activated by fructose-P_2 in addition to fructose-6-P and pyruvate, and thus represent a class distinct from those observed in other photosynthetic bacteria (1). The ADPglucose pyrophosphorylase from *Rps. viridis* is unusual in that it is not activated by pyruvate but is activated by fructose-6-P and fructose-P_2 (unpublished results). ADPglucose pyrophosphorylases with this activator specificity are also found in non-photosynthetic organisms such as Aeromonads, gram negative facultative anaerobes or from gram positive aerobic organisms, *Micrococcus luteus* and *Mycobacterium smegmatis* (1).

The microorganisms of the Enterobacteriaceae (*E. coli, E. aurescens, E. aerogenes, E. cloacae, S. typhimurium, C. freundii, S. dysenteriae*) have ADPglucose pyrophosphorylases activated by fructose-P_2 (11-13). Other effective activators are NADPH and pyridoxal-P. The enteric organisms, *S.*

TABLE 1
ACTIVATORS OF ADP-GLUCOSE PYROPHOSPHORYLASES FROM VARIOUS
BACTERIAL AND PLANT SOURCES

Photosynthetic group	Activators	Non-photosynthetic group
Rhodospirillum rubrum *Rhodospirillum tenue* *Rhodospirillum molischianum* *Rhodospirillum fulvum* *Rhodospirillum photometricum* *Rhodocyclus purpureus*	Pyruvate	--
Rhodopseudomonas acidophila *Rhodopseudomonas capsulata* *Rhodopseudomonas palustris* *Rhodomicrobium vanniellii* *Chromatium vinosum* *Chlorobium limicola*	Pyruvate Fructose-6-P	*Arthrobacter viscosus* *Agrobacterium tumefaciens*
Rhodopseudomonas spheroides *Rhodopseudomonas gelatinosa* *Rhodopseudomonas globiformis*	Pyruvate Fructose-6-P Fructose-P_2	--
Rhodopseudomonas viridis	Fructose-6-P Fructose-P_2	*Aeromonas hydrophila* *Aeromonas formicans* *Myobacterium smegmatis* *Micrococcus luteus*
--	Fructose-P_2 Pyridoxal-P NADPH	*Escherichia aurescens* *Escherichia coli* *Enterobacter aerogenes* *Enterobacter cloacae* *Salmonella typhimurium* *Citrobacter freundii* *Shigella dysenteriae* *Klebsiella pneumoniae* *Salmonella enteritidis*
--	None	*Serratia liquefaciens* *Serratia marcescens* *Clostridium pasteurianum* *Enterobacter hafniae*
Blue green bacteria Green algae Plant leaves	3-P-glycerate	--

liquefaciens, *S. marcescens* and *Enterobacter hafniae* contain however, an enzyme which is not effectively activated by any of the metabolites tested (14). Another ADPglucose pyrophosphorylase with no apparent activator is that isolated from *C. pasteurianum* (15). The unicellular cyanobacteria, *Synechococcus 6301* and *Aphanocapsa 6308* have ADPglucose pyrophosphorylases activated by 3-P-glycerate (16) and in this respect are similar to the ADPglucose pyrophosphorylases found in green algae (17) and in the leaves of higher plants (18-20).

Table 1 also suggests an overlapping of specificity of the activators of the different classes of ADPglucose pyrophosphorylase. The four prevalent activators seen in most classes are pyruvate, fructose-6-P, 3-P-glycerate and fructose-P_2. Almost all the ADPglucose pyrophosphorylases so far isolated from the photosynthetic anaerobic bacteria are activated by pyruvate. Some are also activated by fructose-6-P, and a few are activated by a third metabolite, fructose-P_2. Fructose-P_2 is an activator of the enzyme from enteric organisms, as well as those from the Aeromonads, *M. luteus* and *M. smegmatis*.

However, fructose-6-P, an effective activator of the ADPglucose pyrophosphorylases in the three last mentioned organisms, is not an activator for the enteric enzymes. 3-P-Glycerate, a very potent activator for the enzyme from blue green bacteria, green algae, and plant tissues, can also stimulate the enteric enzymes, although poorly. Conversely, fructose-P_2 is able to activate the plant leaf enzymes, but not as effectively as 3-P-glycerate. Whereas 28 μM fructose-P_2 is needed for 50% of maximal stimulation of the spinach leaf enzyme, only 10 μM 3-P-glycerate is required (19). The maximum stimulation of V_{max} effected by fructose-P_2 (16-fold) is much less than that observed for 3-P-glycerate (58-fold) at pH 8.5 (19).

This overlapping of specificity for the activators in the above ADPglucose pyrophosphorylase classes suggests that the activator sites for the different groups are similar or related to each other. One may therefore hypothesize mutation of that part of the gene specifying the activator site of the ADP-glucose pyrophosphorylase occurring during evolution. Possibly the pressure for change has been the coordination or compatibility of the metabolite(s) acting as the activator with the metabolic activities of the organism.

Most ADPglucose pyrophosphorylases are inhibited by metabolites associated with energy metabolism; namely AMP, ADP or P_i. The enteric ADPglucose pyrophosphorylase (11,21,22), including those found in *Serratia* (14) and *Enterobacter hafniae* (unpublished results), are very sensitive to AMP inhibition, while the plant, algal and blue green bacterial enzymes are most sensitive to P_i (16-20). Other ADPglucose pyrophosphorylases may be sensitive to either P_i, ADP or AMP (1,24). However, no or little inhibition by the above three compounds at moderate concentrations (5 mM or less) is observed for the enzyme isolated from Aeromonads (1,24), *Mycobacterium smegmatis* (25), *R. rubrum* (26) or *R. molischianum* (unpublished results).

<u>Rationale for the Different Activator Specificities of the Bacterial and Plant ADPglucose Pyrophosphorylases</u>. As suggested before, the variation of activator specificity observed for the different ADPglucose pyrophosphorylases is due to evolutionary pressures to coordinate glucan synthesis with the carbohydrate metabolism occurring in the cell. Thus it would be of interest to relate the various metabolite activators to the pathways known to occur in the organism.

Pyruvate is the only glycolytic intermediate capable of activating the ADPglucose pyrophosphorylases of *R. rubrum*, *R. tenue* and *R. molischianum* (1). These photosynthetic organisms are capable of growth under heterotrophic conditions in the light or dark, and autotrophic conditions in the light. Generally they grow very well on pyruvate and TCA cycle intermediates as carbon sources and photosynthetic electron donors.

These organisms cannot utilize glucose or fructose for these purposes. Some *R. rubrum* strains, however, may utilize fructose poorly for growth. Activation of ADPglucose pyrophosphorylase activity by pyruvate in *R. rubrum* is seen whether the cells are grown aerobically in the dark with malate, or anaerobically in the light with either malate, acetate, acetate + CO_2, or CO_2 + H_2 (27). Stanier *et al.* (28) have pointed out the relation of pyruvate formation with glycogen accumulation and this has been reviewed elsewhere (1,26,29).

Since several mechanisms for the synthesis of pyruvate are available when these organisms are grown under various nutritional conditions giving rise to accumulation of glycogen, pyruvate may be considered as the first glycolytic intermediate in gluconeogenesis in *Rhodospirillum*.

The rationale for fructose-6-P or fructose-P_2 being activators is that some of the organisms can metabolize sugars either via Entner-Doudoroff or glycolytic pathways. The enterics are known to catabolize their sugars via glycolysis. However, the metabolism of glucose by *Rps. spheroides* has been shown to occur via the Entner-Doudoroff pathway while degradation of fructose occurs mainly via the Entner-Doudoroff pathway in dark-aerobic conditions and under anaerobic phototrophic conditions, via the glycolytic pathway (30). In *Rps. capsulata* glucose is metabolized via the Entner-Doudoroff pathway while fructose is metabolized via glycolysis (10,31).

The ADPglucose pyrophosphorylases of these organisms may therefore exhibit multiple activation specificity because of the utilization of carbohydrates via a number of metabolic pathways. Pyruvate would be the principle activator for phototrophic or dark aerobic growth of organisms on pyruvate and TCA cycle intermediates. Fructose-6-P and fructose-P_2 would be most important for the time where the Entner-Doudoroff pathway or glycolysis are the predominant carbon assimilatory pathways. However, the predominant metabolic pathways occurring under different nutritional and physiological conditions for many of the organisms have not been fully explored. Thus the above remains as a tentative hypothesis.

The finding that 3-P-glycerate is an activator for the ADPglucose pyrophosphorylases of those tissues or microorganisms undergoing aerobic photosynthesis is consistent with the argument that the activator is related with the type of carbon metabolism occurring in the cell. 3-P-Glycerate is the primary CO_2 fixation product of carbon metabolism in these systems. It is formed by the ribulose-bis-P carboxylase reaction. Thus, the primary CO_2 fixation product of photosynthesis is the activator for an enzyme involved in formation of an end product of photosynthesis, starch.

Kinetic Functions of Activator and Inhibitors. Kinetic studies of partially purified ADPglucose pyrophosphorylases show that the presence of activator in reaction mixtures usually lowers the concentration of the substrates (ATP, glucose-1-P, pyrophosphate, ADPglucose and the cationic activator, Mg^{2+}) required to give 50% of maximal velocity (K_m or $S_{0.5}$) about 2- to 15-fold; i.e., the apparent affinity of the enzyme for the ligand molecule is increased. In addition, the activator may stimulate maximal velocity 2- to 80-fold. The magnitude of stimulation is dependent on the enzyme system studied and pH. Undoubtedly, the prime function of the allosteric activator is to modulate the sensitivity of the enzyme to inhibition by the inhibitors AMP, ADP or P_i. The inhibitors are generally non-competitive with the substrate. For most ADPglucose pyrophosphorylases, the activator at relatively high concentrations can completely reverse the inhibition caused by either AMP, P_i or ADP. A well studied system has been the *E. coli* B enzyme where fructose-P_2 modulates the sensitivity to AMP inhibition (21,22). The $I_{0.5}$ (concentration giving 50% inhibition) for 5'-AMP is about 70 μM at 1.7 mM fructose-P_2. At lower concentrations of activator however lesser concentrations of AMP are required for 50% inhibition. At 60 μM fructose-P_2 only 3.4 μM AMP is needed for 50% inhibition.

From these *in vitro* kinetic studies, it has been postulated that the relative concentrations of the allosteric inhibitors and activators modulate the rate of ADPglucose synthesis and therefore the rate of glycogen or starch synthesis. Under physiological conditions where ADPglucose and glycogen synthetic rates are low, the concentrations of inhibitor and activator are in a ratio that maintains the ADPglucose pyrophosphorylase to a large extent in the inhibited state. Conditions where active glycogen synthesis ensues are those where either a decrease of inhibitor and/or an increase of activator has occurred, thus allowing ADPglucose pyrophosphorylase activity to increase.

Evidence for the Regulatory Effects on ADPglucose Pyrophosphorylase Being Functional *in vivo* in the Regulation of α1,4-Glucan Synthesis. Many experiments in plant systems have shown a direct relationship between the levels of 3-P-glycerate and of P_i and the rate of starch synthesis in the chloroplast. Succinctly, they show that at high 3-P-glycerate to P_i ratios starch accumulates while at low ratios, little starch accumulation occurs (32-37). Similarly it has been shown by Dietzler et al. (38,39) that the rate of glycogen accumulation in stationary phase cultures of *E. coli* could be varied 10-fold by varying either the nitrogen or carbon source. The different rates of glycogen accumulation could be linearly related to the square of the fructose-P_2 concentration found in the bacteria. Fructose-P_2 was shown previously to be the activator of the *E. coli* ADPglucose pyrophosphorylase. The *in vivo* data could be plotted according to the Hill equation and it was found that a Hill slope value of $\bar{n}$ of 2.08 could be obtained (39). These data agreed very well with the *in vitro* kinetic studies previously done with the *E. coli* B ADPglucose pyrophosphorylase where a Hill plot for the fructose-P_2 saturation curve gave a value of about 2 (12,40).

Thus in a good number of systems a direct relationship between activator level and α1,4-glucan synthetic rates have been shown suggesting strongly that the activator effects observed in *in vitro* studies are operative *in vivo*.

More definitive evidence has been obtained with the isolation of a class of mutants of *E. coli* (40-48) and of *S. typhimurium* LT-2 affected in their ability to accumulate glycogen-excess mutants, SG5 and CL1136, with the parent strain *E. coli* B. In minimal media containing 0.75% glucose, the amount of glycogen accumulated in SG5 and CL1136 is two- and four-fold greater, respectively, than that found in *E. coli* B. The rates of glycogen synthesis in both media are about 2- and 3.5-fold greater for the mutants SG5 and CL1136, respectively (Table 2). The levels of activity of the glycogen biosynthetic enzymes (ADPglucose pyrophosphorylase, glycogen synthase and branching enzyme) in the mutants and in the parent strain are equivalent and, therefore, cannot account for the increased rate of accumulation of glycogen present in the mutants. Furthermore, the mutant ADPglucose pyrophosphorylases have approximately the same apparent affinities as the parent enzyme for

TABLE 2
GLYCOGEN ACCUMULATION IN *E. COLI* B AND GLYCOGEN EXCESS MUTANTS

Strains	Glycogen	
	Rate of accumulation ($\mu mol \cdot g^{-1} \cdot h^{-1}$)	Maximal accumulation (mg/g^{-1})
E. coli B	32	20
Mutant SG5	59	35
Mutant CL1136	114	74

The minimal media with glucose as a carbon source and the assays for glycogen accumulation are described in (41,45).

Accumulation of glycogen is expressed as milligrams of anhydroglucose per gram (wet weight) of cells and the value given is the maximal amount accumulated in stationary phase. The rate of glycogen accumulation is expressed as the change of μmoles of anhydroglucose per gram of cell (wet weight) per hour.

the substrates and for the divalent cation activator Mg^{2+} (40, 45). The modified properties are the apparent affinities of the mutant enzymes for the allosteric effectors. The concentration of fructose-P_2 required for 50% of maximal activation ($A_{0.5}$) is about 3-fold less for the SG5 ADPglucose pyrophosphorylase and 12-fold less for the CL1136 enzyme (Table 3). The enzyme from the parent strain is also more dependent on the allosteric activator for maximal activity than are the enzymes obtained from the mutants. Saturating fructose-P_2 concentrations stimulate the V_{max} of ADPglucose synthesis catalyzed by the *E. coli* B enzyme 5-fold but only stimulate 3.3-fold and 1.5-fold the SG5 and CL1136 ADPglucose pyrophosphorylases, respectively. The apparent affinities for the activators NADPH and pyridoxal-5'-P are also greater with the mutant ADPglucose pyrophosphorylases than with the *E. coli* B enzyme (40,45).

The mutant ADPglucose pyrophosphorylases are also less sensitive to the allosteric inhibitor AMP (40,45). Higher concentrations of AMP are required to give 50% inhibition of the mutant ADPglucose pyrophosphorylases than are required for inhibition of the *E. coli* B enzyme (Table 3). Both mutant and *E. coli* B enzymes become more sensitive to AMP inhibition at lower concentrations of fructose-P_2. Lower concentrations of the inhibitors are needed to achieve 50% inhibition. At all concentrations of fructose-P_2, the mutant enzymes are still less sensitive than the parent enzyme to inhibition.

As would be expected, the CL1136 and SG5 ADPglucose pyrophosphorylases have more activity than the *E. coli* B wild type enzyme when assayed under equivalent conditions and energy charge values (40,45). At an energy charge of 0.7 the SG5 enzyme has about five times more activity than the *E. coli* B with 1.5 mM fructose-P_2 as the activator (40). At a range of energy charge of 0.85 to 0.9 and with 0.75 mM fructose-P_2, the SG5 enzyme shows about twice as much activity as the *E. coli* B enzyme (40).

TABLE 3

KINETIC CONSTANTS FOR ALLOSTERIC EFFECTORS OF *E. COLI* B, SG5 AND CL1136 ADP-GLUCOSE PYROPHOSPHORYLASES

Enzyme Source	$A_{0.5}$ fructose-P_2 μM	V_{max} stimulation -fold	$I_{0.5}$ (AMP) fructose-P_2		
			1.5 mM	0.5 mM	0.15 mM μM
E. coli B	68	5	87	41	16
SG5	22	3.3	170	74	29
CL1136	5.2	1.5	680	380	142

The CL1136 ADPglucose pyrophosphorylase is almost completely active in the energy charge range of 0.75 to 1.0 with fructose-P_2 concentrations of 0.75 mM and higher (45). Significant activity (10 to 25% of maximal activity) is even seen at energy charge of 0.4 with fructose-P_2 concentrations of 0.75 mM or higher. In the absence of fructose-P_2 and at an energy charge value of 0.75, the activity of the CL1136 enzyme is almost 30% of the maximal activity.

In contrast, the *E. coli* B ADPglucose pyrophosphorylase activity is less than 4% of the maximal activity at an energy charge of 0.65 even in the presence of 3 mM fructose-P_2. At an energy charge level of 0.75 and with 3.0 mM fructose-P_2, the *E. coli* B enzyme exhibits only 11% of its maximal activity. The CL1136 enzyme shows 98% of its maximal activity under these conditions. The physiological energy charge range in various *E. coli* strains is found to be between 0.74 to 0.9 (49,50). The physiological concentrations of fructose-P_2 ranges between 0.71 to 3.2 mM (49,51). Under these conditions the CL1136 enzyme shows maximal activity and is not sensitive either to change in energy charge or to fluctuations of fructose-P_2. In contrast, the *E. coli* B enzyme is very sensitive to the above concentration range of fructose-P_2 and energy charge range (40,45).

These studies strongly suggest that the increased accumulation of glycogen in the mutants SG5 and CL1136 is due to alterations of the ADPglucose pyrophosphorylases which cause a greater affinity for the activators and a lower affinity for the inhibitor. The correlation of the relative sensitivities of *E. coli* B, SG5 and CL1136 ADPglucose pyrophosphorylases to AMP inhibition and fructose-P_2 activation with the increased rates of glycogen accumulation is in agreement with the expressed view that the cellular levels of the allosteric activators and inhibitors of ADPglucose pyrophosphorylase modulate the rate of ADPglucose synthesis and thereby regulate the rate of synthesis and accumulation of glycogen in the cell (12,29, 40,45).

A similar mutant of *S. typhimurium* LT-2 has been isolated (52) and its properties is shown in Table 4. Mutant JP-51 accumulates almost twice as much as glycogen as the parent LT-2 strain. Examination of the kinetic properties of the mutant and wild type ADPglucose pyrophosphorylases (Table 4) shows that both enzymes have the same $A_{0.5}$ values for fructose-P_2. However, the JP-51 enzyme has about a 5-fold higher $I_{0.5}$ value for AMP than the parent enzyme thus suggesting that the lesser sensitivity to inhibition by AMP is the cause for greater accumulation of glycogen in mutant JP-51. The JP-51 enzyme is of interest in that its kinetic properties appear to differ with the parent enzyme only in respect with sensitivity to AMP inhibition. This is different than that observed for

TABLE 4

CHARACTERIZATION OF *S. TYPHIMURIUM* LT-2 AND ITS MUTANT JP-51

Strain	Maximum glycogen accumulation mg/gram	ADPglucose pyrophosphorylase	
		$A_{0.5}$ (fructose-P_2) μM	$I_{0.5}$ (AMP) μM
S. typhimurium LT-2	12	95	108
Mutant JP-51	20	84	485

Accumulation of glycogen is expressed as milligrams of anhydroglucose per gram (wet weight) of cells. The media is described in (52). The $I_{0.5}$ values for AMP were determined in the presence of 1.0 mM fructose-P_2.

the *E. coli* SG5 and CL1136 mutant ADPglucose pyrophosphorylases where mutation effects the enzyme's sensitivity to activator as well as to inhibition.

E. coli Mutant SG14 ADPglucose Pyrophosphorylase. Pyridoxal-P and NADPH in addition to fructose-P_2 are effective activators of the *E. coli* enzyme. The question, which activator is most important physiologically may be posed. The view that fructose-P_2 may be the most important physiological activator is supported by studies with mutant SG14 (42,44). Mutant SG14 accumulates glycogen at 28% the rate of *E. coli* B and contains about 23 to 25% of the ADPglucose synthesizing activity of *E. coli* B (44). Yet the activity present is still 6-fold greater than that required for the observed rate of glycogen accumulation in SG14. The concentrations of ATP and Mg^{2+} required for 50% of maximal activity ($S_{0.5}$) are 4- to 5-fold higher for the SG14 enzyme than the *E. coli* B enzyme. Whereas, the $S_{0.5}$ values for ATP and Mg^{2+} are 0.38 and 2.3 mM, respectively, for *E. coli* B enzyme in the presence of 1.5 mM fructose-P_2, the $S_{0.5}$ values for ATP and Mg^{2+} are 1.6 and 10.8 mM, respectively, for the SG14 ADPglucose pyrophosphorylase in the presence of saturating fructose-P_2 concentration (4.0 mM). Reports in the literature indicate that the ATP level in growing *E. coli* ranges about 2.4 mM (38,49,53) and the Mg^{2+} level is about 25 to 40 mM (54-56). Therefore, the SG14 ADPglucose pyrophosphorylase would be essentially saturated with respect to these substrates. The apparent affinities ($S_{0.5}$) for glucose-1-P for the *E. coli* B and SG14 enzymes are about the same (44).

The major differences between the SG14 and *E. coli* B ADPglucose pyrophosphorylases appear to be their sensitivities toward activation and inhibition (42,44). Table 5 shows that about 12-fold more fructose-P_2 is needed for 50% maximal stim-

TABLE 5
COMPARISON OF THE REGULATORY PROPERTIES OF MUTANT SG14 ADP-GLUCOSE PYROPHOSPHORYLASE WITH *E. COLI* B ADP-GLUCOSE PYROPHOSPHORYLASE

Strain	$A_{0.5}$ activator			$I_{0.5}$ AMP fructose-P_2 conc.		
	Fru-P_2 µM	NADPH	Pyridoxal-P	4.0 mM	1.5 mM mM	1.0 mM
E. coli B	68	60	16	0.25	0.11	0.08
Mutant SG14	820	--	440	1.0	0.5	0.45

ulation of the SG14 ADPglucose pyrophosphorylase ($A_{0.5}$ = 0.82 mM) than for half-maximal stimulation of the *E. coli* B enzyme, while the $A_{0.5}$ for pyridoxal-P for the SG14 enzyme (0.44 mM) is about 25-fold higher than the $A_{0.5}$ observed for the *E. coli* B enzyme. Pyridoxal-P and fructose-P_2 stimulate ADPglucose synthesis catalyzed by the *E. coli* B enzyme to about the same extent. However, the stimulation of the SG14 ADPglucose pyrophosphorylase seen with pyridoxal-P is only one-half that elicited by fructose-P_2 (42). A notable difference is that NADPH does not stimulate the SG14 enzyme. Compounds similar to NADPH in structure, such as 1-pyrophosphoryl-ribose-5-P and 2'-PADPR, that are capable of activating the *E. coli* B ADPglucose pyrophosphorylase do not activate the SG14 enzyme.

Since the apparent affinity of the SG14 enzyme for its activators is considerably lower than that observed for the *E. coli* B ADPglucose pyrophosphorylase it was an unexpected finding that SG14 is capable of accumulating glycogen even at 28% the rate observed for the parent strain. This rate is accounted for by the relative insensitivity of the SG14 enzyme to inhibition by AMP as seen in Table 5 (24,42). The SG14 ADP-glucose pyrophosphorylase is much less sensitive to AMP inhibition in the concentration range of 0-0.2 mM than is the parent strain enzyme. At a saturating concentration of fructose-P_2 for the SG14 enzyme (4.0 mM) only 7% inhibition of the SG14 enzyme is observed at 0.2 mM AMP. The same concentration of AMP gives 40% inhibition of the *E. coli* B enzyme. At a concentration of fructose-P_2 (1.5 mM) which gives 80% of maximal velocity for the SG14 enzyme, 0.2 mM 5'-AMP causes 80% and 33% inhibition with the *E. coli* B and SG14 enzymes, respectively. A decrease in fructose-P_2 to 1.0 or 0.5 mM further increases the sensitivity of the *E. coli* B ADPglucose pyrophosphorylase activity to inhibition. However, at these concentrations of fructose-P_2 the sensitivity of the SG14 ADPglucose pyrophosphorylase to AMP remains the same or becomes less than that seen at 1.5 mM fructose-P_2. At concentrations of 0.5-1.0 mM of fructose-P_2 the $E.$ $coli$ B enzyme is inhibited 90% or more by 0.2 mM AMP while the inhibition of the SG14 enzyme ranges from 12 to 30%. Although the modification of the SG14 enzyme causes it to have a lower apparent affinity for its activators, it also renders the enzyme less sensitive to AMP inhibition. The insensitivity to AMP renders the SG14 enzyme much less responsive to energy charge than the wild type $E.$ $coli$ B enzyme (44). The two effects on activation and inhibition appear to compensate for each other and allow SG14 to accumulate glycogen at about 28% the rate of the parent strain (44).

The data obtained from the kinetic studies of the SG14 ADPglucose pyrophosphorylase suggest that fructose-P_2 is the most important physiological activator of the $E.$ $coli$ ADPglucose pyrophosphorylase. This is based on the observation that NADPH is not an activator of the SG14 enzyme and that the concentration of pyridoxal-P needed for activation of the enzyme ($A_{0.5}$ = 0.44 mM) is considerably higher than that reported to be present in $E.$ $coli$ B. The concentration (57) of pyridoxal-P is 24 to 48 μM, and most of this metabolite is probably bound to protein in the cell and unavailable for activation of the ADPglucose pyrophosphorylase (58).

The concentration of fructose-P_2 in $E.$ $coli$ is about 0.71-3.2 mM (38,49,51,53) and the $A_{0.5}$ of SG14 ADPglucose pyrophosphorylase is 0.82 mM. The concentration of fructose-P_2 in the $E.$ $coli$ cell is therefore sufficient for activation of ADPglucose synthesis at the rates required for the observed glycogen accumulation rate in SG14.

Chemical and Physical Properties of ADPglucose Pyrophosphorylase. The kinetic studies reviewed in the previous sections describe many interesting and unique properties of the ADPglucose pyrophosphorylases of all classes, in their activator and inhibitor specificities and in the interaction of their activators with the substrates and inhibitors. Further studies attempting to correlate these complex kinetic properties with the physical and chemical properties of the enzymes would be of obvious interest. Studies of this nature on the allosterically modified ADPglucose pyrophosphorylases of SG5, CL1136 and SG14 could provide valuable information on the nature of the allosteric mechanism of the $E.$ $coli$ ADPglucose pyrophosphorylase. Furthermore, a study of the possible immunological relationships on the ADPglucose pyrophosphorylase in various species and genera of bacteria may also provide insight into the evolution and relationships of several bacterial groups.

The $E.$ $coli$ B ADPglucose pyrophosphorylase has been studied in greatest detail (59-61). Gel electrophoresis of the enzyme in SDS suggests a molecular weight of 50,000. The molecular weight of the native enzyme is about 185,000 ± 12,000 indicating that the enzyme is composed of four similar subunits.

Another finding suggesting the similarity of subunits is that peptide mapping of tryptic digests show only enough peptides to account for a single polypeptide unit. Separable peptides that could be distinguished numbered 43 and were 10 fewer than would be expected on the basis of the 52-53 lysines plus arginines found in the amino acid analysis (60). Furthermore terminal amino acid analysis indicates that four valine residues and four arginine residues per molecule of native enzyme are the N-terminal and C-terminal amino acids, respectively (60). Binding studies reveal that ATP binds first and then glucose-1-P (62).

Chromium adenosine triphosphate (CrATP), a potent inhibitor of many enzymes utilizing Mg:ATP as a substrate (63,64), is a potent competitive inhibitor of the *E. coli* ADPglucose pyrophosphorylase (62). If this inactive Mg:ATP analogue is used in the glucose-1-P binding equilibrium dialysis experiments it is found that one mole of glucose-1-P binds per mole of pyrophosphorylase subunit. Only two moles of MgATP or CrATP bind to the tetrameric protein. Thus MgATP sites appear to exhibit half-site reactivity (65). This is in contrast to equilibrium dialysis experiments with ADPglucose where four moles of this substrate bind to one mole of the tetramer protein (62). In the presence of glucose-1-P, however, four moles of CrATP bind to the tetrameric protein. It therefore appears that, in the pyrophosphorylase reaction mechanism in the synthesis direction, two moles of MgATP initially bind to the tetrameric protein. This permits the binding of glucose-1-P to its four binding sites on the tetrameric protein. Further binding of the next two moles of MgATP may then ensue with concomitant catalysis occurring (65).

Preliminary binding experiments also indicate half-site reactivity for the binding sites of the activator fructose-P_2 in contrast to the activator pyridoxal-P (62).

<u>Modification of the Allosteric Site with Pyridoxal-P and Sodium Borohydride</u>. Pyridoxal-P can be covalently linked to an ε-amino group of a lysyl residue in *E. coli* B ADPglucose pyrophosphorylase by reduction with $NaBH_4$ (66,67). Figure 1 and Table 6 show that after incorporation of 0.5 mol of pyridoxyl-P per mol subunit the enzyme no longer requires the presence of allosteric activator in reaction mixtures for high activity. The modified enzyme increases in specific activity, 7- to 9-fold in ADPglucose synthesis, in the absence of activator. The further stimulation by fructose-P_2 is only 1.8- to 4-fold (Fig. 1 and Table 6). The pyridoxyl-P is most likely covalently bound at the allosteric binding site and in its bound form

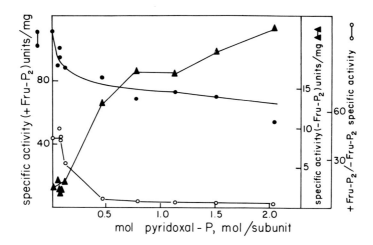

FIGURE 1. Effect of [^{3}H]pyridoxal-P incorporation into *E. coli* B ADPglucose pyrophosphorylase kinetics. Reductive pyridoxylation was carried out in the presence of 0.8 mM ADPglucose and 8 mM MgCl$_2$. Enzyme assays were done in the presence and absence of 1.5 mM fructose-P$_2$.

keeps the enzyme in its high activity conformational form. Further evidence for the pyridoxal-P being bound to the allosteric site is seen in Table 6. Both 1,6-hexanediol-P$_2$ and fructose-P$_2$ can protect the enzyme from being modified with pyridoxal-P and NaBH$_4$. Concomitant with the decrease of pyridoxal-P incorporation the increase in specific activity of the enzyme when assayed in the absence of fructose-P$_2$ is prevented and the enzyme remains significantly activated by the presence of fructose-P$_2$ in reaction mixtures (Table 6).

If ADPglucose plus Mg^{2+} are present during the reductive pyridoxylation the activity in the presence of 1.5 mM fructose-P$_2$ is protected suggesting that in addition to binding to the allosteric site, pyridoxal-P is being covalently attached to a lysyl residue necessary for catalytic action and this site is protected by the presence of the substrate, ADPglucose and the divalent cation cofactor, Mg^{2+}. The presence of ADPglucose plus Mg^{2+} still permits the sharp increase in the unactivated activity as pyridoxal-P still binds to the allosteric activator site.

TABLE 6
COVALENT BINDING OF PYRIDOXAL-P TO *E. COLI* ADP-GLUCOSE
PYROPHOSPHORYLASE: PROTECTION BY 1,6-HEXANEDIOL-P_2
AND FRUCTOSE-P_2

Conditions	P-pyridoxyl enzyme subunit	μmol ADPglucose formed min-mg		
		+fru-P_2	-fru-P_2	+fru-P/-fru-P_2
Unreacted enzyme	--	87.5	2.0	44
Unprotected enzyme	1.2	30	11.1	2.5
+ 1.0 mM hexanediol-P_2	0.68	29	1.8	15.8
+ 1.0 mM fructose-P_2	0.72	43	4.5	9.6
0.8 mM ADPglucose + 8 mM $MgCl_2$	1.2	70	17	4.1

The enzyme was assayed at pH 8.0 with 5 mM $MgCl_2$, 1.5 mM ATP, 0.5 mM [^{14}C]glucose-1-P and 1.5 mM fructose-P_2 where indicated. The enzyme, 0.24 mg, was reacted with 700 nmol pyridoxal-P and 2.1 mM $NaBH_4$ in the presence or absence of the other activators for 30 minutes and then dialyzed overnight before assay.

Since covalent binding to the allosteric site by pyridoxal-P could be achieved it was then possible to determine whether AMP bound to the same site as the activator. Thus ADPglucose pyrophosphorylase was incubated in the presence of ADPglucose and $MgCl_2$, and four increasing concentrations of [^{3}H]pyridoxal-P (0, 20 μM, 59 μM, and 290 μM) before reduction with $NaBH_4$. Table 7 shows the incorporation of pyridoxal-P per subunit: 0, 0.2, 0.4 and 0.8 ml [^{3}H]pyridoxal-P/mol of subunit, respectively. With increasing incorporation of pyridoxal-P, the unactivated specific activity increased from 1.7 to 17.2 units/mg causing the activated specific activity/unactivated specific activity ratio to decrease from 43.5 to 4.4.

ADPglucose pyrophosphorylase in the absence of fructose-P_2, is inhibited only about 60% even by 5.0 mM 5'-AMP (21,22, 40), whereas, in the presence of 1.5 mM fructose-P_2, the enzyme is greater than 90% inhibited. Table 7 shows the 5'-AMP inhibition characteristics of the different pyridoxyl-P enzyme preparations, assayed in the presence of 1.5 mM fructose-P_2. For all degrees of pyridoxylation, the maximum inhibition by 0.4 mM

TABLE 7
5'-AMP INHIBITION OF COVALENTLY MODIFIED AC70R1 [^{3}H]-
PYRIDOXYL-P-ADP-GLUCOSE PYROPHOSPHORYLASE

	Mol [^{3}H]pyridoxal-P incorporated/ mol subunit			
	0	0.2	0.4	0.8
Specific activity + fru-P$_2$ (units/mg)	74.1	66.7	59.7	75.4
Specific activity - fru-P$_2$ (units/mg)	1.7	4.0	11.8	17.2
+Fru-P$_2$ specific activity/ -fru-P$_2$ specific activity	43.5	16.7	5.0	4.4
5'-AMP inhibition in absence of fructose-P$_2$				
I$_{0.5}$ (mM)	1.46	0.18	0.044	0.009
Hill slope, n̄	0.29	0.36	0.54	0.77
% maximum inhibition	40%	60%	73%	89%
5'-AMP inhibition in presence of 1.5 mM fructose-P$_2$				
I$_{0.5}$ (mM)	0.13	0.11	0.09	0.14
Hill slope, n̄	1.6	1.8	1.3	1.6
% maximal inhibition	91%	92%	88%	90%

The enzymes were assayed in the direction of ADPglucose synthesis at pH 7.0. Specific activity was measured in activated conditions in the presence of 1.5 mM fructose-P$_2$, and in unactivated conditions in the absence of fructose-P$_2$. Inhibition by 5'-AMP was carried out in the presence and absence of 1.5 mM activator fructose-P$_2$. I$_{0.5}$ is the concentration of inhibitor giving 50% of the maximal inhibition under conditions of the experiment. The activity in the absence of 5'-AMP is set at 100%.

5'-AMP was about 90%, the I$_{0.5}$ value was about 0.12 mM, and the Hill plot slope value, n̄, was about 1.6.

Thus, pyridoxylation in the presence of protector ADPglucose + MgCl$_2$ produces modified enzyme that still exhibits 5'-AMP inhibition kinetics similar to the nonpyridoxylated enzyme in the presence of fructose-P$_2$ (21,22,40). Pyridoxal-P and 5'-AMP must therefore be binding at different sites, suggesting that the allosteric activator and inhibitor sites are distinct sites.

Table 7 shows the 5'-AMP inhibition characteristics of the different pyridoxyl-P enzyme preparations assayed in the absence of fructose-P_2. With increasing amounts of pyridoxal-P incorporation, the maximum inhibition by 0.4 mM 5'-AMP increased from 40 to 89%, the $I_{0.5}$ value decreased from 1.45 mM to 9 μM and the ñ value increased from 0.29 to 0.77. Thus with increasing pyridoxal-P incorporation into the allosteric activator site, the maximum inhibition increases to about 90%, the modified enzyme is sensitized to 5'-AMP inhibition, and the binding of 5'-AMP becomes less anticooperative.

E. coli B ADPglucose synthase in the presence of low and subsaturating levels of fructose-P_2 exhibits these same kinetic characteristics of 5'-AMP inhibition (21,22,40). In the presence of increasing amounts of subsaturating fructose-P_2, the maximum percent of inhibition increases, the enzyme is very sensitive to inhibition ($I_{0.5}$ is about 7 μM), and the 5'-AMP binding interaction becomes less anticooperative.

Pyridoxylation in the presence of protector ADPglucose + $MgCl_2$, therefore, produces modified enzyme exhibiting 5'-AMP inhibition kinetics when assayed in the absence of fructose-P_2, similar to non-pyridoxylated enzyme when assayed in the presence of fructose-P_2. Therefore, the covalent linkage of pyridoxal-P to enzyme in presence of ADPglucose + $MgCl_2$ is mainly at the allosteric activator site. The modified enzyme is in a high activity form, and exhibits inhibition kinetics similar to that of the fructose-P_2-activated native enzyme. These data suggest that the site of pyridoxylation is in fact the fructose-P_2 binding site.

Primary Sequence of the Allosteric Activator Site of *E. coli* B ADPglucose Pyrophosphorylase. The labeling of the allosteric activator site has enabled us to determine its primary sequence (68). After incorporation of [^{3}H]pyridoxal-P in the presence of ADPglucose + $MgCl_2$, the ADPglucose pyrophosphorylase was degraded into peptides with CNBr. The CNBr pyridoxyl-P peptide that contained over 70% of the total pyridoxal-P incorporated was isolated and purified via gel filtration and CM-cellulose chromatography. Amino acid sequence determination of the CNBr peptide was done on the intact peptide as well as on peptides derived from it via tryptic digestion using automated Edman degradation techniques in a Beckman Model 890C sequenator (68).

The CNBr P-pyridoxyl peptide derived from the enzyme consists of 57 amino residues with a molecular weight of about 6600. The NH_2-terminal of the peptide as shown in Figure 2 has a 16 residue sequence overlap with the previously determined NH_2-terminal sequence of the native enzyme (1,60). The activator site pyridoxyl-P lysine is identified as residue 38 from the native enzyme's NH_2-terminus. Only 39 of the 57 residues

```
                        5                    10
    NH2-Val-Ser-Leu-Glu-Lys-Asn-Asp-His-Leu-Met-
                       15                    20
         Leu-Ala-Arg-Gln-Leu-Pro-Leu-Lys-Ser-Val-
                       25                    30
         Ala-Leu-Ile-Leu-Ala-Gly-Gly-Arg-Gly-Thr-
                       35                    40
         Arg-Leu-Lys-Asp-Leu-Thr-Asn-Lys*-Arg-Ala-
                       45                    50
         Lys-Pro-Ala-Val-His-Phe-Gly-Gly-Lys-[Phe-
                       55                    60
         Arg-Ile-Ile-Asx-Phe-Ala-Leu-Ser-Asn-CmCys-

         Ile-Asn-Ser-Gly-X-Ile-(Arg)]
```

FIGURE 2. Primary sequence of the allosteric site of the $E.$ $coli$ B and mutant CL1136 ADPglucose pyrophosphorylases. P-Pyridoxyl lysine is indicated as Lys*; X refers to an undetermined amino acid residue; (Arg) is inferred by amino acid analysis. The brackets surrounding residues 50 through 62 are the extension of the sequence found with the mutant CL1136 enzyme. The CNBr peptide (see text) starts at residue 11 and has a 16-residue sequence overlap with a previously determined NH_2-terminal sequence of the native enzyme.

have been sequenced. A notable feature of the sequence is the predominance of lysine and arginine residues, especially in close proximity to the pyridoxylated lysine. The sequence, however, of the ADPglucose pyrophosphorylase allosteric binding site does not bear resemblance to pyridoxal-P binding sites of other enzymes (69-71).

The allosteric activators of $E.$ $coli$ B ADPglucose are either bisphosphates, compounds containing an aldehyde and a phosphate group, or compounds containing a carboxylic acid and a phosphate group (1,24,29). Kinetic experiments suggest that the activators are bound to the same site (12,13). Thus, the requirements for binding are satisfied by one phosphate group plus an additional anionic or aldehydic component (1,24). An important aspect of activation may be neutralization of basic groups on the protein by the activator, which may be achieved by phosphate, carboxyl, or aldehyde groups (1,24).

The allosteric activator site contains six positively charged residues in close proximity: arginine at residues 28 and 31, lysine at residue 33, the pyridoxylated lysine at residue 38, arginine at residue 39 and lysine at residue 41. Since at least two basic groups on the protein are probably needed for activator binding, this sequence around the pyridoxylated lysine has great potential for containing both binding components. The pyridoxylated lysine at residue 38 is most likely one of the basic residues involved in activator binding.

Especially interesting is the predominance of arginine residues in close proximity to the pyridoxylated lysine: arginine residues 28, 31 and 39 of the NH_2-terminal. Arginine residues have been shown to serve as positiviely charged recognition sites for negatively charged substrates and anionic cofactors in several enzymes (72,73). One major biological function of arginine might be to interact with phosphorylated metabolites. Studies have indicated that the guanidinium group, by virtue of its planar structure and its ability to form multiple hydrogen bonds with a phosphate moiety, is ideally suited for ionic interaction with phosphate-containing substances (74). Preliminary studies (75) using [^{14}C]phenylglyoxal, indicate the presence of an essential arginine at the allosteric activator site of ADPglucose pyrophosphorylase. Additional studies should determine the exact location of the modified arginine and whether it is present on the same CNBr peptide containing the pyridoxal-P modified lysine or elsewhere in the primary sequence of the subunit.

The mutant CL1136 ADPglucose pyrophosphorylase has been purified to homogeneity and the amino acid sequence of the allosteric activator site of the mutant CL1136 enzyme has been determined. As seen in Figure 2, the primary sequence up to amino residue 49 is exactly the same as found for the parent enzyme. The sequence of the CL1136 activator site peptide has been extended to residue 67. Presently the identity of amino acid residue 65 is unknown. The results indicate that the amino acid sequence immediately adjacent to the allosteric binding site is identical in the parent and mutant enzymes. Thus the altered kinetics seen for the mutant CL1136 enzyme must be due to an amino acid change elsewhere in the peptide chain. This may not be surprising as the mutation in CL1136 effects both the inhibition and activation kinetics of the CL1136 enzyme and thus the activator binding site of the mutant enzyme may not be directly effected.

In conclusion these results indicate that at least part of the activator binding site is very close to the amino terminus of the *E. coli* B ADPglucose pyrophosphorylase. This finding may facilitate elucidation of the structure-function relationships of particular primary sequences with respect to specificity of the ADPglucose pyrophosphorylase activator binding sites. Primary sequence analyses of the NH_2-terminus of the other ADPglucose pyrophosphorylases could show significant differences if the allosteric activator binding sites are also situated close to the NH_2-terminus for the other activator classes of the ADPglucose pyrophosphorylases. Efforts to isolate and characterize the allosteric binding sites of other ADPglucose pyrophosphorylases are thus continuing.

REFERENCES

1. Preiss, J. (1978). Adv. Enzymol. 45, 317.
2. Krebs, E.G., and Preiss, J. (1975). In "MTP International Review of Science, Biochemistry of Carbohydrates" (W.J. Whelan, ed.), Biochemistry Series 1, Vol. 5, p. 337. Butterworths, London and University Park Press, Baltimore.
3. Atkinson, D.E. (1970). In "The Enzymes" (P.D. Boyer, ed.), 3rd Ed., Vol. 1, p. 461. Academic Press, New York.
4. Shen, L.C., and Atkinson, D.E. (1970). J. Biol. Chem. 245, 3996.
5. Strange, R.E., Dark, F.A., and Ness, A.G. (1961). J. Gen. Microbiol. 25, 61.
6. Holm, T., and Palmstierna, H. (1956). Acta Chem. Scand. 10, 578.
7. Sigal, N., Catteneo, J., and Segel, I.H. (1964). Arch. Biochem. Biophys. 108, 440.
8. Shen, L., and Preiss, J. (1966). Arch. Biochem. Biophys. 116, 374.
9. Eidels, L., Edelmann, P.L., and Preiss, J. (1970). Arch. Biochem. Biophys. 140, 60.
10. Eidels, L., and Preiss, J. (1970). Arch. Biochem. Biophys. 140, 75.
11. Ribereau-Gayon, G., Sabraw, A., Lammel, C., and Preiss, J. (1971). Arch. Biochem. Biophys. 142, 675.
12. Preiss, J., Shen, L., Greenberg, E., and Gentner, N. (1965). Biochemistry 5, 1833.
13. Gentner, N., Greenberg, E., and Preiss, J. (1969). Biochem. Biophys. Res. Commun. 36, 373.
14. Preiss, J., Crawford, K., Downey, J., Lammel, C., and Greenberg, E. (1976). J. Bacteriol. 127, 193.
15. Robson, R.L., Robson, R.M., and Morris, J.G. (1974). Biochem. J. 144, 502.
16. Levi, C., and Preiss, J. (1976). Plant Physiol. 58, 753.
17. Sanwal, G.G., and Preiss, J. (1969). Arch. Biochem. Biophys. 119, 454.
18. Preiss, J., Ghosh, H.P., and Wittkop, J. (1967). In "Biochemistry of Chloroplasts" (T.W. Goodwin, ed.), Vol. 2, pp. 131-153. Academic Press, London-New York.
19. Ghosh, H.P., and Preiss, J. (1966). J. Biol. Chem. 241, 4491.
20. Sanwal, G.G., Greenberg, E., Hardie, J., Cameron, E., and Preiss, J. (1973). Plant Physiol. 43, 417.

21. Gentner, N., and Preiss, J. (1967). Biochem. Biophys. Res. Commun. 27, 417.
22. Gentner, N., and Preiss, J. (1968). J. Biol. Chem. 243, 5882.
23. Preiss, J., Crawford, K., Downey, J., Lammel, C., and Greenberg, E. (1976). J. Bacteriol. 127, 193.
24. Preiss, J. (1973). In "The Enzymes" (P.D. Boyer, ed.), 3rd Ed., Vol. 8, p. 73. Academic Press, New York.
25. Lapp, D., and Elbein, A.D. (1972). J. Bacteriol. 112, 327.
26. Furlong, C.E., and Preiss, J. (1969). J. Biol. Chem. 244, 2539.
27. Furlong, C.E., and Preiss, J. (1969). In "Progress in Photosynthetic Research" (H. Metzner, ed.), Vol. 3, pp. 1604-1617. Tubingen, Germany.
28. Stanier, R.Y., Doudoroff, M., Kunisawa, R., and Contopoulou, R. (1959). Proc. Nat. Acad. Sci. USA 45, 1246.
29. Preiss, J. (1969). In "Current Topics in Cellular Regulation" (B.L. Horecker and E.R. Stadtman, eds.), Vol. 1, pp. 125-160. Academic Press, New York.
30. Conrad, R., and Schlegel, H.G. (1977). J. Gen. Microbiol. 101, 277.
31. Conrad, R., and Schlegel, H.G. (1977). Arch. Microbiol. 112, 39.
32. McDonald, P.W., and Strobel, G.A. (1970). Plant Physiol. 46, 126.
33. Kanazawa, T., Kanazawa, K., Kirk, M.R., and Bassham, J.A. (1972). Biochim. Biophys. Acta 256, 656.
34. Steup, M., Peavey, D.G., and Gibbs, M. (1976). Biochem. Biophys. Res. Commun. 72, 1554.
35. Shen-Hwa, C.-S., Lewis, D.H., and Walker, D.A. (1975). New Phytol. 74, 383.
36. Herold, A., Lewis, D.H., and Walker, D.A. (1976). New Phytol. 76, 397.
37. Heldt, H.W., Chon, C.J., Maronde, D., Herold, A., Stankovic, Z.S., Walker, D.A., Kraminer, A., Kirk, M.R., and Heber, U. (1977). Plant Physiol. 59, 1146.
38. Dietzler, D.N., Leckie, M.P., Lais, C.J., and Magnani, J.L. (1974). Arch. Biochem. Biophys. 162, 602.
39. Dietzler, D.N., Leckie, M.P., Lais, C.J., and Magnani, J.L. (1975). J. Biol. Chem. 250, 2383.
40. Govons, S., Gentner, N., Greenberg, E., and Preiss, J. (1973). J. Biol. Chem. 248, 1731.
41. Govons, S., Vinopal, R., Ingraham, J., and Preiss, J. (1969). J. Bacteriol. 97, 970.
42. Preiss, J., Sabraw, A., and Greenberg, E. (1971). Biochem. Biophys. Res. Commun. 42, 180.
43. Preiss, J. (1972). Intrasci. Chem. Rep. 6, 13.
44. Preiss, J., Greenberg, E., and Sabraw, A. (1975). J. Biol. Chem. 250, 7631.

45. Preiss, J., Lammel, C., and Greenberg, E. (1976). Arch. Biochem. Biophys. 174, 105.
46. Damotte, M., Cattaneo, J., Sigal, N., and Puig, J. (1968). Biochem. Biophys. Res. Commun. 32, 916.
47. Cattaneo, J., Damotte, M., Sigal, N., Sanchez-Medina, G., and Puig, J. (1969). Biochem. Biophys. Res. Commun. 34, 694.
48. Creuzat-Sigal, N., Latil-Damotte, M., Cattaneo, J., and Puig, J. (1972). In "Biochemistry of the Glycosidic Linkage" (R. Piras and H.G. Pontis, eds.), p. 647. Academic Press, New York.
49. Dietzler, D.N., Lais, C.J., and Leckie, M.P. (1974). Arch. Biochem. Biophys. 160, 14.
50. Swedes, J.S., Sedo, R.J., and Atkinson, D.E. (1975). J. Biol. Chem. 250, 6930.
51. Dietzler, D.N., Leckie, M.P., and Lais, C.J. (1973). Arch. Biochem. Biophys. 156, 684.
52. Steiner, K.E., and Preiss, J. (1977). J. Bacteriol. 129, 246.
53. Lowry, O.H., Carter, J., Ward, J.B., and Glaser, L. (1971). J. Biol. Chem. 246, 6511.
54. Lubin, M., and Ennis, H.L. (1964). Biochim. Biophys. Acta 80, 614.
55. Webb, B.M. (1968). J. Gen. Microbiol. 43, 401.
56. Silver, S. (1969). Proc. Nat. Acad. Sci. USA 62, 764.
57. Dempsey, W.B. (1972). Biochim. Biophys. Acta 264, 344.
58. Dempsey, W.B., and Arcement, L.J. (1971). J. Bacteriol. 107, 580.
59. Haugen, T., Ishaque, A., Chatterjee, A.K., and Preiss, J. (1974). FEBS Lett. 42, 205.
60. Haugen, T.H., Ishaque, A., and Preiss, J. (1976). J. Biol. Chem. 251, 7880.
61. Ozaki, H., and Preiss, J. (1972). Methods Enzymol. 28, 406.
62. Haugen, T., and Preiss, J. (1979). J. Biol. Chem. 254, 127.
63. Janson, C.A., and Cleland, W.W. (1974). J. Biol. Chem. 249, 2572.
64. DePamphilis, M.L., and Cleland, W.W. (1973). Biochemistry 122, 3714.
65. Lazdunski, N. (1972). In "Current Topics in Cellular Regulation" (B.L. Horecker and E.R. Stadtman, eds.), p. 267. Academic Press, New York.
66. Haugen, T., Ishaque, A., and Preiss, J. (1976). Biochem. Biophys. Res. Commun. 69, 346.
67. Parsons, T.F., and Preiss, J. (1978). J. Biol. Chem. 253, 6197.
68. Parsons, T.F., and Preiss, J. (1978). J. Biol. Chem. 253, 7638.
69. Kempe, T.D., and Stark, G.R. (1975). J. Biol. Chem. 250, 6861.

70. Piszkiewicz, D., Landon, M., and Smith, E.L. (1970). J. Biol. Chem. 245, 2622.
71. Dayhoff, M.D., Barker, W.C., and Hardman, J.K. (1972). In "Atlas of Protein Sequence and Structure" (M.O. Dayhoff, ed.), pp. 53-66. National Biomedical Research Foundation, Washington, D.C.
72. Riordan, J.F., Elvany, K.D., and Borders, C.L., Jr. (1977). Science 195, 884.
73. Borders, C.L., Jr., and Wilson, B.A. (1976). Biochem. Biophys. Res. Commun. 73, 978.
74. Cotton, F.A., Hazen, E.E., Jr., Day, V.W., Larsen, S., Norman, J.G., Jr., Wong, S.T.K., and Johnson, K.H. (1973). J. Am. Chem. Soc. 95, 2367.
75. Parsons, T.F., Carlson, C.A., and Preiss, J. (1977). Fed. Proc. 36, 856, Abstr. 3080.

COVALENTLY INTERCONVERTIBLE ENZYME CASCADE AND METABOLIC REGULATION

P. B. Chock and E. R. Stadtman

Laboratory of Biochemistry
National Heart, Lung, and Blood Institute
National Institutes of Health
Bethesda, Maryland 20014

INTRODUCTION

The singular importance of covalent modification of enzymes in cellular regulation is evident from the ever increasing number of reports demonstrating that activities of key enzymes in metabolism are modulated by their interconversion between covalently modified and unmodified forms (1, 2). These enzymes/proteins are referred to as interconvertible enzymes/proteins and they can be classified into 5 classes, namely (1) those modulated by phosphorylation of a specific serine residue on the enzyme, (2) those regulated by nucleotidylation of a particular tyrosine residue on the protein, (3) those modified by ADP-ribosylation of an arginine or a yet unidentified residue on the enzyme, (4) those modulated by carboxymethylation of an aspartate or glutamate residue on the proteins, and (5) those modified by peptidylation of the carboxyl terminal of a protein. The reactions which lead to these modifications are:

$$\text{ATP} + \ldots(\underset{\underset{\text{OH}}{|}}{\text{Ser}})\ldots \longrightarrow \text{ADP} + \ldots(\underset{\underset{\text{OPO}_3}{|}}{\text{Ser}})\ldots$$

$$\text{NTP} + \ldots(\underset{\underset{\text{OH}}{|}}{\text{Tyr}})\ldots \longrightarrow \text{PPi} + \ldots(\underset{\underset{\text{NMP-O}}{|}}{\text{Tyr}})\ldots$$

$$\text{DPN} + \ldots(\underset{\underset{\text{NH}}{\|}}{\text{Arg}})\ldots \longrightarrow \text{Nicotinamide} + \ldots(\underset{\underset{\text{ADPR-N}}{\|}}{\text{Arg}})\ldots$$

$$+ \ldots(\underset{\underset{\text{X}}{|}}{})\ldots \longrightarrow \text{Nicotinamide} + \ldots(\underset{\underset{\text{ADPR-X}}{|}}{})\ldots$$

$$\text{SAM} + \ldots \underset{\underset{\text{COOH}}{|}}{(\text{Glu})} \ldots \longrightarrow \text{S-adenosylhomocysteine} + \ldots \underset{\underset{\text{COOCH}_3}{|}}{(\text{Glu})} \ldots$$

$$\underset{\underset{H}{|}}{\overset{\overset{R}{|}}{H_2N-C-COOH}} + \ldots X\text{-COOH} \longrightarrow \ldots X-\overset{\overset{O}{\|}}{C}-\overset{\overset{H}{|}}{N}-\overset{\overset{R}{|}}{\underset{\underset{H}{|}}{C}}-\text{COOH}$$

where NTP, NMP, and SAM are nucleoside triphosphate, nucleoside monophosphate and S-adenosylmethionine, respectively. In addition to the covalent modifications shown here, acylation of protein is also known, but its role in enzymic regulation has yet to be determined. In order to understand the rationale for nature to adopt such a regulatory mechanism, we carried out a theoretical analysis of enzyme interconversion systems to reveal their advantages and disadvantages (3, 4). It should be pointed out that this theoretical model is developed based on the results of many studies on the regulation of glutamine synthetase, particularly from the work of Brown *et al*. (5) which showed that adenylylation is not an all or none process; instead, a steady-state is established and its level is determined by the concentration of the effectors involved.

THEORETICAL ANALYSIS

Monocyclic Cascade. Covalent interconversion of an enzyme is a cyclic process resulting from the coupling two cascade systems as shown in Figure 1. In the forward cascade, the inactive converter enzyme (E_i) is first activated by an allosteric effector e_1, as in the case of c-AMP activation of c-AMP dependent protein kinase. The activated converter enzyme then catalyzes the conversion of an interconvertible enzyme, I, from its unmodified form, o-I, to its modified form, m-I, as, for example, occurs when protein kinase catalyzes the phosphorylation of phosphorylase kinase. The regeneration cascade is triggered by the interaction of an allosteric effector e_2 with the inactive form of converter enzyme R, which leads to the formation of its activated form R_a. This activated R_a catalyzes the demodification of the modified interconvertible enzyme to its original form, o-I. Dynamic coupling of these forward and regeneration cascades results in the cyclic interconversion between o-I and m-I and the concomitant hydrolysis of ATP to ADP and Pi. When ATP concentration is maintained in excess relative

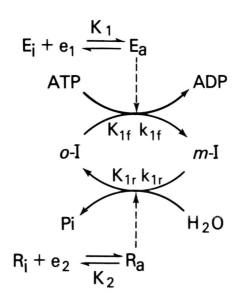

FIGURE 1. *Schematic representation of a monocyclic cascade system.*

to the enzymes involved and at a fairly constant level, then a steady-state will be established in which the rate of m-I formation will be equal to the rate of o-I regeneration. For this steady-state situation, it is possible to derive an equation which shows how the fraction of interconvertible enzyme in the modified form will vary as a function of other cascade parameters. For simplicity (1, 3), ATP dependency is ignored in deriving this steady-state equation because it can be treated as a constant due to the fact that ATP concentration is high relative to the concentrations of converter and interconverter enzymes, and it is fairly constant for any metabolic state. In addition, it is assumed that: (i) the formation of the enzyme-enzyme and enzyme-effector complexes proceed via a rapid equilibrium mechanism; (ii) the concentration of the enzyme-enzyme complexes are negligible compared to the concentrations of the active and inactive enzymes (when this condition is satisfied, $[E] \simeq [E_i] + [E_a]$, $[I] \simeq [o\text{-}I] + [m\text{-}I]$, $[R] = [R_I] + [R_a]$, where $[E]$, $[I]$ and $[R]$ are the total concentration of E, I and R, respectively); and (iii) the concentrations of the allosteric effectors e_1 and e_2 are maintained at constant levels for any metabolic state. With these assumptions, it can be shown that the mole fraction of the modified interconvertible enzyme is given by the equation

$$\frac{[m\text{-}I]}{[I]} = \left[\left(\frac{k_{1r}}{k_{1f}}\right)\left(\frac{K_{1r}}{K_{1f}}\right)\left(\frac{K_2}{K_1}\right)\left(\frac{[R]}{[E]}\right)\left(\frac{[e_2]}{[e_1]}\right) \cdot \frac{(1 + K_1[e_1])}{(1 + K_2[e_2])} + 1 \right]^{-1} \quad (1)$$

in which K_1 and K_2 are the equilibrium constants for the reactions leading to the activated converter enzyme complexes as shown in Figure 1; k_{1f} and k_{1r} are specific rate constants for the reactions which lead to the modification and demodification of the interconvertible enzyme, respectively; K_{1f} and K_{1r} are the association constants for the formation of the enzyme-enzyme complexes, $E_a.o\text{-}I$ and $R_a.m\text{-}I$, respectively, as shown in Figure 1.

Equation 1 shows that the fractional modification of the interconvertible enzyme is modulated by a multiplicative function of 10 parameters. Eight of these parameters (all except e_1 and e_2) can be altered by allosteric interactions of one or more of the cascade enzymes with a single or multiple allosteric effectors. For simplicity, these 8 parameters are treated as 4 parameter ratios ($viz.$, k_{1r}/k_{1f}, K_{1r}/K_{1f} . . . etc.). To demonstrate some properties of the monocyclic cascade, fractional modification of the interconvertible enzyme is plotted as a function of e_1 concentration and variations in one or more of the paratmeter ratios in Equation 1 (Figure 2A). In this figure, all the curves were obtained with K_1, the equilibrium constant for the activation of the converter enzyme E by the effector e_1, being set at 1.0. With this assumption, the dotted line in Figure 2A represents the binding isotherm of e_1 to E_i. When the binding of e_1 to the converter enzyme is linked to the cascade via the catalytic action of E_a in the conversion of $o\text{-}I$ to $m\text{-}I$, the indirect effect of e_1 on the fractional modification of the interconvertible enzyme can vary enormously, depending on the magnitudes of other parameters in Equation 1. For instance, curve 1 shows that, when all parameters in Equation 1 are assigned values of 1.0, the concentration of e_1 required to obtain 50% conversion of $o\text{-}I$ to $m\text{-}I$ (referred to as $[e_1]_{0.5I}$ value) is equal to the value of K_1 which is 1.0; in addition, at saturating levels of e_1 only 67% of the interconvertible enzyme can be converted to its modified form, $m\text{-}I$. In other words, the maximum magnitude of modification,

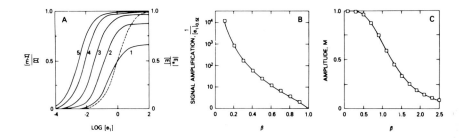

FIGURE 2. Computer simulated curves obtained with Equation 1 showing: (A) The changes with respect to the relationship between fractional modifiction of the interconvertible enzyme and the primary effector e_1 concentration when the parameter ratios are decreased from 1.0 to 0.25 in a stepwise cumulative manner. (B) The relationship between the parameter ratio (β) and signal amplification. (C) The relationship between β and the amplitude of fractional modification of the interconvertible enzyme.

i.e., the amplitude of the modification reaction, is 0.67 under this condition. When each parameter in Equation 1 is varied stepwise by a factor of only 2 in a successively cumulative manner to favor m-I formation, one obtains curves 2 to 5 of Figure 2A. Compared to curve 1, the solid curves in this figure show that as the number of parameter ratios (β) which are varied is increased there is a progressive increase in amplitude of the fractional modification of the interconvertible enzyme, and there is a progressive decrease in e_1 concentration required to convert 50% of the interconvertible enzyme to its modified form. The latter feature is referred to as signal amplification since e_1 is a metabolic signal used to trigger the cyclic cascade system. To facilitate comparisons, this signal amplification is defined as the ratio of the concentration of e_1 required to achieve 50% activation of the converter enzyme E to that required to produce 50% modification of the interconvertible enzyme I (1). With this definition, curve 5 in Figure 2A shows a signal amplification of 320. These two important properties of the cyclic cascade system are better illustrated by Figure 2B and 2C, where the signal amplification and the amplitude are plotted as a function of parameter ratios. Figure 2B shows that when the 4 parameter ratios, β, are varied simultaneously from 1.0 to 0.1, the signal amplification factor

increases from 1 to 10^4 via a complex exponential function.
Figure 2C shows that the amplitude of fractional modification
of the interconvertible enzyme varies from 100% (complete
modification) at $\beta = 0.1$ to almost 0% modification at $\beta = 2.5$.

It should be pointed out that due to signal amplification, the interconvertible enzymes can respond to effector concentrations that are well below the dissociation constant of the effector-converter enzyme complex. In other words, one needs to activate only a relatively small fraction of the converter enzyme to obtain a significant fractional modification of the interconvertible enzyme. This property is illustrated by replots of the data in Figure 2A to show how the fractional modification of interconvertible enzyme varies as a function of the fractional activation of the converter enzyme E. From these replots (Figure 3), it can be seen that only 1% activation of the converter enzyme is needed to obtain a 2%, 7%, 25%, 55%, and 75% modification of the interconvertible enzyme for curves 1, 2, 3, 4, and 5, respectively.

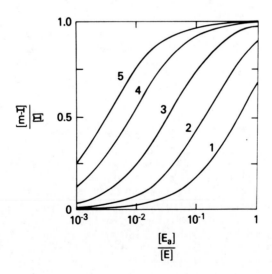

FIGURE 3. *Computer simulated curve showing the relationship between fractional activation of the converter enzyme E and that of the interconvertible enzyme I for the corresponding curves in Figure 2A.*

MODULATION OF PROTEIN FUNCTION 191

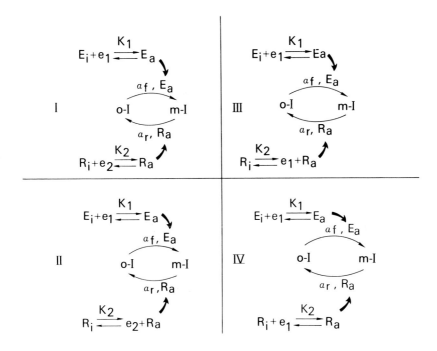

FIGURE 4. *Four regulatory patterns derived for allosteric regulation of monocyclic cascade systems.*

Figure 4 depicts four possible regulatory patterns for monocyclic cascade systems obtained with the assumption that effector e_1 is required to activate the converter enzyme E. Case I is the same as that described in Figure 1, where it is assumed that e_1 activates the converter enzyme E, and e_2 activates the converter enzyme R. In case II, e_1 activates E while e_2 inactivates the R-converter enzyme. In case III, e_1 activates the E converter enzyme and also inactivates the R-converter enzyme. In case IV, e_1 activates both converter enzymes. Numerical analysis of these four regulatory models shows that they can yield a wide variety of patterns for fractional modification of the interconvertible enzyme in response to increasing concentrations of e_1 (1, 3). These patterns differ with respect to their amplitude, signal amplification and sensitivity (1) to changes in e_1 concentration. With the consideration that each cascade enzyme can interact simultaneously with more than one positive and/or negative allosteric effector, it is apparent that a vast number of regulatory patterns can be obtained for the monocyclic cascade systems. In fact, three of the four regulatory patterns shown in Figure 4 have been observed in regulation of the mammalian pyruvate dehydrogenase cascade (6, 7).

Since the properties of the monocyclic cascades systems as shown here are based on the simplified equation, the validity of the simplifying assumptions was checked by analyzing the properties of steady-state equations based on fewer assumptions. For example, without assuming that the concentrations of the activated converter enzyme-interconvertible complexes are negligibly small, a quartic equation containing more than 200 terms was obtained for the expression of modified inconvertible enzyme (3). Analysis of this complex equation yields results that are qualitatively similar to those obtained with the simplified equation. In fact, when the numerical assignment met the simplified conditions, identical fractional modification curves were obtained as a function of e_1 concentration (3). Quantitatively the overall results show that the monocyclic cascade as described by the quartic equation possesses greater signal amplification potential, flexibility, and under certain conditions, it is more sensitive to the change in effector concentration. Thus, the equations derived with the simplifying assumptions are valid for analyzing the fundamental characteristics of the monocyclic cascade systems. In fact, the complex equation confirms the predictions of the simplified equation and demonstrates additionally that monocyclic cascades are potentially more sensitive to effector stimuli and are capable of achieving an even greater signal amplification and flexibility than is predicted by the simplified equations.

It should be emphasized that in order to maintain a steady-state distribution between the modified and unmodified forms of the interconvertible enzyme, nucleoside triphosphate hydrolysis is required. Figure 1 shows that for each complete cycle in a phosphorylation-dephosphorylation cascade, one equivalent of ATP is hydrolyzed to ADP and Pi. However, the role of ATP was disregarded in deriving the steady-state expressions because its concentration is maintained metabolically at a fairly constant level which is several orders of magnitude higher than the concentrations of the enzymes undergoing covalent modification. Nevertheless, it should be pointed out (1) that the rate of this ATP flux is dependent on the concentrations of all enzymes and effectors involved in the cyclic cascade and is dependent on all the constants needed to describe the cascade. Therefore, the rate of ATP consumption can also be regulated by the parameters in Equation 1. This ATP flux is an essential feature for the cyclic cascade regulatory mechanism because it provides the free energy required to maintain the steady-state distribution between the modified and unmodified forms of the interconvertible enzyme at various metabolite-specified levels which are different from that specified by thermodynamic considerations. The ATP consumption is therefore the price the cell must pay to support such an effective control mechanism.

Bicyclic Cascade Systems. When the modified form of an interconvertible enzyme in one cycle catalyzes the covalent modification of an interconvertible enzyme in another cycle, the two cycles are coupled such that the modification of the second inteconvertible enzyme is a function of all the parameters in both cycles (1, 4). Figure 5 shows two types of bicyclic cascades which are known to be involved in metabolic regulation of key enzymes. The so called opened bicyclic cascade (Figure 5A) is involved in the activation of glycogen phosphorylase, and a closed bicyclic cascade (Figure 5B) is utilized in the regulation of glutamine synthetase from gram-negative bacteria. The latter system will be discussed in detail by Dr. E. R. Stadtman. Steady-state analysis of an opened bicyclic cascade (1, 4) revealed that fractional modification of the second interconvertible enzyme, I_2, in the cascade is a multiplicative function of 18 parameters instead of 10 parameters shown in Equation 1 for the monocyclic cascade systems. Therefore, bicyclic cascades are capable of achieving a much greater signal amplification than monocyclic cascade systems. In addition, since 15 of

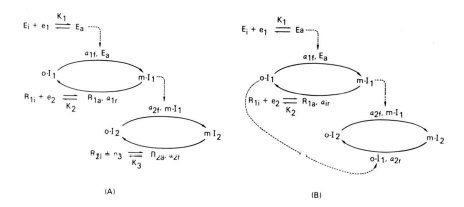

FIGURE 5. Schematic representation of (A) an opened bicyclic cascade system, and (B) a closed bicyclic cascade system.

the 18 parameters in the bicyclic cascade equation are susceptible to modulation by allosteric effectors, bicyclic cascades are endowed with a greater flexibility toward allosteric control. Furthermore, the number of converter enzymes in the bicyclic cascade is greater than in a monocyclic cascade system; therefore, it is possible to obtain a greater number of unique regulatory patterns in response to positive and negative allosteric interactions.

Because each step in a bicyclic cascade (Figure 5) can be regulated by one and the same allosteric effector, it is evident that such cascades are capable of generating a highly "cooperative" response to increasing effector concentration. This is illustrated by the curves in Figure 6A which show how the modification of I varies as a function of e_1 concentration when e_1 is assumed to regulate one or more steps in the cascade, as follows: curve 1, e_1 is assumed to activate the converter enzyme E; curve 2, e_1 is assumed to activate both E and m-I; curve 3, e_1 activates both E and m-I, and inactivates R_{1a}; and in curve 4, e_1 activates both E and m-I and inhibits both R_{1a} and R_{2a}. It is evident from these curves that as the number of steps in the bicyclic cascade being modulated by the effector e_1 increases,

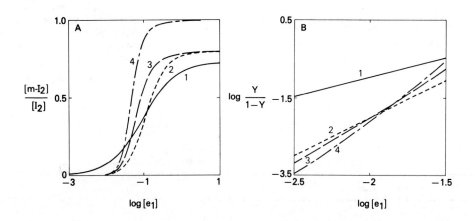

FIGURE 6. Computer simulated curves showing (A) the dependence of the fractional modification of the last interconvertible enzyme in an opened bicyclic cascade on the concentration of allosteric effector e_1, when e_1 is affecting from one (curve 1) to four steps (curve 4) in the cascade. (For detail see text and Reference 1). (B) Hill plots for the corresponding curves in A.

the slopes of nearly linear portions of the curves also
increases. In other words, rectilinear plots of the data
in curves 1 to 4 will yield progressively more sigmoidal
curves. This property is illustrated by the Hill plot in
Figure 6B from which Hill numbers of 1.0, 2.0, 2.5, and 3.0
were obtained for curves 1, 2, 3, and 4, respectively. This
corresponds to the more accurately defined sensitivity index
(1) of 1.0, 3.3, 4.4, and 5.6, respectively. It should be
noted that the Hill number for these curves will vary depending on the values of the parameters used. However, the maximum Hill numbers that can be obtained for situations in which
2, 3, and 4 steps are regulated by the same effector are 2.0,
3.0, and 4.0, respectively. Nevertheless, Figure 6 shows
that when effector e_1 is utilized in more than one reaction
in the cascade in a manner that favors the modification of
interconvertible enzyme I_2, a sensitivity index of greater
than 1.0 would be obtained. In other words, the fractional
modification of I_2 becomes more sensitive to increasing e_1
concentration. In addition, Figure 6 also shows that by increasing the number of reactions in which e_1 plays a direct
role in the cyclic cascade, both the amplitude and the signal
amplification also increased.

Multiplyclic Cascade Systems. The relationship between
the number of cycles in a cascade and its regulatory properties was disclosed by theoretical analysis of a cascade model
containing n cycles as shown in Figure 7. It was assumed
that the cascade is initiated by the activation of the converter enzyme E by an allosteric effector, e_1, and subsequently the modified form of an interconvertible enzyme in
one cycle functions as the converter enzyme in the next
cycle. The steady-state expression for fractional modification of the target enzyme, I_n, is a complex multiplicative
function containing n + 1 terms (1, 4). Analysis of this expression shows that with each additional cycle in the cascade, the modification of the last interconvertible enzyme,
I_n, (the target enzyme) becomes dependent on eight additional
variables which occur together with all the parameters involved in the preceding cycles as a multiplicative function,
and seven of these new parameters can be modulated by allosteric effectors. Therefore, each additional cycle in a cascade results in an enormous increase in signal amplification
potential and allosteric control potential of the system.
The amplification effect is illustrated by the curves in
Figure 8. As a point of reference, curve 0 shows how the
fractional activation of the first converter enzyme E varies
with increasing e_1 concentration when the equilibrium constant K_1 is set at 1.0. When K_1 is held at 1.0 and all

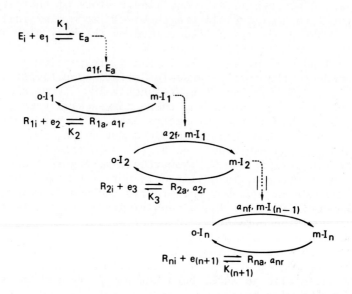

FIGURE 7. *Schematic representation of a multicyclic cascade system.*

parameters that favor modification are assigned values of 2.0, while those favoring demodification are assigned values of 0.5, then the signal amplification is increased progressively with increasing number of cycles in the cascade. From curves 1, 2, 3, and 4 (Figure 8), it can be calculated that the signal amplification factors for one, two, three and four cycle cascades are 3.2×10^2, 1.02×10^5, 3.28×10^7, and 1.05×10^{10}, respectively. Thus, an amplification factor of 10^{10} can be obtained in a four cycle cascade by only 2-fold changes in each parameter. The inset in Figure 8 shows that the log of the amplification factor is proportional to the number of cycles in the cascade.

Kinetic Considerations. So far the properties of the cyclic cascades discussed here are time independent properties. In order for the cell to take advantage of these properties, attainment of the steady-state must be achieved within a reasonable period of time. Kinetic analysis of multicyclic cascade systems (8) shows that the rate of covalent modification of the last interconvertible enzyme in the cas-

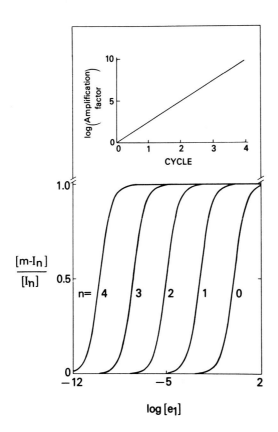

FIGURE 8. *Computer simulated curves to show the dependence of fractional modification of the last interconvertible enzyme on the number of cycles n as a function of log (e_1) when each of the parameters, except K_1 (= 1.0), were varied by a factor of 2 in favor of the modification. The inset depicts the linear relationship between log (signal amplification) and n. (For detail see Reference 1).*

cade is a multiplicative function of the rate constants of all the reactions that lead to the formation of the modified enzyme. Thus, cyclic cascade systems are capable of functioning as rate amplifiers. Figure 9 shows the effect of rate amplification as a function of the number of cycles involved in the cascade. Curve 1 shows that for a monocyclic

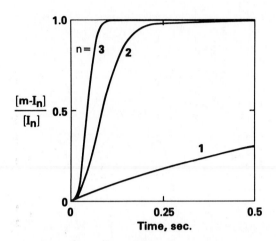

FIGURE 9. Computer simulated time course to show the rate amplification as a function of cycles (n) involved in the cyclic cascade. All the concentration of interconvertible enzymes were set at 10^{-5} M, while the concentration of the converter enzymes and effectors were set at 5×10^{-6} M. All the association constants were equaled to 2×10^5 M^{-1} and forward and reverse rate constants were set at 1,000 sec^{-1} and 10 sec^{-1}, respectively. (For detail see Reference 7).

cascade system, the modification proceeds via a normal exponential function with respect to time. However, for a bicyclic cascade system (curve 2) and a tricyclic cascade system (curve 3), the formation of the modified enzyme in the last cycle proceeds with an intial lag, followed by a burst. This burst is steeper for the tricyclic system than the bicyclic system. With the assumption that the concentrations of all the interconvertible enzymes are the same, and that all forward and reverse rate constants are 1,000 sec^{-1} and 10 sec^{-1}, respectively, Figure 9 shows that within 60 msec, 90% of I_3 is converted to $m-I_3$, while only 35% of I_2, and 6% of I_1 is modified. These differences in time required to achieve certain fractional modifications of interconvertible enzymes for various cycles reflect the multiplier effects with respect to the rate constants in a multicyclic system. The difference in rate can vary enormously depending on the kinetic constants and concentration of both effectors and enzymes involved (8). For example, the dependence of rate amplification on the number of cycles has been shown to be more pronounced than that shown in Figure 9 when the total concentration of interconvertible enzymes is set with $[I_1] < [I_2] < [I_3]$.

From the rate analysis shown in Figure 9 and elsewhere (8), it is evident that with reasonable assumptions for the rate constants, a multicyclic cascade system can achieve steady-state in the millisecond time range. It should be pointed out that with the same kinetic constants, an even greater rate of response could be obtained by proper geographic positioning of the converter enzymes and interconvertible enzymes in multienzyme complexes as occurs in mammalian pyruvate dehydrogenase (9), or on a compact solid support as occurs in the binding of glycogen phosphorylase cascade components on glycogen particles (10). In fact, Cori and his associates (11) have shown that the electrical stimulation of frog sartorius muscle at 30°, phosphoyrlase b is converted to phosphorylase a with a half-time of 700 milliseconds.

CONCLUDING REMARKS

The superiority of a multicyclic cascade relative to a monocyclic cascade system with respect to its signal amplification potential, flexibility, and its rate amplification is evident from the above discussion. It is interesting to note that these advantages are fully utilized by many interconvertible enzymes when they involve the phosphorylation of an enzyme catalyzed by a cyclic nucleoside monophosphate dependent protein kinase. This is illustrated in Figure 10 where a phosphorylation-dephosphorylation cycle is assumed to be involved in the interconversion of o-I and m-I. This monocycle system is extended by the fact that (i) the activation of c-AMP dependent protein kinase, R_2C_2, is initiated by hormone activation of adenyl cyclase which catalyzes the conversion of ATP to c-AMP; (ii) the c-AMP activates the protein kinase; (iii) the activated interconvertible enzyme m-I, catalyzes the conversion of substrate to product. Therefore, with respect to the product formation, this is a five-step cascade, even though only one interconvertible enzyme is involved.

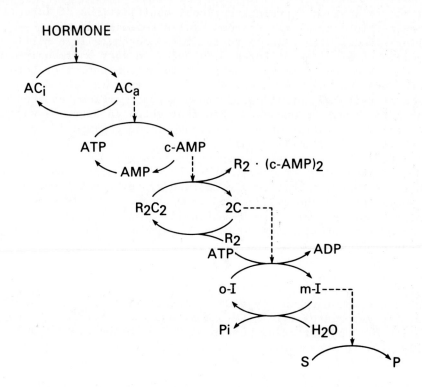

FIGURE 10. *A representation of a multistep cascade which consists of one interconvertible enzyme, I. AC_i and AC_a are inactive and active adenylcyclase, respectively, R_2C_2 is the regulatory (R_2)-catalytic (C) complex of the c-AMP dependent protein kinase; S and P are substrate and product of the activated interconvertible enzyme, m-I.*

It should be emphasized that cyclic cascade systems are fundamentally different from the irreversible, undirectional cascades which are involved in blood clotting and complement fixation. The undirectional cascades once triggered by an appropriate signal respond in an explosive manner to produce an avalanche of product to meet the biological emergency. After the function is fulfilled, the cascades respond to some other regulatory signals which lead to the destruction of the catalytic process. In contrast, the interconvertible enzyme cascades involve the dynamic coupling of two opposing cascades, and thereby establish a steady-state distribution of active and inactive forms, which determine the specific activity of the target enzyme commensurate with metabolic needs. Thus, the specific activity of the target enzyme can vary smoothly and continuously in response to ever changing levels of multiple metabolites which are allosteric effectors of the cascade enzymes. In addition, the interconvertible enzyme cascades possess enormous signal amplification potential, and they can generate a sigmoidal response of interconvertible enzyme activity to increasing concentration of an allosteric effector. Because a minimum of two enzymes, one effector, and 5 additional reaction constants are involved in each cycle, and each enzyme can be a separate target of one or more allosteric effectors, cyclic cascade systems provide high flexibility for metabolic regulation. Finally, interconvertible enzyme cascade can function as rate amplifiers to generate an almost explosive increase in catalytic activity in response to stimuli.

REFERENCES

1. Stadtman, E. R., Chock, P. B. (1978). In "Current Topics in Cellular Regulation", Vol. 13, (B. L. Horecker and E. R. Stadtman, eds.), p. 53, Academic Press, New York.

2. Greengard, P. (1978). Science 199, 146.

3. Stadtman, E. R., and Chock, P. B. (1977). Proc. Natl. Acad. Sci. U.S.A. 74, 2761.

4. Chock, P. B., and Stadtman, E. R. (1977). Proc. Natl. Acad. Sci. U.S.A. 74, 2766.

5. Brown, M. S., Sejal, A., and Stadtman, E. R. (1974). Arch. Biochem. Biophys. 161, 319.

6. Hucho, F., Randall, D. D., Roche, T. E., Burgett, M. W., Pelley, J. W., and Reed, L. J. (1972). Arch. Biochem. Biophys. 151, 328.

7. Pettit, F. H., Pelley, J. W., and Reed, L. J. (1975). Biochem. Biophys. Res. Commun. 65, 575.

8. Stadtman, E. R., and Chock, P. B. (1979). In "The Neurosciences Fourth Study Program", (F. O. Schmitt, ed.), p. 801, MIT Press.

9. Reed, L. J. (1969). In "Current Topics in Cellular Regulation", Vol. 1, (B. L. Horecker and E. R. Stadtman, eds.), p. 233, Academic Press, New York.

10. Meyer, F., Heilmeyer, L. M. G. Jr., Haschke, R. H., and Fischer, E. H. (1970). J. Biol. Chem. 245, 6642.

11. Danforth, W. H., Helmreich, E., and Cori, C. F. (1962). Proc. Natl. Acad. Sci. U.S.A. 48, 1191.

METABOLITE CONTROL OF THE GLUTAMINE SYNTHETASE CASCADE

E. R. Stadtman, P. B. Chock, and S. G. Rhee

Laboratory of Biochemistry
National Heart, Lung, and Blood Institute
National Institutes of Health
Bethesda, Maryland 20014

INTRODUCTION

Glutamine is a compound of central importance in metabolism. It is a primary intermediate in the assimilation of NH_3 on the one hand, and serves as a source of nitrogen in the biosynthesis of purine and pyrimidine nucleotides, of all amino acids, glucosamine-6-P, p-aminobenzoic acid and of nicotinic acid derivatives, on the other hand (1). In order to accommodate these diverse functions, *Escherichia coli* and other gram-negative bacteria have developed a multienzyme cascade system that is able to sense simultaneous changes in the concentrations of many different metabolites and to integrate their combined effects so that the catalytic activity of glutamine synthetase can be adjusted continuously to meet the ever changing demand for glutamine. This cascade system is comprised of two protein nucleotidylation cycles. One involves the cyclic adenylylation and deadenylylation of glutamine synthetase (GS); the other involves the uridylylation and deuridylylation of Shapiro's regulatory protein (P_{II}).

The Adenylylation Cycle. As shown in Figure 1A, the adenylylation and deadenylylation of glutamic synthetase are catalyzed at separate sites (designated AT_a and AT_d, respectively) on a single adenylyltransferase (ATase) (2). The adenylylation of GS at the AT_a site involves the covalent attachment of an adenylyl group from ATP to a unique tyrosyl residue in each subunit (3, 4, 5). Because GS is composed of 12 identical subunits (6, 7), up to 12 adenylyl groups can be coavlently bound to each dodecameric aggregate (3). Moreover, because adenylylated subunits are catalytically inactive under physiological conditions, the specific activity of GS is inversely proportional to the average number ($\bar{n}$) of adenylylated subunits per molecule of enzyme (8).

Figure 1A shows also that the adenylylation of GS at the AT_a site of ATase is opposed by phosphorolysis of the

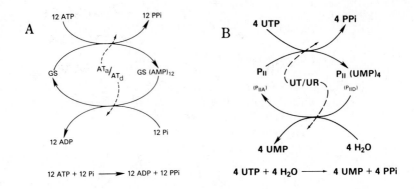

FIGURE 1. (A) Glutamine Synthetase Adenylylation Cycle. (B) Uridylylation Cycle.

adenylyl-tyrosyl bond of adenylylated subunits at the AT_d site of ATase (9); this regenerates unmodified (catalytically active) GS and forms ADP. Since AT_a and AT_d are on the same ATase, it is evident that the two reactions will be coupled. Unless the AT_a and AT_d activities are strictly regulated with respect to one another, GS will undergo senseless cycling between adenylylated and unadenylylated forms, resulting simply in the conversion of ATP and Pi to ADP and PPi. Such futile cycling is avoided by the interaction of the adenylylation cycle with another nucleotidylation cycle in which Shapiro's regulatory protein undergoes interconversion between uridylylated and unmodified forms.

The Uridylylation Cycle. The uridylylation and deuridylylation of the P_{II} protein are catalyzed at separate catalytic centers (designated UT and UR, respectively) on what appears to be a single protein or protein complex (UR/UT) (Figure 1B) (10, 11, 12). Whereas the ratios of UT and UR activities in partially purified preparations can be altered by selective denaturation treatments, it is believed that both activities are contained in a single complex because they copurify, and because point mutations that lead to the loss of one activity, lead simultaneously to a loss of the other (12). As in the adenylylation reaction, the uridylylation of P_{II} involves the covalent attachment of a uridylyl group from UTP to the hydroxyl group of a single tyrosyl residue in each P_{II} subunit. Because P_{II} is composed of 4

MODULATION OF PROTEIN FUNCTION   205

identical subunits (M_r = 11,000), up to 4 uridylyl groups can
be attached per P_{II} molecule (13). Uridylylation of P_{II} at
the UT site of the UT/UR complex is opposed by hydrolytic
cleavage of the uridylyl-tyrosyl bond at the UR (uridylyl
removing) site of the complex to form UMP and unmodified
P_{II} (13). Here too, because the uridylylation and deuridylylation reactions are catalyzed at separate catalytic centers of the same enzyme or enzyme complex, the two steps must
be coupled to yield cyclic interconversion of P_{II} between
uridylylated and unmodified forms, and consequently, the
hydrolysis of UTP to UMP, and PPi.

   The Bicyclic Cascade. The uridylylation cycle and the
adenylylation cycle are linked to one another because the unmodified form of P_{II} (sometimes referred to as P_{IIA}) is required to activate the AT_a site of ATase for the adenylylation of GS, whereas the uridylylated form of P_{II} (sometimes
referred to as P_{IID}) is required to activate the AT_d site for
catalysis of the deadenylylation reaction (10, 11, 13).
Through this reciprocal coupling, a closed type of bicyclic
cascade system is generated as is depicted in Figure 2. In
view of the fact that the nucleotidylation and denucleotidyl-

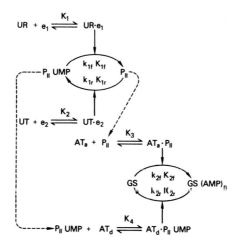

FIGURE 2. *The Glutamine Synthetase Bicyclic Cascade.*

ation reactions in each cycle are tightly coupled, this cascade (except for extreme conditions) is a dynamic processing unit in which the interconvertible proteins undergo continual modification and demodification.

However, for any metabolic condition, a steady-state will be established in which the rates of the forward step and the regeneration step in each cycle will be equal. With reasonable simplifying assumptions (14, 15, 16), it can be shown that in a given steady-state the average number, $\bar{n}$, of adenylylated subunits per GS molecule is given by the expression:

$$\bar{n} = 12 \left[\left(\frac{k_{1r}}{k_{1f}}\right)\left(\frac{k_{2r}}{k_{2f}}\right)\left(\frac{K_{1r}}{K_{1f}}\right)\left(\frac{K_{2r}}{K_{2f}}\right)\left(\frac{K_4}{K_3}\right)\left(\frac{K_2}{K_1}\right)\left(\frac{[UT]}{[UR]}\right)\left(\frac{[AT_d]}{[AT_a]}\right)\left(\frac{[e_2]}{[e_1]}\right) \times \left(\frac{1 + K_1[e_1]}{1 + K_2[e_2]}\right) + 1 \right]^{-1} \quad (1)$$

in which K_1, K_2, K_3, and K_4 are the association constants for the binding of effectors, e_1, e_2, P_{II} and P_{II}(UMP) to the converter enzymes UR, UT, AT_a and AT_d, respectively; k_{1f}, k_{1r}, k_{2f}, and k_{2r} are specific rate constants for the forward and regeneration steps as depicted in Figure 2; K_{1f}, K_{1r}, and K_{2f}, K_{2r} are the association constants for the complexes between the converter enzymes and the interconvertible enzyme substrates, as shown in Figure 2.

It is evident from Equation 1 that the value of $\bar{n}$ is a multiplicative function of 9 different parameter ratios, *i.e.* the ratio of each parameter that governs a step in the regeneration cascade to the corresponding parameter that governs a step in the forward cascade. As noted by P. B. Chock (this symposium, see also References 14-16), only slight fractional activation of a converter enzyme by its positive allosteric effector is needed to obtain substantial covalent modification of an interconvertible enzyme. That is to say, cyclic cascades are endowed with an appreciable <u>signal amplification</u> potential with respect to the effects of primary allosteric stimuli. To facilitate comparisons between different metabolic states and between different cascade systems, signal amplification has been defined as the ratio of the concentration of an allosteric effector required to obtain 50% saturation of the converter enzyme (*viz.*, the UR enzyme) to the concentration of the effector required to support a steady-state in which 50% of the interconvertible enzyme (*viz.*, GS) is covalently modified. The capacity for signal amplification is related to the fact that the fractional modification of an interconvertible enzyme is a multiplicative function of several parameter ratios. The extent of amplification, however, is a function of both the number of parameter ratios and the magnitudes of these ratios. The dependence of these factors on the amplification potential of the GS cascade is illustrated by the data in Figure 3 which shows that as one by one each parameter ratio in Equation 1 is decreased from 1.0 to 0.5 or to 0.25. The signal amplification increases from 1.0 to about 400 or to 100,000, respectively.

It should be emphasized that, if applied to unidirectional (noncyclic) cascades, signal amplification as defined here would be meaningless because in theory any amount of enzyme, no matter how small, if given sufficient time, can catalyze the conversion of any amount of substrate to products. Therefore, one could imagine that even slight fractional activation of a converter enzyme by a positive effector would be sufficient to catalyze the covalent modification of any amount of protein substrate. That is to say, the signal amplification would be infinite. However, when applied to cyclic, interconvertible enzyme, systems signal amplification as defined here is not infinite because the concentration of a primary effector required to support a steady-state

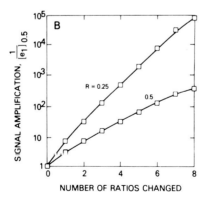

FIGURE 3. *The relationship between the signal amplification and the number of parameter ratios (R) decreased from 1.0 to 0.5 or 0.25 (as indicated).*

in which 50% of the interconvertible enzyme is modified is uniquely defined by the reaction constants that govern the converter enzymes. Therefore, for any given metabolic state, the signal amplification is a unique constant that relates the dissociation constant (Km) of the converter enzyme-effector complex to the reaction constants of the converter enzymes in that metabolic state.

METABOLITE CONTROL

In order to illustrate the responsiveness of $\bar{n}$ to allosteric interactions of metabolites with UR and UT, only the two effectors, e_1 and e_2, are depicted in the GS cascade model shown in Figure 2, and are considered in the formal theoretical analysis which led to Equation 1. It is important to note, however, that the unusual flexibility of the bicyclic cascade with respect to metabolite control derives from the fact that in addition to e_1 and e_2 any one or all of the other 16 parameters that govern the value $\bar{n}$ (see Equation 1 and Figure 2) can be modulated by allosteric or substrate interactions of metabolites with the cascade enzymes. For this reason the bicyclic GS cascade is able to respond to simultaneous changes in the concentrations of a large number of metabolites. In fact, 40 different metabolites have been shown to affect the activities of one or more of the cascade enzymes (14). Therefore, according to Equation 1, the state of adenylylation of GS should reflect the intracellular concentrations of all of these compounds, and a change in the concentration of any one of them should produce a shift in the state of adenylylation. This prediction is supported in principle by the *in vitro* studies of Segal *et al.* (18) which showed that when unadenylylated GS was incubated together with the catalytic components of the bicyclic cascade and a mixture of 8 different effectors, the fraction of adenylylated subunits increases with time and assumes a steady-state level at $\bar{n}$ = 6.0. This is shown by the heavy line (closed squares) in Figure 4. In other words, under these arbitrary conditions, the specific activity of GS is stabilized at one-half its maximal value. Consistent with the steady-state concept, the other curves in Figure 4 show that a change in the concentration of just any one of the effector compounds in the incubation mixture leads to a shift in the steady-state level of adenylylation. The steady-state concept is supported also by *in vivo* experiments of Senior (19) showing that under glucose-limited continuous culture conditions, the level of GS adenylylation is inversely proportional to the specific growth rate, and cor-

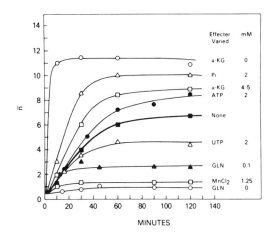

FIGURE 4. Effect of metabolite concentrations on the steady-state level of adenylylated subunits. The heavy line (closed squares) shows the change in $\bar{n}$ with time when 95 μg of GS was incubated in a mixture containing 20 mM $MgCl_2$, 20 mM Pi, 1 mM ATP, 1 mM UTP, 15 mM α-ketoglutarate, 0.3 mM glutamine, and partially purified preparations of P_{II}, AT(containing also UR, UT) as previously described (18). The other curves illustrate the effect of changing the concentration of only one metabolite in the mixture, as indicated. The curves are derived from data in reference (16).

relates well with changes in the intracellular concentration of α-ketoglutarate over the range of 0.5 to 0.9 mM. Whereas these studies confirm in principle the steady-state functions of the GS cascade, further studies to evaluate the kinetic constants of the cascade enzymes are needed before a quantitative description of the regulatory characteristics can be made.

Substitution of P_{IIA} and P_{IID} for the Uridylylation Cycle. Unfortunately, a comprehensive analysis of bicyclic cascade is not yet feasible because instability of the UR/UT complex has thwarted all efforts to obtain the complex in a relatively pure state. For this reason, the interrelationship between GS adenylylation and allosteric regulation of enzymes in the uridylylation cycle cannot be examined directly. It is noteworthy, however, that the only function of the uridylylation cycle is to modulate the proportions of P_{IIA} and P_{IID} in response to variations in the concentrations of effectors that govern the UR and UT activities (see Figure 2). It follows, therefore, that the manifold effects of metabolites on the uridylylation cycle can be simulated by replacement of the uridylylation cycle with mixtures of P_{IIA} and P_{IID} in various proportions. The net effect is therefore to convert the bicyclic cascade into a monocyclic cascade (Figure 5) in which the adenylylation and deadenylylation reactions are governed by the relative concentrations of P_{IIA} and P_{IID}, respectively, and also by the concentrations of α-ketoglutarate and glutamine which exhibit opposite, reciprocal effects on the adenylylation and deadenylylation reactions. From theoretical considerations, it can be shown (16) that the steady-state level of adenylylation of GS in such a cycle is given by the expression:

$$\bar{n} = \left[\frac{1}{12} + \frac{\alpha_{2r}}{12\alpha_{2f}} \left(\frac{1}{(P_{IIA})_{mf}} - 1 \right) \right]^{-1} \quad (2)$$

in which $(P_{IIA})_{mf}$ is the mole fraction of P_{IIA} (*i.e.*,

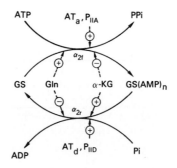

FIGURE 5. Modified monocyclic cascade illustrating the regulatory roles of P_{IIA}, P_{IID}, α-ketoglutarate and glutamine on the adenylylation of GS.

$\dfrac{[P_{IIA}]}{[P_{IIA}] + [P_{IID}]}$, and α_{2f} and α_{2r} are each products of 4 different reaction parameters that govern the adenylylation and deadenylylation reactions, respectively.

The curves in Figure 6A illustrate how, according to Equation (2) the value of $\bar{n}$ will vary as a function of the mole fraction of P_{IIA} and the parameter ratio, α_{2f}/α_{2r}. It is evident from the family of curves that as $(P_{IIA})_{mf}$ is increased from 0 to 1.0, the value of $\bar{n}$ will increase in either a linear, hyperbolic, or a parabolic manner when α_{2f}/α_{2r} = 1.0, > 1.0, or < 1.0, respectively. Since the parameter ratio will vary as a function of allosteric and substrate interactions of metabolites with the cascade proteins, the curves illustrate the remarkable flexibility of the cascade system to metabolite control.

Regulation by α-Ketoglutarate and Glutamine. To test

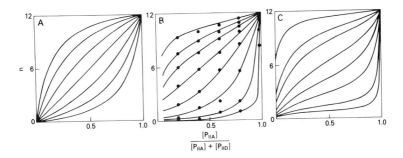

FIGURE 6. *The interdependence of the steady-state level of adenylylation, the mole fraction of P_{IIA} and the ratio α_{2f}/α_{2r}. (A) Computer simulated curves with α_{2f}/α_{2r} set at 8, 4, 2, 1, 0.5, 0.25 and 0.125, from top to bottom, respectively. (B) Experimental data from studies in which $(P_{IIA})_{mf}$ was varied by varying the proportions of pure P_{IIA} and P_{IID}, and the ratio α_{2f}/α_{2r} was varied by varying the relative concentration of α-ketoglutarate and glutamine as described in Reference (2). (C) Computer simulated curves based on the steady-state equation describes the theoretical model in Figure 7. For details see Reference (2) from which this figure was taken.*

the validity of the cascade model depicted in Figure 5, the
steady-state level of adenylylation was measured following
the incubation of unadenylylated GS with a purified preparation of ATase in mixtures containing variable proportions
of pure P_{IIA} and P_{IID} (to vary $(P_{IIA})_{mf}$) and various concentrations of α-ketoglutarate and glutamine (to vary the
α_{2f}/α_{2r} ratio).

As shown in Figure 6B, a family of curves similar to but
not identical to those predicted by Equation 2 was obtained.
The discrepancy between the experimental and theoretical
analysis is due in part to the fact that both α-ketoglutarate
and glutamine exhibit reciprocal effects on the adenylylation
and deadenylylation reactions, and partly because the adenylylation and deadenylylation reactions catalyzed at the AT_a
and AT_d sites, respectively, are each catalyzed by three different enzyme-effector complexes. On the basis of detailed
kinetic studies (2), it is evident that the adenylylation
and deadenylylation cycle is more accurately described by
the model depicted in Figure 7. The steady-state function

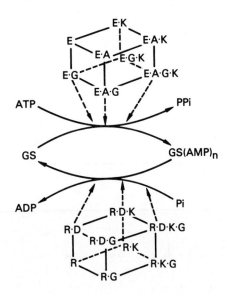

FIGURE 7. *Mechanistic scheme for the adenylylation-deadenylylation cycle. The notations E, R, A, D, K, and G represent AT_a, AT_d, P_{IIA}, P_{IID}, α-ketoglutarate and glutamine, respectively. From Reference (2).*

for this model is described by a complex equation (2) comprised of numerous terms including 10 different binding constants, 12 synergestic or antagonistic coefficients, and 6 rate constants. Of the 28 constants, 22 have been determined directly by kinetic measurements, 4 were determined by computer simulation analysis with rigorous constraints imposed by the other constants, and 2 were determined solely by computer curve fitting (2). By means of these 28 constants and the steady-state equation that describes the model in Figure 7 (2), the steady-state level of $\bar{n}$ was computed as a function of $(P_{IIA})_{mf}$ for the experimental conditions used in obtaining the data summarized in Figure 6B. The theoretical curves generated from these calculations (Figure 6C) are in good agreement with the experimental curves in Figure 6B, showing that the models depicted in Figures 5 and 7 can indeed explain the experimental findings. The data show that for any value of $(P_{IIA})_{mf}$ greater than 0 and less than 1.0, the state of adenylylation can be shifted from nearly 0 to almost 12 by varying the ratio of α-ketoglutarate to glutamine over a relatively narrow range. Conversely, for any given ratio of allosteric effectors, the value of $\bar{n}$ can be shifted from nearly 0 to 12 by changes in the proportions of P_{IIA} and P_{IID}, which in the coupled bicyclic system is achieved by metabolite effects on the UR and UT activities. In other words, the specific activity of GS can be varied continuously from almost zero ($\bar{n}$ = 12) to maximal values ($\bar{n}$ = 0) merely by changes in the relative concentrations of positive and negative allosteric effectors.

<u>Signal Amplification and Apparent Cooperativity</u>. The curves (solid lines) in Figure 8A show how the value of $\bar{n}$ varies when the $(P_{IIA})_{mf}$ is held constant at 0.6 and the concentration of glutamine is varied at several different levels of α-ketoglutarate. Because in all cases the glutamine reacts directly with AT only, the dashed line in Figure 8A is given as a point of reference to show how the fractional saturation of AT with glutamine varies as a function of glutamine concentration in the absence of other metabolite effectors. It is apparent from the other curves in Figure 8A that as the concentration of α-ketoglutarate is decreased from 1.0 mM to 3 μM in order to modulate the parameter ratio, α_{2f}/α_{2r} (see Equation (2) and Figure 5), the concentration of glutamine that is required to stabilize a given state of adenylylation decreases. Note especially, that the concentration of glutamine required to obtain 50% adenylylation of GS is in all instances considerably lower than the concentration required to obtain 50% saturation of AT. This illustrates that, as predicted by the steady-state

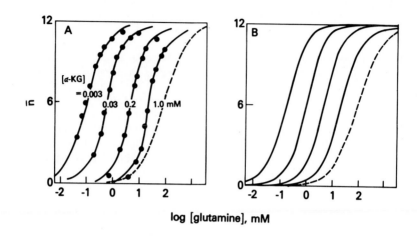

FIGURE 8. (A) Steady-state levels of $\bar{n}$ as a function of glutamine concentration. $(P_{IIA})_{mf}$ was 0.6 and the concentration of α-ketoglutarate for curves from left to right were 3 μM, 0.17 mM and 1.0 mM, respectively. The dashed curve is a calculated saturation curve based on the experimentally determined K_d of 80 mM for the binding of glutamine to AT in the absence of effectors and substrates. (B) Computer simulated curves for the corresponding curves in (A) using the steady-state equation and the experimentally determined reaction constants. For details see Reference (2) from which this figure was taken.

model (14, 15), such cascades are endowed with extraorindary signal amplification potentials with respect to primary allosteric stimuli. For the condition described in Figure 8A, the signal amplification with respect to glutamine concentration varies from 800 to 4 as the concentration of α-ketoglutarate is increased from 3 μM to 1.0 mM.

It is also evident from Figure 8A that the slopes of the nearly linear portions of the $\bar{n}$ versus glutamine concentration curves (solid lines) are considerably steeper than that of the dashed line which describes the noncooperative binding of glutamine to AT in the absence of substrates and other effectors. These greater slopes indicate that the value of $\bar{n}$ is a sigmoidal function of the glutamine concentration, and reflects the fact that glutamine not only stimulates the adenylylation reaction, but it inhibits the deadenylylation reaction (see Figure 5). The experimental results are therefore in accord with the theoretical analysis of a bicyclic cascade (14, 15), which shows that a sigmoidal response (apparent cooperativity) will be obtained when the same effector stimulates the forward step and inhibits the regeneration step in one cycle, or when the same effector stimulates the forward steps in both cycles. From the more conventional Hill-type plots of the data in Figure 8A, a Hill coefficient of 1.5 is obtained. It should be emphasized, however, that this describes the unique situation in which adenylylation of GS is catalyzed by the modification monocyclic cascade illustration in Figure 5, and when the $(P_{IIA})_{mf}$ is 0.6. Under physiological conditions where the adenylylation cycle is coupled to the uridylylation cycle, Hill coefficients of 2.5 or greater (14, 15) may be achieved because in addition to its effects on the adenylylation cycle, glutamine also stimulates the deuridylylation of P_{IID} and perhaps inhibits the uridylylation of P_{IIA} (19).

Utilizing the more rigorous steady-state equation (not shown) and the 28 experimentally determined reaction constants that describe the monocyclic adenylylation cycle depicted in Figure 5, computer simulated curves were derived for the conditions used in the experiments summarized in Figure 8A. The excellent agreement between the simulated data (Figure 8B) and the experimental data (Figure 8A) offer additional support for the steady-state cascade hypothesis.

DISCUSSION

In view of the fact that the covalent modification of an interconvertible enzyme can lead to either complete inactivation of the enzyme (as occurs with the adenylylation of GS), or to activation (as occurs with the phosphorylation of glycogen phosphorylase), interconvertible enzymes have been considered as metabolic switches which can be used to turn ON or OFF critical enzyme activities in response to metabolic demands. However, such an "all or none" concept is not supported by the studies with GS summarized here which show that the interconversion between adenylylated and unadenylylated forms of GS is a dynamic process, and that for any given metabolic condition a steady-state is established in which the rate of adenylylation of glutamine synthetase and the rate of deadenylylation are equal. The fraction of adenylylated subunits in this steady-state determines the specific activity of glutamine synthetase, and is specified by the magnitudes of 16 parameters that govern the activities of the converter enzymes that catalyze the covalent modification reactions. Through multiple interactions with allosteric and substrate effectors, the four converter enzyme activities are programmed to sense changes in the concentrations of a multitude of metabolites. By causing changes in the catalytic constants of the cascade enzymes, these multiple interactions lead continuously to rapid shifts in the steady-state distribution of adenylylated and unadenylylated GS subunits, and consequently to changes in the specific activity of glutamine synthetase commensurate with the ever changing metabolic need for glutamine. In addition to its role as a metabolic integration system, the cascade is endowed with an enormous signal amplification potential, making it possible to obtain large shifts in the specific activity of GS in response to changes in primary effector concentrations that are orders of magnitude lower than the dissociation constants of the effector-converter enzyme complexes. Furthermore, the cascade exhibits unusual flexibility with respect to allosteric control patterns. Depending upon the metabolic situation, it can not only respond differently to changes in the concentration of a given metabolite, but it can also modulate the amplitude of the metabolite effect, and can generate a cooperative-type of response to one particular allosteric effector and not another. These unique characteristics provide the versatility and sensitivity needed in the regulation of one of the most important enzymes in metabolism.

REFERENCES

1. Stadtman, E. R. (1973). In "The Enzymes of Glutamine metabolism" (S. Prusiner and E. R. Stadtman, eds.), p. 1, Academic Press, New York.

2. Rhee, S. G., Chock, P. B., and Stadtman, E. R. (1978). Proc. Natl. Acad. Sci., U.S.A. 75, 3138.

3. Kingdon, H. S., Shapiro, B. M., and Stadtman, E. R. (1967). Proc. Natl. Acad. Sci., U.S.A. 58, 1703.

4. Wulff, K., Mecke, D., and Holzer, H. (1967). Biochem. Biophys. Res. Commun. 28, 740.

5. Shapiro, B. M., and Stadtman, E. R. (1968). J. Biol. Chem. 243, 3769.

6. Shapiro, B. M., and Ginsburg, A. (1968). Biochemistry 7, 2153.

7. Valentine, R. C., Shapiro, B. M., and Stadtman, E. R. (1968). Biochemistry 7, 2143.

8. Stadtman, E. R., Shapiro, B. M., Ginsburg, A., Kingdon, H. S., and Denton, M. D. (1968). Brookhaven Symp. Biol. 21, 378.

9. Anderson, W. B., and Stadtman, E. R. (1970). Biochem. Biophys. Res. Commun. 41, 704.

10. Brown, M. S., Segal, A., and Stadtman, E. R. (1971). Proc. Natl. Acad. Sci. U.S.A. 68, 2949.

11. Mangum, J. H., Magni, G., and Stadtman, E. R. (1973). Arch. Biochem. Biophys. 158, 514.

12. Bancroft, S., Rhee, S. G., Neumann, C., and Kustu, S. (1978) J. Bacteriol. 134, 1046.

13. Adler, S. P., Purich, D., and Stadtman, E. R. (1975). J. Biol. Chem. 250, 6264.

14. Stadtman, E.R., and Chock, P. B. (1978). In "Current Topics in Cellular Regulation", Vol. 13, (B. L. Horecker and E. R. Stadtman, eds.), p. 53, Academic Press, New York.

15. Chock, P. B., and Stadtman, E. R. (1977). Proc. Natl. Acad. Sci. U.S.A. 74, 2766.

16. Stadtman, E. R., Chock, P. B., and Rhee, S. G. (1979). In 11th Miami Winter Symposium, "From Gene to Protein: Information Transfer in Normal and Abnormal Cells", (T. R. Russell, K. Brew, J. Schultz, and H. Faber, eds.) in press, Academic Press, New York.

17. Stadtman, E. R., and Chock, P. B. (1979). In "The Neurosciences Fourth Study Program", (F. O. Schmitt, ed.), p. 801, MIT Press.

18. Segal, A., Brown, M. S., and Stadtman, E. R. (1974). Arch. Biochem. Biophys. 161, 319.

19. Senior, P. J. (1975). J. Bacteriol. 123, 407.

20. Engleman, E. G., and Francis, S. H. (1979). Arch. Biochem. Biophys. 191, 602.

Modulation of Protein Function

THE MOLECULAR DYNAMICS AND BIOCHEMISTRY OF COMPLEMENT[1]

Hans J. Müller-Eberhard[2]

Department of Molecular Immunology, Research Institute of Scripps Clinic, La Jolla, California 92037

INTRODUCTION

Complement consists of a set of proteins that occurs in plasma in inactive, but activatable form. These proteins have the potential of generating biological activity by entering into protein-protein interactions. In the course of these interactions, indigenous complement enzymes are assembled and activated, and fission and fusion products of complement proteins are formed. One of the fusion products is the membrane attack complex (MAC) which is responsible for complement dependent cytolysis. Through fission products arising from enzymatic cleavage, complement molecules express different biological activities. Cells responding to the stimuli of complement reaction products include polymorphonuclear leukocytes, monocytes, lymphocytes, macrophages, mast cells, smooth muscle cells and platelets. In vivo, complement participates together with antibodies and various cellular elements in host defense mechanisms against infections. It also participates in inflammation and immunologic tissue injury. These seemingly disparate biological functions of complement may be reduced to the same basic reactions on the molecular and cellular level. Thus, complement is increasingly emerging as a humoral system capable of generating molecular effectors of cellular functions. Background information on the molecular biology and biochemistry of complement may be obtained from recent reviews (1-3).

COMPONENTS AND REGULATORS

To date, 21 distinct proteins have been recognized as constituents of the complement (C) system: 13 may be considered components proper

[1]This is publication number 1735 from the Research Institute of Scripps Clinic. This work was supported by U.S. Public Health Service Grants AI 07007 and HL 16411.
[2]Cecil H. and Ida M. Green Investigator in Medical Research, Research Institute of Scripps Clinic.

and eight regulatory proteins. The components include five proteolytic enzymes: C1r, C1s, C2, Factor B and Factor D. These are serine proteases inasmuch as they are inhibitable by DFP. The eight non-enzyme components include C1q, the recognition protein of the classical pathway, C3, the most versatile component of the C system, C4, the modulator of C2, and C5, C6, C7, C8 and C9, the precursors of the membrane attack complex.

Some of the known properties of the components are listed in Tables 1-3. The proteins of the classical pathway of activation are described in Table 1. C1q has the largest molecular size and is the most complex. It is a collagen-like protein that is composed of 18 polypeptide chains containing a large number of glycine, hydroxyproline and hydroxylysine residues with a galactose-glucose disaccharide unit linked to many of the hydroxylysine residues (4). Three similar, but distinct chains (A, B, C) of 20,000-22,000 dalton occur six times in the molecule, forming six structural and functional subunits (5). This is apparently accomplished by the non-covalent interaction of nine pairs of S-S linked chains: six A-B pairs and three C-C pairs. The amino acid sequence of the chains as well as the ultrastructure of the protein have been elucidated (6,7). C1r consists of two apparently identical, non-covalently linked chains (8). C1s is a single chain protein that dimerizes in presence of Ca^{++} (8). Both C1r and C1s have a similar amino acid composition and as far as investigated, considerable homology in primary structure. C2 is a single chain protein with two free sulfhydryl groups that upon oxidation with iodine form an intramolecular disulfide bond. As a result, C2 activity in the C reaction increases approximately 20-fold (1). C3, a two chain protein, is the precursor of several biologically active fragments (1) (Fig. 1). It supplies one of the two anaphylatoxins (C3a) (3), one of the subunits of three C enzymes (C3b) (2),

TABLE 1
PROTEINS OF THE CLASSICAL PATHWAY
OF COMPLEMENT ACTIVATION

Protein	Molecular Weight	Number of Chains	Electroph. Mobility	Conc. (µg/ml)
C1q	400,000	6x3	γ_2	65
C1r	190,000	2	β	50
C1s	88,000	1	α	40
C2	117,000	1	β_1	25
C3	180,000	2	β_2	1600
C4	206,000	3	β_1	640

the major C derived opsonin (C3b)(9), and a leukocytosis inducing factor (C3e). At least three of its physiological fragments, C3a, C3b and C3d, address specific cell surface receptors. C4 is composed of three chains (1).

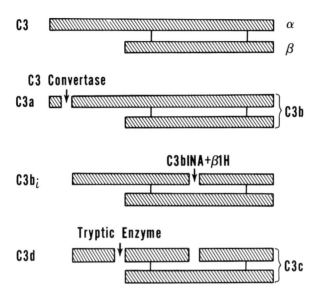

FIGURE 1. Schematic description of the physiological activation and control of C3: In succession, three enzymes cause liberation of C3a anaphylatoxin, generation of C3b and its transient binding site, inactivation of C3b and fragmentation of C3b$_i$ into two immunochemically distinct pieces (1).

The proteins of the alternative pathway of C activation are described in Table 2. Factor B, or proactivator, constitutes the key enzyme of the pathway. It is the precursor of two fragments, Ba and Bb, the latter being the enzymatic site carrying subunit of the alternative C3/C5 convertase (2). Factor D, or proactivator convertase, is the activating enzyme of Factor B and a trace protein in human serum (10). It occurs in plasma and serum only in its potentially active form. Expression of its enzymatic activity depends entirely on modulation of its substrate, Factor B, by C3b. C3 constitutes the major non-enzyme protein of the alternative pathway (11). It is thought to be a subunit of the initial enzyme of the alternative pathway and in the form of C3b it discriminates between activators and non-activators of this pathway (12).

Properdin does not appear to be an essential component of the alternative pathway, while β1H and C3bINA are necessary for its function.

TABLE 2
PROTEINS OF THE ALTERNATIVE PATHWAY
OF COMPLEMENT ACTIVATION

Protein	Symbol	Molecular Weight	No. of Chains	Electroph. Mobility	Conc. (µg/ml)
C3	C3	180,000	2	β	1600
Proactivator	B	93,000	1	β	200
Proactivator Convertase	D	24,000	1	β-γ	1
C3b Inactivator	C3bINA	88,000	2	β	34
β1H	β1H	150,000	1	β	500
Properdin	P	224,000	4	γ	20

The proteins of the membrane attack pathway are described in Table 3. C5 supplies the second anaphylatoxin (C5a) which is also a potent chemotactic factor (3) and C5b which is the nucleus of assembly of the membrane attack complex (MAC) (1). C6 and C7 are single chain proteins of similar physical and chemical properties, including similar amino acid composition. C8 is composed of three chains and may be intimately involved in the interaction of the MAC with the hydrophobic groups of membrane lipids (13). C9 is a one-chain acidic glycoprotein.

TABLE 3
PROTEINS OF THE MEMBRANE ATTACK PATHWAY
OF COMPLEMENT

Protein	Molecular Weight	Number of Chains	Electroph. Mobility	Conc. (µg/ml)
C5	180,000	2	β_1	80
C6	128,000	1	β_2	75
C7	121,000	1	β_2	55
C8	154,000	3	γ_1	55
C9	79,000	1	α	60

Properties of the regulatory proteins may be summarized as follows. Properdin (P), which is stabilizer of the alternative C3/C5 convertase, is composed of four apparently identical subunits which are held together by non-covalent forces (2). Its molecular weight is 180,000. β1H (mol. wt. 150,000) is a single chain glycoprotein which controls the function of the alternative pathway through its affinity for a strategic site on C3b (14-16). The C3b inactivator (C3bINA) (mol. wt. 80,000) is an endopeptidase which is composed of two non-identical chains. It cleaves the α-chain of C3b without causing overt fragmentation of the molecule (17) (Fig. 1). C3bINA also cleaves the α-chain of C4b which leads to fragmentation of the molecule into C4c and C4d (17). β1H functions as cofactor of C3bINA for cleavage of C3b (14,17) and the C4 binding protein (C4-bp) is the cofactor of C3bINA for cleavage of C4b (18). The C1q inhibitor (C1qINH) is a β-globulin which can precipitate C1q in gel diffusion experiments. $\overline{C1}$ inhibitor ($\overline{C1}$ INH) (mol. wt. 105,000) is a single chain acidic glycoprotein capable of blocking all activated enzymic sites in the $\overline{C1}$ complex (8). The membrane attack complex inhibitor (MAC-INH) or S-protein (mol. wt. 70,000) is a newly recognized protein (19). It inhibits the forming MAC by physically blocking its hydrophobic membrane binding site. The serum carboxypeptidase B (SCPB) (mol. wt. 310,000) is responsible for efficient inactivation of the anaphylatoxins, C3a and C5a (1,3).

THE PATHWAYS

Upon contact with activators, the C proteins organize themselves to form two pathways of activation, the classical and the alternative, and the common terminal pathway of membrane attack. While self-assembly may occur in cell-free solution, the C proteins have the unusual ability to transfer themselves from solution to the surface of biological particles and to function as a solid phase enzyme system. The capacity of C to mark and prepare particles for ingestion by phagocytic cells and its potential to attack and lyse cells is based on this ability. Transfer is accomplished through activation of metastable binding sites which are transiently revealed by the respective activating enzymes. Cleavage of a critical peptide bond leads to dissociation of an activation fragment of a given component and to exposure of structures that are concealed in the native molecule. Owing to the revealed site, a molecule can bind to a suitable acceptor and establish a firm association with it. Failing collision with the acceptor within a finite time period after activation, the site decays and the molecule remains unbound in the fluid phase. As such, it cannot be activated again. It has become apparent that through the mechanism of metastable binding sites activated C3, C4 and C5b-7 can bind directly to biological membranes, activated C2, Factor B and C5 to their respective acceptors C4b, C3b and C6 (1).

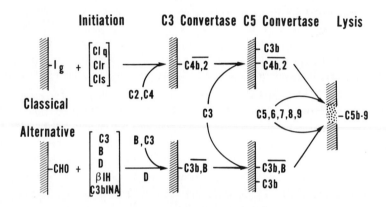

FIGURE 2. Schematic representation of the classical and alternative pathways of complement activation and of the common, terminal pathway of membrane attack. The shaded areas represent biological membranes on the surface of which the sequential reactions proceed. The biologically active by-products C3a, C5a, Ba and Bb have not been indicated (20).

The two pathways of activation have a similar molecular organization (Fig. 2). An initial enzyme catalyzes the formation of the target bound C3 convertase which in turn catalyzes the formation of the target bound C5 convertase. The C5 convertase of either pathway, by cleaving C5, can set in motion the self-assembly of the membrane attack complex, C5b-9. The initial enzymes differ in substrate specificity, the classical enzyme being specific for C2 and C4 and the alternative enzyme for C3. The classical pathway is activated by immune complexes of the IgG and IgM type. The alternative pathway is activated by plant, fungal and bacterial polysaccharides and lipopolysaccharides in particulate form. In addition, certain animal cells are activators, such as human lymphoblastoid cells, rabbit erythrocytes and neuraminidase treated sheep erythrocytes (2).

MODULATION OF PROTEIN FUNCTION

Several biologically important entities arise as by-products of the C reaction. For instance, both pathways generate C3a and C5a, the two anaphylatoxins (3), and C3b, the opsonin of the C system (9).

THE CLASSICAL PATHWAY OF ACTIVATION

Figure 3 depicts a schematic representation of the molecular dynamics of the pathway. The events involve six proteins: C1q, C1r, C1s, C2, C3 and C4. The reaction commences with reversible binding of the C1 complex (C1q, C1r, C1s) (8) to the Fc-regions of two or more antibody molecules in immune complexes. Binding occurs through the six globular regions of the C1q subcomponent and may initiate activation. The precise mechanism of the internal activation of C1 is unknown. However, unlike precursor C1, activated C1 ($\overline{C1}$) contains C1r and C1s in cleaved form (8,21). Instead of two 83,000 dalton chains, $\overline{C1r}$ is composed of two 56,000 and two 27,000 dalton chains. The smaller chains bear the proteolytic site (21,22) and are linked by disulfide bonds to the larger chains. $\overline{C1r}$ activates C1s proenzyme by limited proteolysis which results in a $\overline{C1s}$ chain structure that resembles that of $\overline{C1r}$.

Assembly of the classical C3 convertase from C2 and C4 is catalyzed by the $\overline{C1s}$ subcomponent of $\overline{C1}$. Cleavage of the a-chain of C4 exposes the metastable binding site through which C4b can attach to the target surface (1). Bound C4b forms a Mg^{++} dependent enzymatically inactive complex with native C2. The complex acquires C3 convertase activity upon cleavage of C2 (23). By-products of this reaction are the two activation peptides, C2b and C4a, which have no known biological activity. The proteolytic site of $\overline{C4b,2a}$ resides in the C2a subunit, and by cleavage of the a-chain of C3 it exposes the metastable binding site through which C3b can attach to the target surface (1). By-product of this reaction is the activation peptide C3a which like C5a has anaphylatoxin activity (3). C3 convertase acquires C5 cleaving activity as a result of binding of C3b in its immediate microenvironment. Bound C3b acts by modulating C5 such that it can be acted upon by C2a. To denote the dependence of the C5 cleaving function on C3b, the enzyme is referred to as $\overline{C4b,2a,3b}$. The enzyme cleaves the a-chain of C5 generating C5a and C5b, the latter fragment constituting the first intermediate product of the membrane attack pathway (1).

Three regulatory mechanisms are apparent. First, the metastable binding sites of activated C4 and activated C3 impose spatial constraints on the location of the forming enzymes. Second, decay-dissociation of C2a from both the C3 and the C5 convertase limits their function in time. Third, C3b and C4b can be cleaved by the enzyme C3bINA so that they can no longer serve as enzyme subunits.

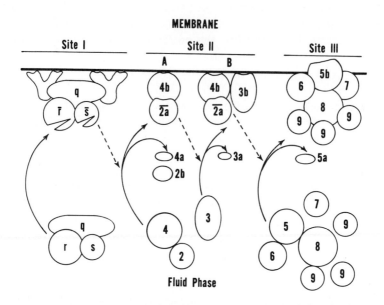

FIGURE 3. Pictorial representation of the molecular events associated with activation of the classical and membrane attack pathways of complement. Site I denotes the immune complex to which C1 binds and undergoes activation. Site II represents the membrane position on which first the C3 convertase and subsequently the C5 convertase are assembled. Cleavage of C5 initiates the self-assembly of the C5b-9 complex which attacks the membrane at a third topographically distinct site (Site III) (1).

THE ALTERNATIVE PATHWAY OF ACTIVATION

The pathway is composed of six proteins: C3, Factor B, Factor D, β1H, C3bINA and properdin. Basically, the molecular organization of the pathway may be understood in terms of one key enzyme, Factor B, functioning at all steps of the pathway and being modulated by different proteins, namely C3, C3b and properdin. In addition, C3b fulfills a similar key function in that it confers upon the pathway the ability to discriminate between activators and non-activators (12).

Fundamental to the understanding of the alternative pathway is the C3b dependent positive feedback mechanism (11). As illustrated in Figure 4, a molecule of C3b and a molecule of Factor B form, in presence of Mg^{++}, the loose bimolecular complex C3b,B. In complex with C3b, Factor B becomes susceptible to cleavage by Factor D, its activat-

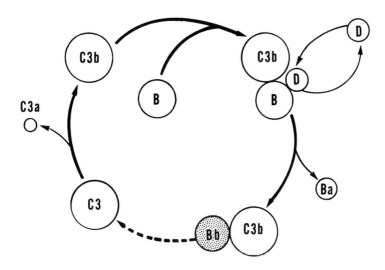

FIGURE 4. The C3b-dependent positive feedback. One molecule of C3b can create many molecules of C3 convertase in presence of an unlimited supply of Factor B. Since Factor D, the activating enzyme of Factor B, is not consumed in the reaction, a chain reaction is created (20).

ing enzyme, which results in formation of the C3 convertase of the alternative pathway, $\overline{C3b,Bb}$ (2). The α-fragment of Factor B is dissociated in the process. In acting upon C3, the enzyme supplies, in a short period of time, many molecules of C3b, each of which is capable of initiating the formation of a molecule of C3 convertase provided the supply of Factor B is not limiting. Since Factor D is not incorporated into the enzyme complex, it can activate many C3b,B complexes (10). In its uncontrolled form, the process resembles a chain reaction. Two questions arise: How is this amplification system set in motion and how is it controlled. Regulation is provided by C3bINA and by β1H. β1H binds to C3b and this binding is competitive with that of Bb (14-16). As a result, β1H either blocks the formation of $\overline{C3b,Bb}$ or disassembles the enzyme, dissociating Bb in inactive form. In complex with β1H, C3b is readily cleaved and inactivated by C3bINA (14, 16). The enzyme responsible for setting in motion the feedback mechanism is generated when native C3, Factors B and D and Mg^{++} interact in free solution at physiological concentration (12). Although not known with certainty, the initial enzyme appears to be $\overline{C3,Bb}$ where native C3 rather than C3b serves as a subunit. The first molecule of C3b produced by this enzyme will then set in motion the feedback reaction. The control of the reaction mixture containing native C3, Factors B, D and Mg^{++} is provided by β1H and C3bINA and may concern only the product of the initial enzyme, C3b and secondary amplification. Examination of the controlled five-protein system at 37° revealed a remarkable stability in that neither C3 nor Factor B consumption could be detected during several hours of incubation (12). However, it must be assumed that a small number of C3b molecules are continually produced by the initial enzyme.

For initiation of the pathway by activators, five proteins are sufficient: C3, Factors B, D, β1H and C3bINA (12). It is thought that upon introduction of biological particles into the stable fluid phase system, the initial enzyme deposits a small number of C3b molecules on the surface of these particles. C3b deposition from the fluid phase is a random process involving activation of the metastable binding site of C3. At the surface of non-activators, bound C3b is subject to the same control that controls fluid phase C3b: interaction with Factor B is blocked by β1H and C3b is cleaved and inactivated by the C3bINA (24, 12).

On the surface of activators, C3b bound through its non-specific, metastable binding site interacts with surface markers through a separate site (12). Engagement of the marker site inactivates the β1H site so that binding of β1H to C3b is reduced by an order of magnitude (16). Since efficient action of C3bINA requires β1H, C3b escapes inactivation and is available for unrestricted formation of C3 convertase. The activator bound C3 convertase also escapes control and consequently each molecule of C3 convertase produces numerous bound C3b molecules and C3b,Bb complexes (Fig. 5) (24, 12).

As the multiplicity of C3b molecules increases around each enzyme complex, three novel activities are generated: C5 convertase activity, properdin binding activity and opsonic activity. Acquisition of C5 cleaving capacity by the C3 convertase is a function of additional C3b molecules (2). Present evidence indicates that C3b is required for the modulation of C5 so that it becomes a substrate for the enzyme. The active site responsible for C3 and C5 cleavage resides in the Bb subunit. The C3/C5 convertase is innately labile and decays at 37° with a half-life of 1.5 min. Native properdin upon collision with the labile C3/C5 convertase becomes physically associated with it and "activated" in the process. Binding-activation of properdin results in stabilization of the enzyme such that its half-life at 37° increases to 10 min. (2). Amplification results in the accumulation of many bound C3b molecules and thus in coating of biological particles with the protein (opsonization). These unengaged C3b molecules are capable of interacting with C3b-specific cell surface receptors that occur on phagocytic and certain other cells (9). Our present concept of the molecular dynamics underlying activation of the alternative pathway is shown in Figure 6.

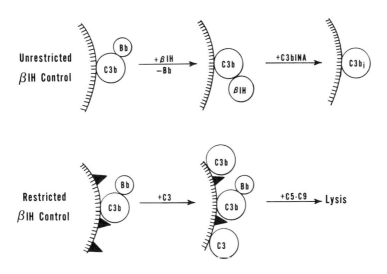

FIGURE 5. Schematic representation of differential control by β1H of bound C3b: dependence on cell surface markers. Non-activator cell surfaces (above) allow unrestricted control of bound C3 convertase. Activator surfaces (below) restrict β1H binding and allow amplification of the alternative pathway (16).

THE MEMBRANE ATTACK PATHWAY

C-dependent cytolysis is entirely a function of the activated membrane attack complex (MAC). The five precursor proteins of MAC form a loose protein-protein complex which upon cleavage of C5, by either the classical or the alternative C5 convertase, self-assembles into an exceedingly firm complex (1,25). After dissociation of C5a, nascent C5b forms a stable bimolecular complex with C6 (26). Then, C5b,6 and C7 form a trimolecular complex which for less than 9 msec. has the ability to bind to the surface of biological membranes (27). Subsequently, C8 is adsorbed to C5b-7 and C9 to C5b-8. Leakage of cell membranes ensues at the C8 stage of the assembly, indicating a major role of C8 in the interaction of MAC with the interior of a membrane. Present evidence suggests that C5b-7 overcomes the charge barrier and the fully assembled MAC the hydrophobic barrier of a biological membrane (28).

Recently, two developments have shed light on the mechanism of action of the MAC as well as on the relationship of its ultrastructure to the morphology of C produced membrane lesions. First, during assembly of the MAC, high affinity phospholipid binding sites are generated (29). Whereas the five precursor proteins exhibit no affinity for phospholipids,

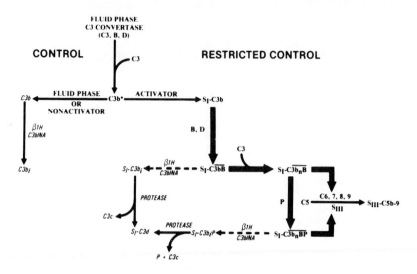

FIGURE 6. Schematic representation of the molecular dynamics of the alternative pathway. "Control" represents the situation that exists in the fluid phase, or upon introduction of non-activator particles. "Restricted control" occurs upon introduction of activator particles. C3b* denotes nascent C3b which is endowed with a labile binding site. S_I denotes the site of initial C3b deposition and its secondary interaction with structures of an activator particle. S_{III} denotes the attachment site of the membrane attack complex, C5b-9. Control reactions are italicized; amplification reactions are in bold type (12).

lipid binding capacity appears at the C5b-7 stage of assembly, where 400 phospholipid molecules can be bound per complex and increases to 1500 molecules per fully assembled MAC. Second, the MAC extracted from membranes of C lysed cells has been found to constitute a dimer of C5b-9 with a molecular weight of 1.7 million (30). Its probable structural formula is $C5b_2, 6_2, 7_2, 8_2, 9_6$. Visualization of the dimer by electron microscopy showed that its morphological appearance is identical to that of C produced membrane lesions (30). Both have a ring-like appearance with an inner and outer diameter of 100 and 200 Å, respectively.

Two alternative models are being considered to explain MAC action on the molecular basis: it either forms a transmembrane protein channel (31,32), or it causes reorganization of the lipid bilayer with formation of transmembrane lipid channels (28,29). The demonstrated unusual lipid binding capacity of the MAC strongly suggests that upon its insertion into membranes, it causes increased permeability primarily by phospholipid reorganization (29).

CONCLUSION

The past twenty years have witnessed a gradual emergence of C research from serology and its incorporation into molecular biology and biochemistry. Many proteins had to be discovered, isolated and studied before the ongoing investigations of the entire C system could commence. It became apparent that biological activities ascribed to C are generated by complex groups of proteins acting in concert. As a result of the past decade's work, most of the activities attributable to C can now be related to defined molecular entities, the chemical structure of some of which has been elucidated. Since most of the work has been conducted on human C, molecular, immunochemical and biochemical insights became immediately applicable to basic and clinical investigations of human disease.

REFERENCES

1. Müller-Eberhard, H.J. (1975). Ann. Rev. Biochem. 44, 697.
2. Götze, O., and Müller-Eberhard, H.J. (1976). Adv. Immunol. 24, 1.
3. Hugli, T.E., and Müller-Eberhard, H.J. (1978). Adv. Immunol. 26, 1.
4. Calcott, M.A., and Müller-Eberhard, H.J. (1972). Biochem. 11, 3443.
5. Porter, R.R. (1977). Fed. Proc. 36, 2191.
6. Reid, K.B.M., and Porter, R.R. (1976). Biochem. J. 155, 19.
7. Knobel, H.R., Villegar, W., and Isliker, H. (1975). Europ. J. Immunol. 5, 78.
8. Cooper, N.R., and Ziccardi, R.J. (1976). In "Proteolysis and Physiological Regulations" (D.W. Ribbons and K. Breed, eds.), p. 167. Academic Press, New York.
9. Gigli, I., and Nelson, R.A., Jr. (1968). Exp. Cell Res. 51, 45.
10. Lesavre, P., and Müller-Eberhard, H.J. (1978). J. Exp. Med. 148, 1498.
11. Müller-Eberhard, H.J., and Götze, O. (1972). J. Exp. Med. 135, 1003.
12. Schreiber, R.D., Pangburn, M.K., Lesavre, P., and Müller-Eberhard, H.J. (1978). Proc. Nat. Acad. Sci. USA 75, 3948.
13. Kolb, W.P., and Müller-Eberhard, H.J. (1976). J. Exp. Med. 143, 1131.
14. Whaley, K., and Ruddy, S. (1976). J. Exp. Med. 144, 1147.

15. Weiler, J.M., Daha, M.R., Austen, K.F., and Fearon, D.T. (1976). Proc. Nat. Acad. Sci. USA 73, 3268.
16. Pangburn, M.K., and Müller-Eberhard, H.J. (1978). Proc. Nat. Acad. Sci. USA 75, 2416.
17. Pangburn, M.K., Schreiber, R.D., and Müller-Eberhard, H.J. (1977). J. Exp. Med. 146, 1977.
18. Fujita, T., Gigli, I., and Nussenzweig, V. (1978). J. Exp. Med. 148, 1044.
19. Podack, E.R., Kolb, W.P., and Müller-Eberhard, H.J. (1977). Fed. Proc. 36, 1209.
20. Müller-Eberhard, H.J. (1978). In "Molecular Basis of Biological Degradative Processes" (R. Berlin, H. Hermann, I. Lepow and J. Tanzer, eds.), p. 65. Academic Press, New York.
21. Cooper, N.R., and Ziccardi, R.J. (1977). J. Immunol. 119, 1664.
22. Sim, R.B., Porter, R.R., Reid, K.B.M., and Gigli, I. (1977). Biochem. J. 163, 219.
23. Müller-Eberhard, H.J., Polley, M.J., and Calcott, M.A. (1967). J. Exp. Med. 125, 359.
24. Fearon, D.T., and Austen, K.F. (1977). Proc. Nat. Acad. Sci. USA 74, 1683.
25. Kolb, W.P., and Müller-Eberhard, H.J. (1973). J. Exp. Med. 138, 438.
26. Lachmann, P.J., and Thompson, R.A. (1970). J. Exp. Med. 131, 643.
27. Podack, E.R., Biesecker, G., Kolb, W.P., and Müller-Eberhard, H.J. (1978). J. Immunol. 121, 484.
28. Esser, A.F., Kolb, W.P., Podack, E.R., and Müller-Eberhard, H.J. (1979). Proc. Nat. Acad. Sci. USA 76, March Issue.
29. Podack, E.R., Biesecker, G., and Müller-Eberhard, H.J. (1979). Proc. Nat. Acad. Sci. USA 76, February Issue.
30. Biesecker, G., Podack, E.R., Halverson, C.A., and Müller-Eberhard, H.J. (1979). J. Exp. Med. 149, February Issue.
31. Mayer, M.M. (1973). Scientific American 54, 229.
32. Bhakdi, S., and Tranum-Jensen, J. (1978). Proc. Nat. Acad. Sci. USA 75, 5655.

CYCLIC NUCLEOTIDE-INDEPENDENT PROTEIN KINASES FROM RABBIT RETICULOCYTES AND PHOSPHORYLATION OF TRANSLATIONAL COMPONENTS[1]

Jolinda A. Traugh, Gary M. Hathaway, Polygena T. Tuazon, Stanley M. Tahara, Georgia A. Floyd, Robert W. Del Grande, and Tina S. Lundak

Department of Biochemistry, University of California Riverside, California 92521

Five different cyclic nucleotide-independent protein kinases have been isolated from rabbit reticulocytes. The enzymes are distinct from the type I and type II cAMP-dependent protein kinases and the free catalytic subunit of these enzymes. Three of the cyclic nucleotide-independent protein kinases, casein kinase I, casein kinase II and the hemin controlled repressor, have been obtained in highly purified form. The physical and chemical properties of these enzymes have been studied and the possible modes of regulation examined. In addition, two protease-activated kinases have also been observed in reticulocytes. These five cyclic nucleotide-independent protein kinases have different substrate specificities with respect to histone and casein. All of the protein kinases, cyclic nucleotide-independent and cAMP-dependent, differentially phosphorylate components of the protein synthesizing system including 40S ribosomal subunits and initiation factors 2, 3, 4B and 5. Each of the protein kinases phosphorylates two or more of these components and the initiation factors and ribosomal subunits are in turn modified by at least two different protein kinases. The result is a multiply phosphorylated system. Alterations in the phosphorylation state of these components can be observed under conditions which activate or inhibit specific protein kinase activities.

INTRODUCTION

In 1970, Kabat (1) and Loeb and Blat (2) showed that phosphate was covalently bound to ribosomal proteins. The majority of the phosphate was associated with a single 40S ribosomal protein which was subsequently identified in rat

---

[1]This research was supported by Grant GM 21424 from the U. S. Public Health Service.

liver as S6 (3) and as S13 in reticulocytes (4). Both in intact reticulocytes (5) and in vitro (4) phosphorylation of S13 was regulated, at least in part, by cAMP. Although considerable effort has been expended to elucidate the role of this phosphorylation event in protein synthesis, the function of this modification has remained obscure. However, in the intervening years several translational initiation factors have been shown to be phosphorylated both in reticulocytes (6) and in reticulocyte lysates (7). These include initiation factor 2 (eIF-2), eIF-3 and eIF-4B. Although not observed in the experiments with reticulocytes and lysates, eIF-5 has been shown to be phosphorylated in vitro by purified protein kinases (8).

In order to determine the role of these multiple phosphorylation events in the control of protein synthesis, we have identified and purified the protein kinases from reticulocytes. The enzymes have been examined with respect to their physical and chemical properties and mode of regulation in order to approach the question of control of protein synthesis by phosphorylation in a systematic manner. In addition, we have added effectors of protein synthesis to reticulocytes and reticulocyte lysates and monitored alterations in the incorporation of radioactive phosphate into the various translational components.

RESULTS AND DISCUSSION

A number of substrate-specific protein kinases have been identified in the post-ribosomal supernatant fraction from rabbit reticulocytes. These enzymes are listed in Table I. The first enzymes to be described, the type I and type II cAMP-dependent protein kinases, are separable by chromatography on DEAE-cellulose (9,10). During the purification of these enzymes, mixed histone (Sigma IIA) has been used as substrate. The enzymes follow the pattern of the classical cAMP-dependent protein kinases in which the inactive holoenzyme is readily dissociated in the presence of cAMP releasing active catalytic subunits. In addition, some free catalytic subunit is observed in reticulocyte lysates which does not adhere to DEAE-cellulose (9). The activated type I and type II enzymes as well as the free catalytic subunit are inhibited by the addition of isolated regulatory subunit and by the heat-stable inhibitor protein (9).

Using casein as substrate, two cyclic nucleotide-independent protein kinases have been resolved by chromatography

TABLE I

MODE OF REGULATION OF PROTEIN KINASES
FROM RABBIT RETICULOCYTES

Protein Kinase	Regulatory Compound	Mode of Regulation
cAMP-dependent (type I and II)	cAMP	Activation
Casein kinase I	Not known	--
Casein kinase II	5S RNA	Inhibition
Hemin controlled repressor	Hemin	Inhibition
Protease activated kinase I	Trypsin; chymotrypsin	Activation
Protease activated kinase II	Trypsin; Ca^{2+} stimulated protease	Activation

on DEAE-cellulose (11). They have been specified as casein kinase I and II in order of elution from the resin. These enzymes are not activated by either cAMP or cGMP and are not inhibited by either the regulatory subunit of the cAMP-dependent protein kinase or by the heat-stable inhibitor protein. Unlike the cAMP-dependent protein kinases and the other cyclic nucleotide-independent enzymes, these enzymes are in an activated state in a reticulocyte lysate and no increase in activity is observed either upon purification or by addition of exogenous compounds. Although numerous substances have been examined, no compound has been identified which alters the enzymatic activity of casein kinase I. Recently, we have isolated a 5S RNA from the post-ribosomal supernatant fraction of reticulocytes which specifically inhibits casein kinase II. This RNA has no effect on the cAMP-dependent protein kinases or upon the other cyclic nucleotide-independent enzymes.

A third cyclic nucleotide-independent protein kinase, the hemin controlled repressor, has been identified in reticulocytes (12-16). This enzyme does not readily phosphorylate either casein or mixed histone, but exhibits a high degree of substrate specificity. To date the only substrates which have been identified for the enzyme are the α-subunit of eIF-2 and the protein kinase itself. This protein kinase is in an inactive state in reticulocytes that are rapidly synthesizing globin. However, under conditions where hemin is limiting, the protein kinase is activated with a consequent inhibition of globin synthesis.

Recently we have discovered two protease activated kinases which are separable by DEAE-cellulose chromatography (17). These have been termed protease activated kinases I and II. These enzymes are distinct from the cAMP-dependent and the other cyclic nucleotide-independent protein kinases. These protein kinases are present in an inactive form in the reticulocyte supernatant fraction and are activated in vitro by trypsin and chymotrypsin. Protease activated kinase II is also converted by a calcium activated protease from reticulocytes. The enzymes have been partially purified using mixed histone as substrate. Initial studies on the protease activated kinases show that they use only ATP and are inhibited by increasing concentrations of monovalent cations.

Three of the cyclic nucleotide-independent protein kinases have been obtained in a highly purified form by ion-exchange chromatography and by gel filtration. A summary of some of the properties of casein kinase I, casein kinase II and the hemin controlled repressor are shown in Table II. Casein kinase I is a monomer with a molecular weight of 37,000 and a sedimentation coefficient of 3.2 S (11). The enzyme uses ATP in the phosphotransferase reaction with an apparent K_m of 13 µM for ATP and 900 µM for GTP. The protein kinase activity is stimulated by monovalent cations with optimal activity between 140 and 300 mM KCl (18).

Casein kinase II has a sedimentation coefficient of 7.5 S and is composed of multiple subunits of 42,000, 38,000 and 24,000 daltons (11). The enzyme is a tetramer with an α, α', β_2 composition. Both ATP and GTP are utilized in the phosphotransferase reaction with an apparent K_m of 10 µM for ATP and 40 µM for GTP. As observed with casein kinase I, casein kinase II is also stimulated by monovalent cations with optimal activity between 100 and 250 mM KCl (18).

TABLE II

CHARACTERISTICS OF THREE CYCLIC-NUCLEOTIDE INDEPENDENT PROTEIN KINASES[a]

Protein kinase	Subunit Molecular Weight	Subunit Structure	Nucleotide Specificity	Monovalent Cation Effect	Self-Phosphorylation
Casein kinase I	37,000	Monomer	ATP	Stimulation	Yes
Casein kinase II	42,000 38,000 24,000	α, α', β_2	ATP/GTP	Stimulation	β subunit
Hemin controlled repressor	90,000	n.d.	ATP/GTP	Inhibition	Yes

n.d. = not determined.

[a] Casein kinases were purified by ion-exchange chromatography as described by Hathaway and Traugh (11). The hemin controlled repressor was purified by chromatography on DEAE-cellulose, phosphocellulose, Sephadex G-150 and Biogel A-1.5 (19). Casein kinase I and II were greater than 85% and 80% pure respectively. The hemin controlled repressor was greater than 95% pure. The subunit molecular weights were determined by polyacrylamide gel electrophoresis in sodium dodecyl sulfate as described elsewhere (25). The subunit structure was determined from the sedimentation coefficient and quantitation of the Coomassie Brilliant Blue bound to each subunit (11). Self-phosphorylation was determined by incubating the purified protein kinases with $(\gamma-{}^{32}P)ATP$ and analyzing the mixture by gel elecrophoresis followed by autoradiography (11).

The hemin controlled repressor has a subunit molecular weight of 90,000 (19) and preferentially utilizes ATP in the phosphotransferase reaction, although GTP will also react (12). In contrast to the results obtained with the casein kinases, monovalent cations inhibit the hemin controlled repressor with 50% inhibition observed at around 100 mM KCl.

It is important to note that all three of the purified protein kinases are self-phosphorylated. With casein kinase II, phosphate is incorporated into the 24,000 dalton subunit (11). Up to one mole of radioactive phosphate is incorporated per mole of β subunit; a total of two phosphates per enzyme. When casein kinase II is maximally self-phosphorylated and compared with the native enzyme, no differences are observed in the rate of phosphorylation of casein (20). However, a notable decrease in the phosphorylation of eIF-2α is observed when the hemin controlled repressor is self-phosphorylated (19). This decrease occurs only when the enzyme is preincubated with ATP. GTP, CTP, UTP or ADPCP have little or no effect on the activity.

Casein kinase I and II and the hemin controlled repressor can be added to the list of self-phosphorylated protein kinases which already include the cAMP-dependent (21) and cGMP-dependent enzymes (22). We would like to suggest that self-phosphorylation may be a common feature of all protein kinases and it will be of interest to determine whether additional enzymes including the protease activated kinases substantiate this theory.

We have shown that the casein kinases are structurally different. To establish that the enzymes are unique, it is necessary to show that they modify different substrates or amino acid residues. This has been accomplished by examining chymotryptic digests of purified β-casein B phosphorylated by casein kinase I, casein kinase II and the cAMP-dependent protein kinases. It has been determined that the phosphoryl groups added by each of the enzymes are in different chymotryptic peptides (23). In further studies the phosphorylated amino acids in β-casein A^2 and $α_{s1}$-casein A have been determined and the amino acid composition of the phosphorylated chymotryptic peptides has been analyzed (18). The results of these studies are summarized in Table III. Casein kinase I phosphorylates serine residues and a similar sequence is adjacent to the modifiable serine in both $α_{s1}$- and β-caseins. We have concluded from these studies that the recognition determinants for casein kinase I are Glu-X-Ser. Casein kinase II modifies threonine residues and phosphorylates the same sequence, Thr-Glu-Asp,

TABLE III

RECOGNITION DETERMINANTS OF CASEIN KINASE
I AND II WITH PURIFIED CASEINS[a]

Protein Kinase	Recognition Determinants	
	α_{s1}-Casein A	β-Casein A^2
Casein kinase I	Glu-Leu-<u>Ser</u>	Glu-Glu-<u>Ser</u>
Casein kinase II	<u>Thr</u>-Glu-Asp	<u>Thr</u>-Glu-Asp

[a] The casein variants were phosphorylated with casein kinase I and II and the phosphorylated chymotryptic peptides were separated by two-dimensional fingerprinting (23). Phosphoserine and phosphothreonine was determined and the amino acid composition of the phosphorylated chymotryptic peptides was analyzed (18). The phosphorylated sites were assigned on the basis of the amino acid analysis of the peptides and phosphoserine and phosphothreonine determinations in conjunction with the primary sequences of the individual variants.

in both α_{s1}- and β-caseins. However, threonine does not appear to be an absolute requirement for the enzyme since casein kinase II modifies serine residues in several of the initiation factors.

As mentioned previously, casein kinase II is inhibited by a 5S RNA. Highly purified inhibitor RNA has been prepared from the post-ribosomal supernatant fraction of reticulocytes by phenol extraction followed by gel filtration on Sephadex G-100 (20). The inhibitor activity is reversed by incubation with pancreatic ribonuclease A or micrococcal nuclease but not by deoxyribonuclease I, trypsin or pronase. Assuming a molecular weight of approximately 50,000, the inhibitor appears to act stoichiometrically with casein kinase II.

The ultimate goal of this research is to determine the relationship between phosphorylation and the regulation of protein synthesis. Thus, although histone and casein have been used to purify the protein kinases (with the exception of the hemin controlled repressor), the enzymes have been shown to modify components of the protein synthesizing system. As shown in Table IV, the cAMP-dependent protein kinases phosphorylate the 130,000 dalton subunit of eIF-3 (a large molecular weight complex composed of approximately ten different subunits) and one of the 40S ribosomal proteins,

TABLE IV

PHOSPHORYLATION OF TRANSLATIONAL COMPONENTS[a]

Protein Kinase	eIF-2	eIF-3	eIF-4B	eIF-5	40S Ribosomal Subunit
cAMP-dependent (type I and II)	-	130,000	-	-	32,500 (S13)
Casein kinase I	-	-	+	+	-
Casein kinase II	53,000	130,000 69,000 35,000 (?)	+	+	-
Hemin controlled repressor	38,000	-	-	-	-
Protease activated kinase I	-	130,000	+	+	17,500 (S15)
Protease activated kinase II	53,000	-	?	?	32,500 (S13)

[a] Initiation factors were individually phosphorylated and analyzed by polyacrylamide gel electrophoresis in sodium dodecyl sulfate followed by autoradiography as described previously (25).

S13 (4,24). As stated previously, the hemin controlled repressor phosphorylates only the α-subunit (M_r 38,000) of eIF-2 (a factor composed of three different subunits). Casein kinase I phosphorylates two single subunit factors, eIF-4B and eIF-5 (25). These four initiation factors are also phosphorylated by casein kinase II. In addition, the two protease activated kinases differentially phosphorylate the same four initiation factors (17) as well as two proteins, S13 and S15, in 40S ribosomal subunits (26).

It is clear that casein kinase II, the cAMP-dependent enzymes and the hemin controlled repressor phosphorylate different sites on the initiation factors since different subunits of eIF-2 and eIF-3 are phosphorylated. From the work with the casein variants, it is probable that different sites on eIF-4B and eIF-5 are modified by casein kinases I and II. Since the protease activated kinases have still different substrate specificities with regard to the initiation factors and ribosomal proteins, it is suggested that they also have unique recognition determinants.

It is of interest that each of the protein kinases phosphorylates two or more components of the protein synthesizing system. In addition, each of the four initiation factors and 40S ribosomal subunits are phosphorylated by at least two different protein kinases. Thus the potential exists for a highly regulated system. These phosphorylation events, except for the phosphorylation of S15, eIF-5 and the 35,000 dalton subunit of eIF-3, have been observed when intact reticulocytes are incubated with a nutritional medium containing $^{32}P_i$ (6,27). Using a radioactive creatine phosphate·ATP·GTP regenerating system coupled with a cell-free protein synthesizing system, all of the proteins except eIF-5 and S15 have been shown to be phosphorylated under conditions of optimal globin synthesis (7). From the studies in vivo, in cell lysates and with the purified components, it can be concluded that two of the protein kinases, casein kinase I and II, are in an activated state. In addition, a portion of the cAMP-dependent protein kinases are active. We have not observed the activated form of the protease activated kinases in fresh reticulocyte lysates and S15 is not phosphorylated either in vivo or in cell lysates.

Under conditions of optimal hemin availability, the hemin controlled repressor is essentially inactive; little or no phosphorylation of eIF-2α is observed in hemin supplemented lysates. However, under conditions of hemin deprivation, increased phosphate incorporation into eIF-2 and a 55,000-60,000 dalton protein in the high salt-wash

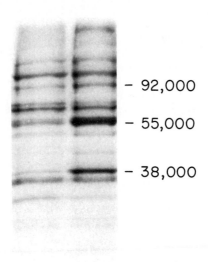

+H   −H
3 minutes

Figure 1. Reticulocyte lysate was prepared and incubated with the radioactive phosphate exchange system described by Floyd and Traugh (28) in the presence and absence of 25 µM hemin. Incubation was terminated at three min. The lysates were fractionated by centrifugation and the high salt wash fraction from ribosomes was examined by gel electrophoresis followed by autoradiography (28). The autoradiographs from hemin supplemented (+) and hemin deprived (−) lysates are shown.

fraction from ribosomes is observed (Figure 1). A three to four fold increase in phosphate incorporation into these two proteins is observed after three minutes incubation in the absence of hemin, just prior to inhibition of protein synthesis at five minutes. The unidentified phosphorylated protein is not an established initiation factor and the function is as yet unknown (27). These studies indicate a correlation between specific protein phosphorylation, activation of the protein kinse activity associated with the hemin controlled repressor, and inhibition of protein synthesis. Hemin appears to act directly on the protein kinase and 6 μM hemin inhibits the activity of the purified enzyme (greater than 95% pure) by 50% (19). Thus under conditions where hemin is limiting, the enzyme is activated. The hemin controlled repressor is also regulated by self-phosphorylation. We postulate that the hemin controlled repressor preferentially phosphorylates eIF-2, and when this substrate has been depleted, the enzyme self-phosphorylates resulting in inhibition of the repressor activity.

The phosphorylation of S13 remains enigmatic. Although S13 may contain up to five phosphoryl groups in vivo, only two phosphoryl groups can be added to the salt-washed subunits in vitro with the cAMP-dependent enzymes (26). The 40S ribosomal subunits obtained by centrifugation in 0.5 M KCl contain approximately one phosphate on S13. Thus it appears that the addition of other phosphoryl groups is dependent either upon protease activated kinase II or upon the structural properties of the ribosome.

While one of these phosphorylation events is clearly involved in the control of protein synthesis, the role of the others remains to be identified. One of the problems in the pursuit of this goal has been the ubiquitousness of the protein kinases. This causes difficulties in examining the action of one enzyme without interference by the others. Using the studies described in this article as a basis, it is now feasible to attempt to elicit the functional role of these events in the regulation of protein synthesis. It is possible that some of the phosphorylation reactions may be involved in control of protein synthesis during the early stages of differentiation when the cell is initiating globin synthesis. Alternatively, specific phosphorylation or dephosphorylation events may be a signal for destruction of the protein synthesizing system as the reticulocyte matures. The latter possibilities will be difficult to examine. However, the complexity of the multiple phosphorylation events make these studies intriguing.

The fact that five different cyclic-nucleotide independent protein kinases have been identified in the cytoplasm of such a highly differentiated cell type as the reticulocyte is worthy of further consideration. These protein kinases are regulated by different mechanisms such as hemin, RNA and proteolytic cleavage. Even the two protease activated kinases are not converted by the same endogenous protease activity. It is interesting to ponder the function of these enzymes in more complex cell types. For instance, many of the alterations occurring in embryonic development in the absence of transcription may be due to activation or inhibition of one or more protein kinases. The protein kinases may also play a role in the determination of differentiation if the regulatory agents for these enzymes are altered. Hormonal control of one or more of these cyclic nucleotide-independent protein kinases cannot be eliminated since the levels of the compounds mediating the enzymatic activities can be readily elevated or diminished.

ACKNOWLEDGMENTS

We would like to thank Dr. William C. Merrick, Department of Biochemistry, Case Western Reserve University for supplying the purified initiation factors and Elizabeth W. Bingham, Eastern Regional Research Center, United States Department of Agriculture for supplying the casein variants used in these studies.

REFERENCES

1. Kabat, D. (1970). Biochemistry 9, 4160-4175.
2. Loeb, J. E., and Blat, C. (1970). FEBS Lett. 10, 105-108.
3. Gressner, A. M., and Wool, I. G. (1974). J. Biol. Chem. 249, 6917-6825.
4. Traugh, J. A., and Porter, G. G. (1976). Biochemistry 15, 610-616.
5. Cawthon, M. L., Bitte, L. F., Krystosek, A., and Kabat, D. (1974). J. Biol. Chem. 249, 275-278.
6. Benne, R., Edman, J., Traut, R. R., and Hershey, J. W. B. (1978). Proc. Natl. Acad. Sci. USA 75, 108-112.
7. Floyd, G. A., Merrick, W. C., and Traugh, J. A. (1979) Eur. J. Biochem. (in press).
8. Traugh, J. A., Tahara, S. M., Sharp, S. B., Safer, B., and Merrick, W. C. (1976). Nature 263, 163-165.
9. Traugh, J. A., and Traut, R. R. (1974). J. Biol. Chem. 249, 1207-1212.
10. Tao, M., and Hackett, P. (1973) J. Biol. Chem. 248, 5324-5332.
11. Hathaway, G. M., and Traugh, J. A. (1979). J. Biol. Chem. 254, 762-768.
12. Tahara, S. M., Traugh, J. A., Sharp, S. B., Lundak, T. S., Safer, B., and Merrick, W. C. (1978) Proc. Natl. Acad. Sci. USA 75, 789-793.
13. Kramer, G., Cimadevilla, J. M., and Hardesty, B. (1976). Proc. Natl. Acad. Sci. USA 73, 3078-3082.
14. Farrell, P. J., Balkow, K., Hunt, T., Jackson, R. J., and Trachsel, H. (1977). Cell 11, 187-200.

15. Gross, M., and Mendelewski, J. (1977). Biochem. Biophys. Res. Commun. 74, 559-569.
16. Levin, D. H., Ranu, R. S., Ernst, V., and London, I. M. (1976). Proc. Natl. Acad. Sci. USA 73, 3112-3116.
17. Tahara, S. M. (1979). Ph. D. Thesis, University of California, Riverside.
18. Tuazon, P. T., Bingham, E. W., and Traugh, J. A. (1979). Eur. J. Biochem. (in press).
19. Lundak, T. S., and Traugh, J. A. (in preparation).
20. Hathaway, G. M., and Traugh, J. A. (in preparation).
21. Erlichman, J., Rosenfeld, R., and Rosen, O. M. (1974). J. Biol. Chem. 249, 5000-5003.
22. de Jonge, H. R., and Rosen, O. M. (1977). J. Biol. Chem. 252, 2780-2783.
23. Tuazon, P. T., and Traugh, J. A. (1978). J. Biol. Chem. 253, 1746-1748.
24. Traugh, J. A., and Lundak, T. S. (1978). Biochem. Biophys. Res. Commun. 83, 379-384.
25. Hathaway, G. M., Lundak, T. S., Tahara, S. M., and Traugh, J. A (1979). "Methods in Enzymology" 60, 495-511.
26. Del Grande, R. W., and Traugh, J. A. (in preparation).
27. Floyd, G. A. (1979). Ph.D. Thesis, University of California, Riverside.
28. Floyd, G. A., and Traugh, J. A. (1979). "Methods in Enzymology" 60, 511-521.

AN APPROACH TO THE STUDY OF PHOSPHOPROTEIN AND CYCLIC NUCLEOTIDE METABOLISM IN CULTURED CELL LINES WITH DIFFERENTIATED PROPERTIES[1]

Ora M. Rosen[*,+], Chen K. Chou[*,#], Jeanne Piscitello[*], Barry R. Bloom[o,o], Charles Smith[*], Peter J. Wejksnora[+], Rameshwar Sidhu[*], Charles S. Rubin[*,¢]

Departments of Molecular Pharmacology[*], Medicine[+], Molecular Biology[#], Cell Biology[o], Microbiology and Immunology[o], Biochemistry[+], and Neuroscience[¢], Albert Einstein College of Medicine, Bronx, New York 10461

INTRODUCTION

The observation that the catalytic component of the cAMP-dependent protein kinase purified from bovine cardiac muscle could catalyze the phosphorylation of its own cAMP-binding protein in an intramolecular reaction (1,2) led to two questions: (1) What are the biochemical consequences of phosphorylation *in vitro* and (2), does phosphorylation occur *in vivo*. The first could be approached in standard fashion by designing methods to assess the functions of the cAMP-binding protein in its phosphorylated and non-phosphorylated states. The second question is less easily handled. Its solution depends upon the ability to resolve the two forms of binding protein in unpurified cell extracts and the availability of cells whose hormonal and/or physiological state can be conveniently manipulated.

In this report we will first review our work on self-phosphorylation of the protein kinase purified from bovine heart: its potential for regulation *in vivo* and some methods for detecting its phosphorylation state in unpurified cell extracts. We will then present a progress report on two cell culture systems whose differentiated functions are sensitive to hormones and in which we are beginning to study the role of protein phosphorylation under physiological conditions. These cells are a murine macrophage-like cell line, J774.2, in which we have developed stable variants in cyclic nucleotide metabolism and the murine 3T3-L1 cell line that differentiates into adipocytes *in vitro*.

---

1 This work was supported by National Institutes of Health Grants AM09038 (O.M.R.), GM22792 (C.S.R.), AM21248 (O.M.R. and C.S.R.), AI-07118 and 10702 (B.R.B.); by American Cancer Society Grant BC-121 (O.M.R.).

RESULTS

Phosphorylation of Bovine Heart cAMP-dependent Protein Kinase. The principal cAMP-dependent protein kinase in bovine cardiac muscle has been purified (3) and classified on the basis of its chromatographic behavior, immunological properties (4), affinity for cAMP and subunit interactions (5,6) as a type II kinase. The enzyme is a tetramer, composed of a cAMP-binding protein dimer (subunit Mr = 55,000) and two catalytic subunits (Mr, 38,000). The inactive holoenzyme is activated by dissociation and release of its catalytic subunits, a process resulting from the interaction of cAMP with the binding protein component of the enzyme. The holoenzyme incorporates 1 mole of phosphate from ATP into each of two seryl residues in its cAMP-binding protein dimer (1). The dissociated phosphorylated binding protein but not the phosphorylated binding protein residing in the holoenzyme is a substrate for the phosphoprotein phosphatases of cardiac muscle (7). The larger of these phosphoprotein phosphatases has been purified about 150-fold from bovine cardiac muscle (8). It has a molecular weight of 180,000 and contains, as one of its subunits, the low molecular weight phosphatase (Mr = 31,000) previously purified to homogeneity from this tissue (7). This holophosphatase (phosphatase H) can be converted into its more active low molecular weight form (phosphatase S) by the addition of 4.0 $\underline{M}$ urea, limited proteolysis or heat (8). Both forms of the enzyme act only on the dissociated phosphorylated cAMP-binding protein (among other phosphoproteins), the phosphatase S having a 10-fold higher specific activity on this substrate than phosphatase H (8). Phosphatases S and H are easily distinguishable, even in unpurified extracts, by their chromatographic and electrophoretic behavior (8). It is thus feasible to look for possible interconversion in vivo.

Both phospho- and dephospho-holoenzyme can be completely dissociated by cAMP (9). The dephospho-binding protein, however, reassociates more readily with the catalytic subunit than its phosphorylated homolog (9). Following complete dissociation of protein kinase by cAMP, the dephosphorylated enzyme can attain 50% reassociation at a concentration of cAMP 10-fold higher than that required to observe reassociation of the phosphorylated form (10). There is reason to believe, therefore, that activation of phosphoprotein phosphatase (e.g., by conversion of phosphatase H to phosphatase S) could hasten the inactivation of protein kinase by promoting its reassociation. The first step in an analysis of this construct was to establish methods to distinguish phospho- and dephospho-binding proteins in cell-free extracts. To this end we employed a photoaffinity analog of cAMP, 8-azido cyclic [^{32}P]AMP, previously shown to react specifically and in stoichiometric amounts with cAMP-binding proteins (11-14). Following labelling, purified phospho- and dephospho-cAMP binding proteins could be clearly resolved by one dimensional analysis in 10% SDS PAGE[1]/ the phosphorylated protein migrating as if it had a molecular weight of 56,000 and the dephosprotein behaving as a protein of molecular weight 54,000 (14). An electrophoretic analysis of bovine cardiac muscle homogenates showed that most of the protein that incorporated 8-azido-cyclic [^{32}P]AMP moved to the position of the phospho enzyme and could be converted to the apparent molecular weight of the dephosphoprotein by treatment with phosphatase (14). Confirmation of this conclusion and a more complete resolution of the phospho- and dephospho-binding proteins was achieved by isoelectric focusing in 8 M urea. Both the 8-azido-cyclic [^{32}P]AMP and the ^{32}P-labelled proteins had the same pI of 5.35 which shifted to 5.40 upon dephosphorylation. Two dimensional electrophoretograms of the cAMP-binding proteins present in soluble extracts of bovine cardiac muscle demonstrated that this protein is predominantly in the phosphorylated form (14). In other bovine and murine tissues, variable ratios of phospho- and dephospho-type II kinases were observed[2]/.

Approaches to Studying the Role of Protein Phosphorylation in the Differentiated Functions of a Macrophage-like Cell Line J774.2. J774.2, a cell line derived from a murine reticulum cell sarcoma (15,16) exhibits a number of macrophage-like characteristics including Fc-mediated phagocytosis, lysozyme secretion and plasminogen activator secretion (17). Non-phagocytic variants of J774.2 were selected by exposing cells to IgG-coated sheep erythrocytes that had previously incorporated the adenosine analog, tubericidin (18). Those cells able to phagocytize these erythrocytes were killed by the subsequent intracellular release of tubericidin phosphate. Non-phagocytic cells could then be cloned from the surviving population. Phagocytosis by some of these cloned lines was subsequently shown to be specifically and fully corrected by the addition of 8-Br-cAMP or agents such as cholera toxin, L-isoproterenol or prostaglandin E_1, each of which promotes endogenous synthesis of the nucleotide (19). All of the lines were able to rosette antibody-coated erythrocytes. To evaluate the role of cAMP on Fc-mediated phagocytosis in parental J774.2, it was necessary to first render phagocytosis suboptimal. This was done two ways: (1) by growing the

---

[1]/ The abbreviations are: SDS, sodium dodecyl sulfate; PAGE, polyacrylamide gel electrophoresis.
[2]/ Rangel-Aldao, R. and Rosen, O.M. Unpublished observation.

cells in 'suspension' on non-tissue culture Petri dishes and (2) by treating the cells with insulin (60 ng/ml) for 5 h prior to assay. In both situations, the depression of Fc-mediated phagocytosis could be restored to control levels by raising the intracellular concentration of cAMP (19). The evidence that (1) genetic variants defective in Fc-mediated phagocytosis can be corrected by cAMP, (2) J774.2 grown in suspension can have its phagocytosis restored by cAMP and (3) insulin-mediated inhibition of phagocytosis can be overcome by cAMP has led us to conclude that there is a specific role for cAMP and, pari passu, protein phosphorylation, in Fc-mediated phagocytosis.

Cyclic AMP has two other known effects on J774.2 - it inhibits cell growth and it inhibits plasminogen activator secretion. The first effect was utilized to select for variants defective in either the ability to synthesize cAMP or the ability to respond to cAMP (20). Cells resistant to the growth inhibitory effects of cholera toxin were cloned and found to have defective adenylate cyclases. Their protein kinase activity and sensitivity to exogenous 8-Br-cAMP were the same as for the wild-type J774.2 The basal adenylate cyclase activities of the cholera toxin resistant clones were lower than normal (2-3 pmols cAMP formed/min/mg protein compared to 10-11 pmols/min/mg) and could not be enhanced by any of the agonists effective in stimulating the activity of control cells 10-15-fold (isoproterenol, prostaglandin E_1, cholera toxin, fluoride)(20). Cells with adenylate cyclase defects were in general less phagocytic than control cells. One of the cloned lines, $CTRM_1$, was most defective in phagocytosis (36% of its cells were phagocytically active compared to 80-90% of J774.2) and had the lowest basal cAMP content (0.6 pmols/10^6 cells compared to 1.5 pmols/10^6 cells for J774.2). Its capacity to undergo phagocytosis was fully restored by the addition of 0.1 m$\underline{M}$ 8-Br-cAMP.

The selection of variants resistant to growth inhibition by 8-Br-cAMP, a method based upon a similar selection technique used in S49 lymphoma cells (21), resulted in the acquisition of cloned cell lines with protein kinase defects. The variant studied in most detail thus far, $J_7M_3H_2$, has an aberrant cyclic nucleotide-binding protein(s) requiring 10-fold excess cAMP to saturate and fully activate the kinase (its Ka for activation by cAMP and Km for binding cAMP were approximately 280 nM and 51 nM, respectively, compared to 35 nM and 4 nM for these functions in the parental line) (20). Curiously, this line does not exhibit any defect in phagocytosis. Studies are currently underway to analyze the precise structural defects in the kinase using 8-azido [^{32}P]cyclic AMP to specifically label the cAMP binding protein in cell extracts (22). Preliminary data indicate that there is a low molecular weight cAMP-binding protein in the variant (Mr ~ 38,000) that is not apparent in J774.2. It is not yet known whether this protein derives from the principal type I cAMP-binding protein in J774.2 (Mr 49,000) or is, itself, an abnormal gene product.

The availability of protein kinase- and adenylate cyclase-defective cells has permitted a more detailed analysis of the observation that cAMP inhibits the secretion of plasminogen activator (23). The addition of either cholera toxin (0.5 pM) or 8-Br-cAMP (20 µM) inhibited the secretion of plasminogen activator 85% in less than 90 min whereas the secretion of lysozyme remained unaffected. Plasminogen activator secretion by cells defective in adenylate cyclase was similarly inhibited by 8-Br-cAMP but not by cholera toxin. As predicted for protein kinase-mediated cAMP effects, cells with defective protein kinases were resistant to the inhibitory effects of both cholera toxin and 8-Br-cAMP. Thus, cAMP has opposing effects on two functions of J774.2: Fc-mediated phagocytosis and plasminogen activator secretion.

The existence of defined cAMP-linked functions such as growth inhibition, enhancement of Fc-mediated phagocytosis and inhibition of plasminogen activator secretion in a cell line amenable to the development of specific variants in phagocytosis, cAMP synthesis and cAMP-mediated protein phosphorylation should permit us to unravel the role of protein phosphorylation in these complex functions. It may also provide a system for analyzing the relationship between cyclic nucleotide-dependent and -independent protein phosphorylation reactions and the effect, thereon, of insulin.

Approaches to Studying the Role of Protein Phosphorylation during Adipocyte Conversion in 3T3-L1 Cells. The 3T3-L1 line was developed and cloned by Green and colleagues (24-26). After achieving confluence the fibroblast-like cells differentiate in a focal distribution and acquire the morphological and biochemical properties of adipocytes. Differentiation is accompanied by 10-50-fold increases in the activities of enzymes involved in triglyceride biosynthesis (27-29), the acquisition of ACTH- and insulin-responsiveness (30) and a 30-40-fold increment in cell surface insulin receptors (31,32). Assessment of the biochemical concomitants of differentiation is facilitated when uniform and rapid differentiation is induced by treatment of the monolayers with 1-methyl-3-isobutylxanthine (0.5 mM) and dexamethasone for 48 h (31). Analysis of the proteins synthesized by cells differentiated under these conditions compared to the proteins synthesized by undifferentiated 3T3-L1 cells revealed at least 60 changes (33) in cytoplasmic, non-histone chromosomal, and plasma membrane fractions. One of the proteins altered by differentiation has been identified as actin, the synthesis of which decreased 70-80% during differentiation (33). It is likely that many of the proteins synthesized in higher amounts during differentiation are enzymes involved in lipid and intermediary metabolism. In fact, the number of proteins responsible for the altered metabolism of the adipocytes has been estimated to be about 50 (34).

One of the unique features of the 3T3-L1 cell lines is its developmental acquisition of insulin sensitivity. Concomitant with the increase in high affinity insulin receptors from about 7,500/cell to about 250,000/cell (31), hexose transport becomes insulin-sensitive. Both [^{14}C]2-deoxyglucose uptake and $^{14}CO_2$ production from [^{14}C] glucose are increased 3-8-fold by 0.1-1.0 ng/ml insulin (35). Since the cells are in culture, they can be treated with drugs and radioactive precursors and can be manipulated as single cell suspensions without exposure to proteolytic enzymes. It appeared reasonable, therefore, to determine whether insulin could direct the phosphorylation of one or more proteins at a concentration and within the time period previously established for its effect on hexose transport.

Differentiated cells incubated with ^{32}Pi for 50 min followed by the addition of insulin (1-10 ng) for 10 min showed an insulin-dependent phosphorylation of a protein of molecular weight 31,000 (Fig. 1). This phosphorylation was insulin specific, apparent within 5 min of the addition of as little as 0.5 ng/ml insulin, eliminated by anti-insulin serum, unaffected by protein synthesis inhibition (cycloheximide, 2.0 µg/ml) or by 1 µM L-isoproterenol, an adenylate cyclase agonist (30) and mimicked by the addition of rabbit anti-insulin receptor antiserum[3/].

Preliminary identification of this protein suggests that it is either identical or very similar to ribosomal protein S6, a protein known to undergo reversible phosphorylation in vivo and to be sensitive to changes in intracellular cAMP (37), serum and growth conditions (38). Designation as S6 is based upon sedimentation properties, comigration with S6 on 2-dimensional PAGE systems (39,40) designed to resolve

---

[3/] Antiserum to the insulin receptor was a gift from Drs.S. Jacobs and P. Cuatrecasas.

↳ pH 5
sodium dodecyl sulfate

A

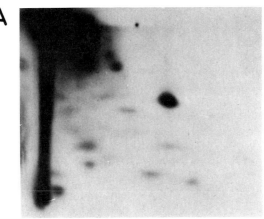

B

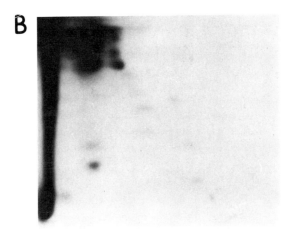

FIGURE 1. Radioautograms of two dimensional polyacrylamide gel electrophoresis of insulin-treated and control cells. Homogenates were prepared from cells treated with ^{32}Pi for 60 min with (A) or without (B) 7.0 ng insulin/ml for 10 min (41). Proteins were then extracted with 2 volumes of glacial acetic acid, dialyzed, lyophilized and finally solubilized in urea-containing sample buffer (19). Non-radioactive mouse L-cell ribosomal protein was added to each sample and 2.5×10^5 cpm of trichloroacetic acid precipitable radioactivity were applied to each gel. Two dimensional electrophoresis was performed according to Gorenstein and Warner (39) following which gels were stained and radioautograms obtained.

ribosomal proteins, and association with the 40S ribosomal subunit following dissociation and sucrose density gradient centrifugation (41).

It is not yet known whether this phosphorylation is directly involved in insulin action or merely reflective of a change in the state of kinases and phosphatases brought about by insulin. However, the magnitude of the effect (> 10-fold), sensitivity and specificity of the response suggests that it may be useful in establishing assays of insulin action in cell-free systems.

CONCLUSION

Elucidation of the role of protein phosphorylation in bioregulation depends upon different kinds of experimental approaches. Ultimately, however, it will be necessary to correlate the results of studies on purified proteins in vitro with changes that occur in vivo. The premise of this presentation is that the genetic and physiological modifications possible with certain differentiated cell lines will, in time, facilitate such a synthesis.

REFERENCES

1. Erlichman, J., Rosenfeld, R., and Rosen, O.M. (1974). J. Biol. Chem. 249, 5000.
2. Rangel-Aldao, R., and Rosen, O.M. (1976). J. Biol. Chem. 251, 7526.
3. Rubin, C.S., Erlichman, J., and Rosen, O.M. (1972). J. Biol. Chem. 247, 26.
4. Fleischer, N., Rosen, O.M., and Reichlin, M. (1976). Proc. Natl. Atad. Sci. U.S.A. 73, 54.
5. Hofmann, R,, Beavo, J.A., Bechtel, P.J., and Krebs, E.G. (1975). J. Biol. Chem. 250, 7795.
6. Corbin, J.D., Keely, S.L., and Park, C.S. (1975). J. Biol. Chem. 250, 218.
7. Chou, C.K., Alfano, J., and Rosen, O.M. (1977). J. Biol. Chem. 252, 2855.
8. Chou, C.K., and Rosen, O.M. Manuscript submitted for publication.
9. Rangel-Aldao, R., and Rosen, O.M. (1976). J. Biol. Chem. 251, 3375.
10. Rangel-Aldao, R., and Rosen, O.M. (1977). J. Biol. Chem. 252, 7140.
11. Walter, U., Uno, T., Liu, A.Y.C., and Greengard, P. (1977). J. Biol. Chem. 252, 6494.
12. Munevama, K., Bauer, R.J., Shulman, D.A., Rubins, R.K., and Simon, L.N. (1971). Biochemistry 10, 2390.

13. Haley, B.E. (1975). Biochemistry 14, 3852.
14. Rangel-Aldao, R., Kupiec, J.W., and Rosen, O.M. (1979). J. Biol. Chem. In Press.
15. Ralph, P., Pritchard, J., and Cohn, M. (1975). J. Immunol. 114, 888.
16. Ralph, P., and Nakoinz, I. (1975). Nature (London) 257, 393.
17. Bloom, B.R., Diamond, B., Muschel, R., Rosen, N., Schneck, J., Damiani, G., Rosen, O., and Scharff, M. (1978). Fed. Proc. 37, 2765.
18. Muschel, R.J., Rosen, N., and Bloom, B.R. (1977). J. Exp. Med. 145, 175.
19. Muschel, R.J., Rosen, N., Rosen, O.M., and Bloom, B.R. (1977). J. Immunol. 119, 1813.
20. Rosen, N., Piscitello, J., Schneck, J., Muschel, R., Bloom, B.R., and Rosen, O.M. (1979). J. Cell Physiol. 98, 125.
21. Bourne, H.R., Coffino, P., and Tomkins, G.M. (1976). J. Cell Physiol. 85, 611.
22. Piscitello, J., and Rosen, O.M. Manuscript in preparation.
23. Rosen, N., Schneck, J., Bloom, B.R., and Rosen, O.M. (1978). J. Cyclic Nuc. Res. 5, 345.
24. Green, H., and Kehinde, O. (1974). Cell 1, 113.
25. Green, H., and Kehinde, O. (1975). Cell 5, 19.
26. Green, H., and Kehinde, O. (1976). Cell 7, 105.
27. Mackall, J.C., Student, A.K., Polakis, S.F., and Lane, M.D. (1976). J. Biol. Chem. 251, 6462.
28. Wise, L.S., and Green, H. (1978). Cell 13, 233.
29. Spooner, P.M., Chernick, S.S., Garrison, M.M., and Scow, R.O. (1978). J. Biol. Chem.
30. Rubin, C.S., Lai, E., and Rosen, O.M. (1977). J. Biol. Chem. 252, 3554.
31. Rubin, C.S., Hirsch, A., Fung, C., and Rosen, O.M. (1978). J. Biol. Chem. 253, 7570.
32. Reed, B.C., Kaufmann, S.H., Mackall, J.C., Student, A.K., and Lane, M.D. (1974). Proc. Natl. Acad. Sci. U.S.A. 74, 4876.
33. Sidhu, R., Rubin, C.S., and Rosen, O.M. Manuscript submitted for publication.
34. Green, H. (1978). In Miami Winter Symposia: "Differentiation and Development" (F. Ahmad, ed.), 15, pp. 13-36. Academic Press, New York.
35. Rosen, O.M., Smith, C.J., Fung, C., and Rubin, C.S. (1978). J. Biol. Chem. 253, 7579.
36. Rosen, O.M., Fung, C., Hua, G., and Rubin, C.S. (1979). J. Cell Physiol. In Press.

37. Gressner, A.M., and Wool, I.G. (1976). J. Biol. Chem. 251, 1500.
38. Lastick, S.M., Nielsen, P.J., and McConkey, E.H. (1977). Molec. Gen. Genet. 152, 223.
39. Gorenstein, C., and Warner, J.R. (1976). Proc. Natl. Acad. Sci. U.S.A. 73, 1547.
40. Wittman, H.G. (1974). Methods Enzymol. 30, 457.
41. Smith, C., Wejksnora, P.J., Rubin, C.S., and Rosen, O.M. Manuscript submitted for publication.

GLYCOGEN SYNTHASE KINASE-2 AND PHOSPHORYLASE KINASE ARE THE SAME ENZYME

Noor Embi, Dennis B. Rylatt and Philip Cohen

Department of Biochemistry, University of Dundee,
Dundee DD1 4HN, Scotland, U.K.

ABSTRACT   Homogeneous preparations of phosphorylase kinase from rabbit skeletal muscle catalyse a calcium dependent phosphorylation of glycogen synthase $\underline{a}$ isolated from the same tissue. The calcium dependent glycogen synthase kinase activity copurifies with phosphorylase kinase throughout the standard procedure for the isolation of the latter enzyme. At the final step of the purification, gel filtration on Sepharose 4B, the elution profiles for glycogen synthase kinase and phosphorylase kinase activities are identical. ICR/IAn mice, which completely lack muscle phosphorylase kinase activity, do not contain detectable calcium dependent glycogen synthase kinase activity. These results indicate that the calcium dependent phosphorylation of glycogen synthase is catalysed by phosphorylase kinase and not by another calcium dependent protein kinase that might be contaminating the preparation.

The phosphorylation of glycogen synthase $\underline{a}$ by phosphorylase kinase reaches a plateau at 0.6-0.8 molecules of phosphate incorporated per subunit and is accompanied by a 2-fold decrease in the activity. The phosphorylation takes place on a unique serine residue located 7 amino acids from the N-terminus of the polypeptide chain. The amino acid sequence surrounding serine-7 shows considerable similarity to the amino acid sequence surrounding the phosphoserine in phosphorylase $\underline{a}$.

Glycogen synthase kinase-2, a protein kinase present as a trace contaminant in highly purified preparations of glycogen synthase also phosphorylates serine-7 exclusively. Evidence is presented which demonstrates that glycogen synthase kinase-2 and phosphorylase kinase are the same enzyme. The reasons why the identity of these proteins was not realised previously are discussed.

The rate of phosphorylation of glycogen synthase $\underline{a}$ by phosphorylase kinase is 2-3 fold slower than the rate of phosphorylation of phosphorylase $\underline{b}$ when identical concentrations of the two protein substrates are used (6 μM). At physiological concentrations of glycogen synthase (0.3 mg/ml) and phosphorylase (8.0 mg/ml), the time required for half maximal phosphorylation of each enzyme by phosphorylase kinase is similar. These results suggest that the phosphorylation of glycogen synthase by phosphorylase kinase is likely to be physiologically significant, and the implications of these findings are considered.

INTRODUCTION

The discovery that glycogen synthase is regulated by a phosphorylation-dephosphorylation mechanism in mammalian skeletal muscle was made while following-up the observation that the degree to which the activity could be stimulated by the allosteric activator glucose-6-phosphate depended on the metabolic state of the cell (1,2). Two forms of the enzyme were isolated subsequently, one of which was almost fully active in the absence of glucose-6-phosphate (termed the a-form) whereas the other (the b-form) was largely dependent on glucose-6-phosphate for activity (3). The conversion of glycogen synthase a to b was found to take place when partially purified preparations of the a-form were incubated with ATP-Mg. The reaction was stimulated by cyclic AMP and accompanied by phosphorylation of the protein (4). Subsequently, homogeneous preparations of glycogen synthase a were shown to be converted to a glucose-6-phosphate dependent 'b' form, by the same cyclic AMP dependent protein kinase which also catalysed the activation of phosphorylase kinase (5,6). Since the activity of cyclic AMP dependent protein kinase in skeletal muscle is determined by the level of adrenaline in the circulation, this result suggested that the hormone promoted the breakdown of glycogenolysis in two ways. First by activation of phosphorylase kinase, phosphorylase and the pathway of glycogenolysis; second, by inactivation of glycogen synthase and the pathway of glycogen synthesis.

The results did not however rule out the possibility that other glycogen synthase kinases might exist that were important in the regulation of glycogen synthesis by other factors, such as the contractile state of the muscle, the level of circulating insulin, or the concentration of tissue glycogen. In 1974 we reported that highly purified preparations of rabbit skeletal muscle glycogen synthase were contaminated by two types of protein kinase that phosphorylated the enzyme. One of these was cyclic AMP dependent protein kinase, but the other was an activity termed glycogen synthase kinase-2 (7). The latter enzyme could be distinguished from cyclic AMP

dependent protein kinase in a number of ways. Its activity
was unaffected by cyclic AMP or the specific protein inhibitor
of cyclic AMP dependent protein kinase. It had a higher Km
for ATP (0.4 versus 0.02 mM) and it could use GTP as phosphoryl donor (Km = 5 mM) whereas cyclic AMP dependent protein
kinase could not.

Following the digestion of the native enzyme with trypsin,
the sites phosphorylated by glycogen synthase kinase-2 were
liberated much more slowly as trichloracetic acid soluble
phosphopeptide material than were the sites phosphorylated by
cyclic AMP dependent protein kinase, and this indicated that
the sites phosphorylated by the two protein kinases were
distinct (7). This supposition was confirmed by amino acid
sequence analyses. The two serine residues phosphorylated by
cyclic AMP dependent protein kinase (termed site 1a and 1b)
are distinct from the serine phosphorylated by the endogenous
glycogen synthase kinase-2 activity (termed site 2) (8-10).
Furthermore, site 2 is located only 7 amino acids from the N-
terminus of the polypeptide chain (10).

The endogenous glycogen synthase kinase-2 activity was
found to be unaffected by the addition of calcium ions or the
presence of EGTA (7), and this initially suggested that it was
distinct from phosphorylase kinase, which is known to be
completely dependent on calcium ions for activity (11).
However, we showed recently that the endogenous glycogen
synthase kinase-2 activity was stimulated considerably by the
addition of the calcium dependent regulator protein (termed
calmodulin) in the presence of calcium ions (12), and also
that calmodulin is a subunit of muscle phosphorylase kinase
(13). Furthermore, Roach et al recently reported that
phosphorylase kinase catalyses the phosphorylation of glycogen
synthase (14). This information led us to make a detailed
appraisal of the relationship between glycogen synthase
kinase-2 and phosphorylase kinase. The results of these
experiments are described in this paper, and they demonstrate
that glycogen synthase kinase-2 and phosphorylase kinase are
indeed one and the same enzyme.

METHODS

All proteins were isolated from rabbit skeletal muscle.
Phosphorylase b (15) and phosphorylase kinase (16) were
purified to homogeneity and the specific protein inhibitor of
cyclic AMP dependent protein kinase (17) was partially
purified. Calmodulin was a homogeneous preparation obtained
by heat treatment of purified phosphorylase kinase (13).
Calmodulin-Sepharose was prepared according to Klee and
Krinks (18). Glycogen synthase a was isolated from the

protein-glycogen complex by chromatography on DEAE-cellulose and fractionation with polyethylene glycol 6000 (19,20). Prior to the ion exchange step, the glycogen was degraded using either the endogenous phosphorylase b and debranching enzyme activity (19,20) or purified salivary amylase. Approximately 80-90% of endogenous calmodulin dependent glycogen synthase kinase-2 activity was removed from purified glycogen synthase a by passing the enzyme through calmodulin-Sepharose equilibrated at 5.0 mM glycerophosphate 0.2 mM EDTA-2 mM magnesium chloride-0.2 mM calcium chloride. The glycogen synthase emerged in the breakthrough fraction of the column. The molecular weights of phosphorylase and glycogen synthase were taken as 100,000 and 88,000 respectively, and their absorbance indices ($A_{280nm}^{1\%}$) as 13.1 and 13.4.

Phosphorylase kinase was assayed at pH 6.8 or 8.2 (16). The incubations contained 0.2 mM calcium chloride or 1.0 mM EGTA. Glycogen synthase was assayed at pH 6.8 in the presence or absence of 10 mM glucose-6-phosphate (21). The phosphorylation of phosphorylase b and glycogen synthase a was carried out in the standard assay for phosphorylase kinase, except that [γ-^{32}P]ATP replaced the 'cold' ATP, the ATP concentration was reduced from 3.0 mM to 1.0 mM, and sufficient protein kinase inhibitor was included to inhibit cyclic AMP dependent protein kinase, which was a trace endogenous contaminant in purified glycogen synthase a. The incorporation of phosphate into protein was measured as described previously (21). Since the rate of phosphorylation of glycogen synthase was not linear above 0.1 molecules of phosphate incorporated per subunit, the initial rates of phosphorylation was measured within this limit.

Muscle extracts were prepared from phosphorylase kinase deficient (ICR/IAn) or control (C3H/He-mg) mice (22), taken to 35% ammonium sulphate, and the precipitates obtained by centrifugation were redissolved and dialysed against 50 mM glycerophosphate-2 mM EDTA, 15 mM mercaptoethanol pH 7.0. Phosphorylase kinase was purified 20-fold in a 70% yield by this treatment in the C3H/He-mg mice, and was concentrated 10-fold over the muscle extracts. No phosphorylase kinase activity remained in the 35% ammonium sulphate supernatant, and no phosphorylase kinase activity was present in either the 35% ammonium sulphate precipitate or supernatant from the ICR/IAn mice.

RESULTS

Phosphorylation of Glycogen Synthase by Purified Phosphorylase Kinase. Phosphorylase kinase was found to catalyse a calcium dependent phosphorylation of glycogen

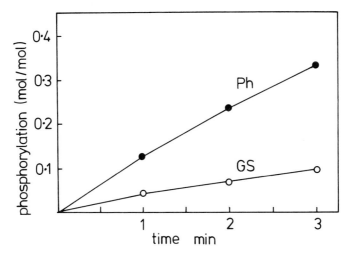

FIGURE 1. Relative rates of phosphorylation of glycogen phosphorylase b (●, Ph) and glycogen synthase a (O, GS) by purified phosphorylase kinase (1.0 µg/ml). The assays were carried out at 6 µM substrate concentration pH 8.2. No significant phosphorylation was observed in the absence of calcium ions.

synthase a confirming the work of Roach et al (14). The initial rate of phosphorylation of glycogen synthase a was 2-3 fold slower than for phosphorylase b, when the equimolar concentrations of the two protein substrates (6 µM) were used (Figure 1). This concentration is well below the Km for phosphorylase b which is above 10 mg/ml (100 µM) under these conditions (23).

Copurification of Phosphorylase Kinase and Calcium Dependent Glycogen Synthase Kinase. The calcium dependent glycogen synthase kinase activity was found to copurify with phosphorylase kinase activity throughout the standard isolation procedure for the latter enzyme (Table 1). At the final gel filtration on Sepharose 4B, which yields essentially pure phosphorylase kinase, the elution profiles for calcium dependent glycogen synthase kinase and phosphorylase kinase activities were virtually superimposable (Figure 2). The activity ratios (pH 6.8/8.2) were very similar with either phosphorylase or synthase as substrates, and were usually 0.06 ± 0.02 using purified phosphorylase kinase (not illustrated).

TABLE 1
COPURIFICATION OF CALCIUM DEPENDENT PHOSPHORYLASE KINASE AND GLYCOGEN SYNTHASE KINASE ACTIVITIES FROM RABBIT SKELETAL MUSCLE

Step	Activity Ratio PhK/GSK
1. Extract	16
2. pH 6.1 precipitate	19
3. 30,000 rpm supernatant	17
4. 30% ammonium sulphate precipitate	18
5. Sepharose 4B eluate	18

Muscle extracts were fractionated according to the standard procedure for the isolation of phosphorylase kinase (16). Assays were carried out at pH 8.2 using 5 mg/ml phosphorylase *b* or 0.5 mg/ml glycogen synthase *a*. Glycogen synthase kinase (GSK) and phosphorylase kinase (PhK) activities were stimulated at least 10-fold by calcium ions at each step under these conditions.

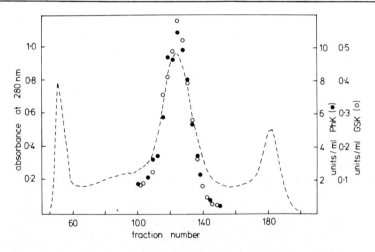

FIGURE 2. Elution of calcium dependent glycogen synthase kinase (O, GSK) and phosphorylase kinase (●, PhK) activities from Sepharose 4B (step 5, Table 1). The broken line shows the absorbance at 280 nm. The column (150x5 cm) was equilibrated in 50 mM sodium glycerophosphate-2.0 mM EDTA-15 mM mercaptoethanol pH 7.0, the flow rate was 60 ml/h and fractions of 12 ml were collected.

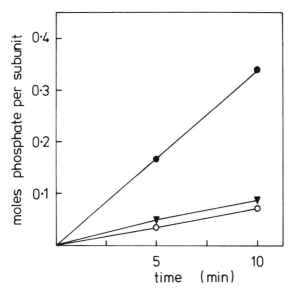

FIGURE 3. Phosphorylation of rabbit muscle glycogen synthase a (0.5 mg/ml) by fractions prepared from phosphorylase kinase deficient (ICR/IAn) mice (▼) and "normal" (C3H/He-mg) mice (●). Muscle extracts were fractionated by a 0-35% ammonium sulphate precipitation and equivalent amounts of each fraction was added in the assays. The open circles show the phosphorylation of glycogen synthase by endogenous glycogen synthase kinase-2 activity. Assays were carried out at pH 8.2 in the presence of calcium ions.

Phosphorylation of Glycogen Synthase in Phosphorylase Kinase Deficient Mice. ICR/IAn mice possess <0.2% of normal phosphorylase kinase activity in their skeletal muscles (23). Muscle extracts from ICR/IAn mice and "normal" (C3H/He-mg) mice were fractionated by a 0-35% ammonium sulphate precipitation (see Methods) and the ability of these fractions to catalyse the phosphorylation of glycogen synthase was examined (Figure 3). The fraction prepared from "normal" mice catalysed a calcium dependent phosphorylation of glycogen synthase, the activity ratio phosphorylase/glycogen synthase being 2.5 ± 0.5 at 6 μM substrate concentration, i.e. a very similar value to that observed with purified phosphorylase kinase from rabbit skeletal muscle (Figure 1). In contrast, the corresponding fractions prepared from phosphorylase kinase deficient mice did not phosphorylate glycogen synthase at a significant rate (Figure 3).

Effect of Phosphorylase Kinase on the Activity of Glycogen Synthase. The phosphorylation of glycogen synthase by purified phosphorylase kinase reached a plateau at 0.6-0.8 molecules of phosphate per subunit from preparation to preparation. In control experiments performed at the same time, the phosphorylation of phosphorylase b reached a plateau at 0.8-0.95 molecules of phosphate per subunit from preparation to preparation. The activity ratio of glycogen synthase ($\mp$ glucose-6P) decreased from 0.7-0.75 to 0.35-40 after maximal phosphorylation. The decrease in the activity ratio was due solely to a decreased activity in the absence of glucose-6-phosphate.

Amino Acid Sequence at the Site on Glycogen Synthase Phosphorylated by Phosphorylase Kinase. We have shown previously that the trace endogenous glycogen synthase kinase-2 activity which contaminates purified glycogen synthase phosphorylates a single serine residue located seven amino acids from the N-terminus of the polypeptide chain (10).

Purified phosphorylase kinase was found to phosphorylate glycogen synthase, at the same residue phosphorylated by the endogenous glycogen synthase kinase-2 (serine-7), using exactly the same methodology described in reference 10. The amino acid sequences surrounding the phosphoserines in phosphorylase and synthase are compared in Table 2.

TABLE 2
AMINO TERMINAL SEQUENCES OF RABBIT MUSCLE
GLYCOGEN PHOSPHORYLASE AND GLYCOGEN SYNTHASE
CONTAINING THE SITES OF PHOSPHORYLATION BY
PHOSPHORYLASE KINASE (10, 24-26)

Phosphorylase

$$\text{glu-lys-arg-}\underline{\text{lys-gln-ile-ser-val}}\overset{P}{\underset{14}{}}\text{-arg-gly-}\underline{\text{leu}}\text{-ala-}\underline{\underline{\text{gly}}}\text{-val-}\underline{\underline{\text{glu}}}$$

Synthase

$$\overset{1}{\text{pro}}\text{-leu-ser-}\underline{\underline{\text{arg}}}\text{-}\underline{\underline{\text{thr}}}\text{-}\underline{\text{leu-}}\overset{P}{\underset{7}{\underline{\text{ser-val}}}}\text{-ser-}\underline{\underline{\text{ser}}}\text{-}\underline{\text{leu}}\text{-pro-}\underline{\underline{\text{gly-leu}}}\text{-}\underline{\underline{\text{glu}}}^{15}$$

Identical residues are shown by a full line and conservative differences by a broken line. The numbers indicate distances from the N-terminus of each enzyme.

Comparison of the Properties of the Endogenous Glycogen Synthase Kinase-2 and the Endogenous Phosphorylase Kinase Activities Present in Purified Glycogen Synthase. The glycogen synthase kinase-2 activity present as a trace endogenous contaminant in purified glycogen synthase was found to be unaffected by the presence or absence of calcium ions at pH 6.8, and this initially suggested that this enzyme was distinct from phosphorylase kinase (7). However, the report of Roach et al (14) followed by the findings described above prompted a re-evaluation of this question.

A study of the properties of the trace phosphorylase kinase activity which also contaminates purified glycogen synthase (7) showed that like the endogenous glycogen synthase kinase-2 activity it was also unaffected by calcium ions at pH 6.8. When the properties of phosphorylase kinase were investigated at each stage of the purification of glycogen synthase, it was found that the ion-exchange chromatography on DEAE-cellulose was largely responsible for the conversion of phosphorylase kinase to a calcium insensitive form (Table 3). If the assays were carried out at pH 8.2 in the presence of calcium ions, 90-95% of the phosphorylase kinase activity was lost by chromatography on DEAE-cellulose. However, as this step was accompanied by both a rise in the activity ratio pH 6.8/8.2, and a loss in calcium sensitivity (Table 3), the loss in activity was much smaller, if the assays were performed at pH 6.8 in the presence of EGTA. Degradation of the glycogen by the action of endogenous phosphorylase, which involves incubation of the protein-glycogen complex with 0.5M phosphate (19,20), also promoted the desensitization of phosphorylase kinase to calcium ions, as compared to degradation with salivary amylase (Table 3).

The dependence of phosphorylase kinase on calcium ions was lost much more rapidly at pH 6.8 than at pH 8.2, and the trace phosphorylase kinase activity in purified glycogen synthase was always unaffected by calcium ions at pH 6.8. However at pH 8.2 slight stimulation by calcium ions (1.2-fold to 1.7-fold) was usually detectable in freshly purified preparations of glycogen synthase. When the glycogen was degraded with salivary amylse, the stimulation by calcium ions was usually ca 2-fold at pH 8.2, but there was still no detectable effect of calcium ions at pH 6.8.

The endogenous glycogen synthase kinase-2 activity behaved in an identical manner to the endogenous phosphorylase kinase activity. No effect of calcium ions was observed at pH 6.8, and the stimulation by calcium ions at pH 8.2 was identical to that determined for the endogenous phosphorylase kinase activity of the same preparation (not illustrated).

TABLE 3

CHANGES IN THE PROPERTIES OF PHOSPHORYLASE KINASE DURING THE ISOLATION OF GLYCOGEN SYNTHASE

Step	Activity Ratio ± Ca^{2+}						Activity Ratio pH 6.8/8.2	
	pH 6.8			pH 8.2				
	1a	1b	2	1a	1b	2	1a	2
1. Protein-glycogen pellet	8.0	8.0	—	30	30	—	0.15	—
2. 30,000 rpm supt. after glycogenolysis	1.6	4.0	2.2	4.5	30	11.0	0.17	0.09
3. DEAE-cellulose pH 7.5	1.0	1.3	1.1	2.0	2.6	1.6	0.30	0.36
4. 5-10% polyethylene glycol pptn. (pure glycogen synthase)	1.0	1.0	1.0	1.5	2.0	1.2	0.35	0.44

In preparation 1, the material was divided into two parts; in one portion the glycogen was degraded using the endogenous phosphorylase and debranching enzyme activities (1a) and in the other portion with purified salivary amylase (1b). In preparation 2 only the degradation with endogenous phosphorylase was employed.

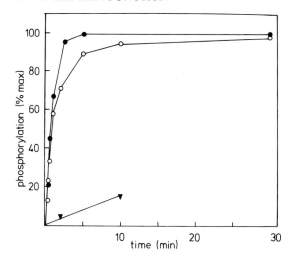

FIGURE 4. Relative rates of phosphorylation of glycogen synthase (O) and phosphorylase (●) by phosphorylase kinase. The assays were carried out at pH 8.2 in the presence of calcium ions using 8 mg/ml phosphorylase b or 0.3 mg/ml synthase a, and 0.02 mg/ml phosphorylase kinase. The closed triangles illustrate endogenous glycogen synthase kinase-2 activity, which was negligible at 30-60 sec, at which time half maximal phosphorylation of glycogen synthase had taken place.

Relative Rates of Phosphorylation of Phosphorylase and Synthase at Physiological Concentrations. The intracellular concentration of phosphorylase (8 mg/ml) is nearly 30-fold higher than glycogen synthase (0.3 mg/ml) (27). At these concentrations, the time required for half maximal phosphorylation of each enzyme was very similar (Figure 4). The time for half maximal phosphorylation of both substrates was 30-60 sec at pH 8.2 using 0.02 mg/ml phosphorylase kinase (Figure 4). However, it should be noted that the intracellular concentration of phosphorylase kinase is 0.7-0.8 mg/ml in rabbit skeletal muscle (27).

DISCUSSION

Several lines of evidence demonstrated that the calcium dependent glycogen synthase kinase in mammalian skeletal muscle is identical to phosphorylase kinase. These included the copurification of the two activities (Figure 2, Table 1), similar rates of phosphorylation (Figures 1, 4), similar amino acid sequences at the sites of phosphorylation (Table 2), similar pH 6.8/8.2 activity ratios, and absence of calcium dependent glycogen synthase kinase activity in phosphorylase kinase deficient mice (Figure 3).

The cyclic AMP <u>independent</u> protein kinase which contaminates purified glycogen synthase, termed glycogen synthase kinase-2, was originally thought to be distinct from phosphorylase kinase because its activity at pH 6.8 was completely unaffected by calcium ions, whereas purified phosphorylase kinase is almost totally dependent on calcium ions for activity. However the present results show that the standard procedure for the purification of glycogen synthase in this laboratory converts phosphorylase kinase to a form which is no longer regulated by calcium ions at pH 6.8, and which has a 5-10 times higher pH 6.8/8.2 activity ratio (Table 3). The properties of the endogenous phosphorylase kinase activity present in purified glycogen synthase (<u>ca</u> 1-2 µg per mg synthase) are therefore identical to the endogenous glycogen synthase kinase-2 activity, and the endogenous phosphorylase kinase is sufficient to account for the observed rate of phosphorylation of glycogen synthase. Further evidence for the identity of glycogen synthase kinase-2 and phosphorylase kinase, was obtained from the finding that both these activities phosphorylate the same serine residue (serine-7) on glycogen synthase.

We have reported previously, that the endogenous glycogen synthase kinase-2 activity which was unaffected by calcium ions at pH 6.8, could nevertheless be activated by calmodulin in the presence of calcium ions (12). The activation was variable and ranged from less than 2-fold to more than 10-fold from preparation to preparation (12). Phosphorylase kinase purified by the standard procedure (16) contains calmodulin in stoichiometric amounts with the other three subunits of the enzyme (13,28), and is almost certainly the subunit responsible for conferring calcium sensitivity to the enzyme. However the further addition of calmodulin to purified phosphorylase kinase produces additional activation of the enzyme and this activation is also variable, ranging from 1.5-fold to 10-fold from preparation to preparation (24 and unpublished work). This appears to result from the binding of a second molecule of calmodulin, since the calmodulin stimulated activity can be inhibited completely by the anti-psychotic drug trifluorperzine, or by troponin-I, whereas the calcium dependent activity in the absence of added calmodulin is unaffected by these agents (28). If this interpretation is correct, it would seem that the procedure used for the isolation of glycogen synthase causes phosphorylase kinase to lose its regulation by the tightly bound molecule of calmodulin, without losing its regulation by the second molecule of calmodulin. This idea is supported by the finding that activation of endogenous glycogen synthase kinase-2 by calmodulin can be blocked completely by trifluoperazine or troponin-I (unpublished work).

It is of interest that the sites on glycogen synthase and phosphorylase phosphorylated by phosphorylase kinase are both very near the N-termini of these proteins, and also that the amino acid sequences at the sites of phosphorylation are very similar. This suggests that specificity of phosphorylase kinase may either lie in the recognition of a specific feature of the primary structure, or a particular secondary structure determined by a local region of primary structure. This idea has been suggested by Graves and coworkers, but they have also suggested that the arginine residue at residue 16 in phosphorylase is critical to the specificity of phosphorylase kinase (29,30). Since this arginine residue is replaced by serine in glycogen synthase (Table 2) and glycogen synthase a and phosphorylase b are phosphorylated at rather similar rates (Figures 1, 4), the basis for the specificity of phosphorylase kinase would seem to require further investigation.

When the phosphorylation of glycogen synthase and phosphorylase by phosphorylase kinase was studied at physiological substrate concentrations, the time required for half maximal phosphorylation of each enzyme was very similar (Figure 4). This suggests that the phosphorylation of glycogen synthase by phosphorylase kinase may well be physiologically significant. It is an attractive idea, that the simultaneous phosphorylation of glycogen synthase and phosphorylase by phosphorylase kinase is a mechanism for achieving synchronous inhibition of glycogen synthesis and activation of glycogenolysis during muscle contraction. Furthermore, since phosphorylase kinase is itself activated by cyclic AMP dependent protein kinase, the inhibition of glycogen synthase in response to adrenaline may involve phosphorylation by two different protein kinases. Complete phosphorylation of glycogen synthase a by phosphorylase kinase only decreased the activity 2-fold in the absence of glucose-6-phosphate, but larger effects of this phosphorylation on the activity may well be found when the phosphorylation by phosphorylase kinase is combined with phosphorylations catalysed by other glycogen synthase kinases. Direct proof of these ideas will however require the demonstration that serine-7 becomes phosphorylated in vivo in response to nervous and hormonal stimulation of skeletal muscle.

Following our original identification of glycogen synthase kinase-2 (7), we subsequently purified an enzyme 4000-fold from skeletal muscle which was unaffected by cyclic AMP or calcium ions, and which catalysed the phosphorylation and inactivation of glycogen synthase (19). Although this enzyme was thought to be identical to the endogenous glycogen synthase kinase-2 which contaminated purified glycogen synthase, it is now clear that it is a distinct enzyme termed glycogen synthase kinase-3. This enzyme phosphorylates a site distinct from those phosphorylated by either cyclic AMP dependent protein kinase or phosphorylase kinase, and evidence in support of these statements will be presented in a subsequent publication (D.B. Rylatt and P. Cohen, in preparation).

ACKNOWLEDGMENTS

This work was supported by grants from the Medical Research Council, London and the British Diabetic Association. Noor Embi acknowledges a postgraduate scholarship from the National University of Malaysia. Philip Cohen was the recipient of a Wellcome Trust Special Fellowship from January 1976 to December 1978.

REFERENCES

1. Villar-Palasi, C. and Larner, J. (1961). Biochim. Biophys. Acta, 39, 173.
2. Villar-Palasi, C. and Larner, J. (1961). Arch. Biochem. Biophys. 94, 436.
3. Friedman, D.L. and Larner, J. (1963). Biochem. 2, 669.
4. Rosell-Perez, N. and Larner, J. (1964). Biochem. 3, 773.
5. Schlender, K.K., Wei, S.H. and Villar-Palasi, C. (1969). Biochim. Biophys. Acta, 191, 272.
6. Soderling, T.R., Hickinbottom, J.P., Reimann, E.M., Hunkeler, F.L., Walsh, D.A. and Krebs, E.G. (1970). J. Biol. Chem. 245, 6617.
7. Nimmo, H.G. and Cohen, P. (1974). FEBS Lett. $\underline{47}$, 162.
8. Huang, T.S. and Krebs, E.G. (1977) Biochem. Biophys. Res. Commun. 75, 643.
9. Proud, C.G., Rylatt, D.B., Yeaman, S.J. and Cohen, P. (1977). FEBS Lett. 80, 435.
10. Rylatt, D.B. and Cohen, P. (1979). FEBS Lett. in the press.
11. Brostrom, C.O., Hunkeler, F.L. and Krebs, E.G. (1971). J. Biol. Chem. 246, 1961.
12. Rylatt, D.B., Embi, N. and Cohen, P. (1979). FEBS Lett. in the press.
13. Cohen, P., Burchell, A., Foulkes, J.G., Cohen, P.T.W., Vanaman, T.C. and Nairn, A.C. (1978). FEBS Lett. 87, 287.
14. Roach, P.J., DePaoli-Roach, A.A. and Larner, J. (1978). J. Cyc. Nuc. Res. 4, 245.
15. Fischer, E.H. and Krebs, E.G. (1958) J. Biol. Chem. 231, 65.
16. Cohen, P. (1973). Eur. J. Biochem. 34, 1.
17. Nimmo, G.A. and Cohen, P. (1978). Eur. J. Biochem. 87, 341.

18. Klee, C.B. and Krinks, M.H. (1978). Biochm. 17, 120.
19. Nimmo, H.G., Proud, C.G. and Cohen, P. (1976). Eur. J. Biochem. 68, 21.
20. Caudwell, F.B., Antoniw, J.F. and Cohen, P. (1978). Eur. J. Biochem. 86, 511.
21. Nimmo, H.G., Proud, C.G. and Cohen, P. (1976) Eur. J. Biochem. 68, 21.
22. Cohen, P.T.W., Burchell, A. and Cohen, P. (1976). Eur. J. Biochem. 66, 347.
23. Cohen, P.T.W. and Cohen, P. (1973). FEBS Lett. 29, 113.
24. Titani, K., Cohen, P., Walsh, K.A. and Neurath, H. (1975). FEBS Lett. 55, 120.
25. Titani, K., Koide, A., Hermann, J., Ericsson, L.H., Kumor, S., Wade, R.D., Walsh, K.A., Neurath, H. and Fischer, E.H. (1977). Proc. Natl. Acad. Sci. USA, 74, 4762.
26. Huang, T.S. and Krebs, E.G. (1979). FEBS Lett. in the press.
27. Cohen, P., Antoniw, J.F., Nimmo, H.G. and Proud, C.G. (1975). Biochem. Soc. Trans. 3, 849.
28. Cohen, P., Embi, N., Foulkes, J.G., Hardie, D.G., Nimmo, G.A., Rylatt, D.B. and Shenolikar, S. (1979). XIth Miami Winter Symposium, in the press. (Academic Press, New York).
29. Tessmer, G.W., Skuster, J.R., Tabatabai, L.B. and Graves, D.J. (1977). J. Biol. Chem. 252, 5666.
30. Graves, D.J., Uhing, R.J., Janski, A.M. and Viriya, J. (1978). J. Biol. Chem. 253, 8010.

Modulation of Protein Function

METHYLATION AND DEMETHYLATION IN THE BACTERIAL CHEMOTACTIC SYSTEM[1]

Sharon M. Panasenko and Daniel E. Koshland, Jr.

Department of Biochemistry, University of California, Berkeley, California 94720

ABSTRACT A novel type of protein modification, the reversible formation of carboxy methyl esters, is an essential feature of sensory transduction during bacterial chemotaxis. It has been demonstrated that methylation of glutamate residues in a class of membrane proteins in *E. coli* and *S. typhimurium* occurs during the chemotactic responses of these organisms. Physiological studies of the responses of cells in which this methylation reaction is blocked indicate that the methylation process plays a role in information processing and adaptation. The methylation reaction is catalyzed by a substrate-specific, SAM-dependent methyl transferase. The hydrolysis of the methyl esters is catalyzed by a protein methyl esterase. Both the methyl transferase and methyl esterase are found in the soluble fraction of extracts derived from wild type *Salmonella* or *E. coli* cells. Analysis of mutant strains indicate that the *che*R gene product of *Salmonella* (*che*X in *E. coli*) is necessary for the expression of the methyl transferase activity. The methyl transferase activity is associated with the *che*X gene product of *Salmonella* (*che*B in *E. coli*). The development of an *in vitro* methylation reaction as well as the isolation of mutants in all the components of the methylation system has allowed us to dissect the processes into its components. Thus, it was possible to correlate changes in the pattern of protein methylation with alterations in physiological behavior, and to relate these changes to the time course of the response.

---

[1] This work was supported by NIH Grant #AM09765, and by an NIH Fellowship #F32 NS05456 to SMP.

INTRODUCTION

Bacteria have a well developed sensory system which allows them to swim towards attractants, which are usually nutrients, and away from repellents, which are usually indicators of toxic conditions. The chemotactic system which mediates this behavior can be divided into three parts as is schematically depicted in Figure 1.

The receptors which detect the signals from the external environment are located either in the membrane or in the periplasmic space just outside of the inner membrane. There are approximately 20 or 30 receptors known which can detect signals to give the chemotactic response. Several of these have been isolated and characterized (for a review see ref.1). Each receptor binds a specific class of chemoeffector molecules and the formation of the ligand-receptor complex leads to a conformational change in the protein. This conformational change is presumably the initial signal that is then processed by a common signal transmission apparatus. The signal transmission system has been shown to involve approximately 9 gene products. The output from the signal processing system, the motor response, provides a signal to the bacterial flagella which control the swimming behavior of the cell.

A methylation reaction is an integral part of the signal processing system. It is known that glutamate residues of a class of membrane proteins of approximately 65,000 molecular weight are methylated during chemotaxis (2,3). The following relationships between the methylation of these proteins and the chemotactic response have been established: *(i)* The methylation process requires S-adenosylmethionine, cells in

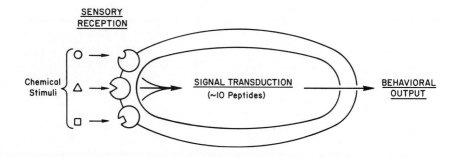

FIGURE 1. Chemotaxis in *Salmonella typhimurium* and *Escherichia coli*, a typical sensory-response phenomenon.

which production of S-adenosylmethionine is blocked cannot respond normally to stimuli (4,5,6). *(ii)* Mutant strains deficient in methylation are defective in chemotaxis (7,8). *(iii)* The level of methylation of the 65,000 molecular weight proteins changes in response to attractants and repellents in the medium (9).

Understanding the details of the role of methylation in the complicated process of signal transmission has awaited elucidation of the biochemistry of the system as well as the behavioral patterns of mutants defective in methylation. In this report we describe the use of *in vitro* methylation assays to characterize the enzymatic activities involved in methylation and in addition we present the results of various studies of the behavior of methylation mutants.

RESULTS

The 65,000 molecular weight proteins can be methylated *in vitro* using S-adenosylmethionine as a methyl donor. Membranes are prepared from sonicated *S. typhimurium* cells by centrifugation and incubated with soluble extracts from either *Salmonella* or *E. coli* in the presence of S-adenosyl(^{3}H-methyl)-methionine. The methylated products are then analyzed either by assaying the radioactivity in the 65,000 molecular weight region of sodium dodecyl sulfate polyacrilamide gels (8) or by assaying for protein carboxyl groups by the method of Axelrod and Daly (10). By comparing the results of these two methods it has been established that under our incubation conditions the 65,000 dalton proteins are the predominant carboxylmethylated species.

The protein carboxymethyltransferase that catalyzes the methylation reaction has been partially purified (8,11). A comparison of the properties of this enzyme with the methyltransferases from mammalian systems is shown in Table 1. Although the enzymes are similar in several respects, *e.g.* molecular weight and pH optimum, the K_m for S-adenosylmethionine is 5-10 fold greater for the bacterial enzyme. An additional and very striking difference, however, is the high degree of substrate specificity characteristic of the bacterial enzyme. As indicated in Table 2, of all the protein substrates tested, only the 65,000 molecular weight proteins can serve as methyl acceptors in the bacterial system. This is in marked contrast to the mammalian carboxymethyltransferases for which a wide variety of polypeptides serve as methyl acceptors.

It should be noted that the high degree of specificity exhibited by the chemotactic enzyme is not a general property of prokaryotic methyltransferases. Kim *et al.* (12) have

TABLE 1
COMPARISON OF BACTERIAL AND MAMMALIAN PROTEIN CARBOXYLMETHYLTRANSFERASES

Source	MW	pH optimum	S-adenosyl-methionine K_m (μM)	Activity in crude cell extracts (pmoles/min/mg protein)	Specificity	Ref.
S. typhimurium	40,000	5.7-7.2	10	10	narrow	8,11
Calf thymus	35,200	6.0; 7.5	1	3.0	broad	20
Rat erythrocytes	26,000	6-7	2	0.9	broad	21
Cow pituitary	—	5.5	1.5	4.2	broad	22
Ox brain	35,000	6.2	2.7	3.2	broad	23

TABLE 2
SPECIFICITY OF THE *SALMONELLA* CARBOXYLMETHYLTRANSFERASE

Protein tested	Methyl-acceptor activity [a]
S. typhimurium membranes	100
E. coli membranes	103
Erythrocyte membranes	3.1
HeLa membranes	0.0
Bovine γ-globulin	4.5
Histones (calf thymus)	4.2
Bovine serum albumin	3.6
Soybean trypsin inhibitor	3.0
Bovine insulin	2.4
Ovalbumin	1.8
Actin	1.5
Myosin	0.0
Cytochrome c	0.8
Flagellin (*S. typhimurium*)	0.6
Ribonuclease A	0.1

[a] The indicated proteins (2 mg/ml) were incubated in the presence of methyltransferase plus S-adenosyl-(^{3}H-methyl)methionine. The degree of methylation (methyl-acceptor activity) is given as percentage of the level of methylation of the 65,000-dalton *Salmonella* proteins after incubation under similar conditions.

reported a methyltransferase from *E. coli* which is similar to the mammalian enzymes in its specificity.

The rapid changes in the levels of methylation that occur upon addition of chemoeffectors to bacterial cells (9) would suggest that a mechanism exists for the hydrolysis of these protein carboxylmethyl esters under the appropriate conditions. Because of the extreme lability of these esters it was supposed that demethylation was a spontaneous event. Unlike proteins methylated by other carboxylmethyltransferases, the methylated 65,000 molecular weight proteins are stable at neutral pH. The ability to prepare membranes containing methylated 65,000 molecular weight protein has enabled us to assay demethylation *in vitro*.
When methylated membranes were incubated with soluble extracts methanol was produced with concomitant reduction in the level of methylation of the 65,000 protein (cf. Figure 2). This

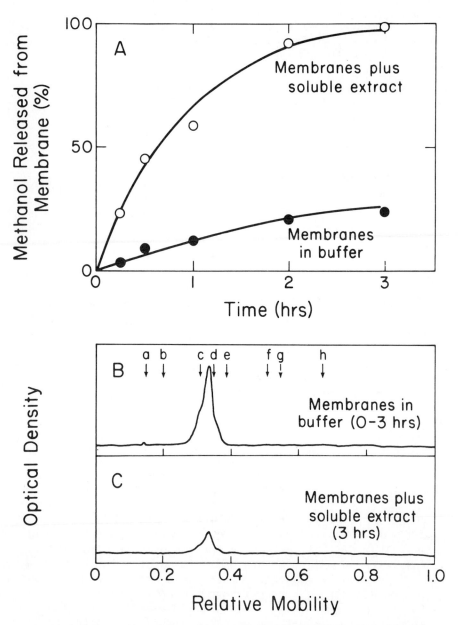

FIGURE 2. Hydrolysis of methylated proteins by extracts of *S. typhimurium*. Demethylation was followed either by measuring the release of ^{3}H methanol from ^{3}H-methyl-labeled membranes (A) or by measuring the level of radiolabel in the 65,000 dalton region of SDS polyacrylamide gels before (B) and after (C) incubation with soluble cell extracts.

methylesterase activity was heat labile and non-dialyzable.

Further insight into the role of the methylation system in bacterial chemotaxis came from the study of mutants defective in these components of the system (*i.e.* methyltransferase and methylesterase). Several hundred *Salmonella* chemotaxis mutants have been isolated. These have been shown to fall into nine complementation classes. These genes and their location on the *Salmonella* genetic map are shown in Figure 3. Six genes have been shown to lie in two adjacent operons. Similar results have been obtained in *E. coli* (13). Mutant genes in *E. coli* are complemented by the wild type gene from *Salmonella* and *vice versa*. The results of complementation analysis indicate a direct correspondence between the genes in the two bacteria (14).

In Table 3 is shown the results of a survey of the various nonchemotactic mutants that were assayed for methyltransferase activity *in vitro* (8). Only one class of che⁻ strains lacked activity, those designated *che*R. From this result we conclude that *che*R gene product is necessary for methyltransferase activity. Similar studies were conducted to determine the genes responsible for methylesterase activity (15). The results of these studies are shown in Table 4. Whereas most of the mutants, including the *che*R methyltransferase mutant had significant levels of methylesterase, one class lacked activity in both *Salmonella* and *E. coli*. These were the *che*X mutants. In *Salmonella* one other class, the *che*T mutants, lacked activity. Mutants in the gene corresponding to *che*T (*che*Z) in *E. coli* show normal activity, however. Parkinson has found that in *E. coli* these two genes, *che*B and *che*Z, complement poorly and has postulated that the gene products interact (13). Our results support this idea although the *che*Z gene product may not be essential for activity in *E. coli*. The simplest explanation of these data

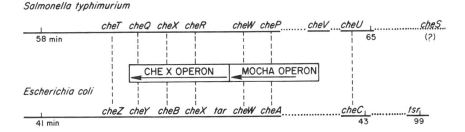

FIGURE 3. Location of chemotaxis genes on bacterial genome.

TABLE 3

METHYLTRANSFERASE ACTIVITY IN SOLUBLE EXTRACTS
OF *SALMONELLA* CHE MUTANT STRAINS

Complementation class [a]	% of wild-type methylase activity
*che*P	178
*che*Q	187
*che*R	< 5
*che*S	90
*che*T	105
*che*U	113
*che*V	33
*che*W	183
*che*X	55

[a] One mutant strain from each complementation class was tested.

is that the *E. coli che*B and *Salmonella che*X genes code for the methylesterase, and the *che*T and *che*Z gene products function to regulate the activity.

The availability of these mutants has allowed us to study the effects of aberrant methylation on the behavioral responses of the bacteria and, in fact, it is the behavioral phenotypes of some of these mutants that have given us the first clues to the biological function of the methylation system. It was found that the swimming behavior of methylation mutants was often altered drastically. As shown in Table 5, absence of methyltransferase produces smooth swimming whereas the absence of methylesterase produces tumbling. In addition these mutants are capable of responding to chemotactic stimuli. The response times are altered, however, and in the case of the methyltransferase mutant, adaptation appears to take place much more slowly than in wild type. Indeed, a methyltransferase mutant has been reported in *E. coli* which appears to be unable to adapt and which therefore responds permanently to chemical stimuli (16).

TABLE 4

METHYLESTERASE ACTIVITY IN SOLUBLE EXTRACTS OF *SALMONELLA* AND *E. COLI* CHEMOTAXIS MUTANTS

Complementation class [a]		% of wild-type activity	
Salmonella	*E. coli*	*Salmonella*	*E. coli*
cheP (1)	cheA (1)	60	102
cheQ (1)	cheY (1)	57	59
cheR (1)	cheX (1)	45	55
cheS (1)	—	125	—
cheT (4)	cheZ (4)	< 5	79-138
cheU (1)	—	66	—
cheV (1)	—	74	—
cheW (1)	cheW (1)	87	64
cheX (2)	cheB (1)	< 5	< 5
—	tar (1)	—	277
—	tsr (1)	—	317
—	tar/tsr (1)	—	187

[a]Number of strains tested from each complementation class is given in parentheses.

TABLE 5

BEHAVIORAL RESPONSES OF MUTANT STRAINS OF *SALMONELLA*

Strain	Unstimulated swimming behavior	Stimulus	Response duration [a]
ST1 (che⁺ control)	random	1 mM aspartate	2 min
		4.5 mM phenol	1.5 min
ST4 (cheX)	tumbly	1 mM aspartate	1 min
ST1038 (cheR)	smooth	4.5 mM phenol	4 min

[a]Responses measured as described in Ref. 17.

DISCUSSION

The picture that emerges from the study of the biochemistry and genetics of the methylation process during bacterial chemotaxis is one of a set of proteins in the cytoplasmic membrane where they might be in close proximity to both chemoreceptors and flagella. The membrane proteins are reversibly modified by carboxylmethylation. A highly specific mechanism exists for the methylation and demethylation of these proteins and the level of methylation is correlated with the presence of attractants and repellents. These are the sort of properties one would predict for a system which transmits information from receptors to flagella, the reversible methylation being ideally suited to the transient changes which occur during response and adaptation to chemoeffectors.

The unstimulated swimming behavior observed in methylation mutants (smooth swimming in methyltransferase deficient and tumbling in demethylase deficient strains) would be expected if the level of methylation somehow directly or indirectly controlled the tumble frequency of the cell. Mutants in the 65,000 dalton proteins have been isolated in *E. coli* and their behavior is also enlightening (18,19). Mutants in one class, *tsr*, are defective in response to a specific set of chemoeffectors including serine, alanine and some repellents. Mutants in a second and third class are unable to respond to yet others.

The relationship between the genetics of chemotaxis and protein carboxy methylation is shown schematically in Figure 4. During chemotaxis, chemoeffectors bind to receptor proteins located at the cell periphery. This results in a conformational change in the receptor which in some way is transmitted to the 65,000 dalton proteins. We envision at

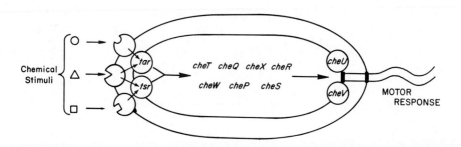

FIGURE 4. Roles of chemotaxis gene products in protein carboxy methylation reactions.

least two parallel pathways in this initial stage of signal transmission. Some classes of chemoeffectors utilize a pathway involving the *tsr* gene product, others utilize a pathway involving the *tar* gene product, still others may use both. Thus the function of the methyl-accepting proteins may be as channeling elements in the signal transmission process. The methyltransferase and demethylase might then serve as key regulatory elements in controlling the level of methylation and thus the response of the cells.

REFERENCES

1. Koshland, D.E., Jr. (1977). In "Advances in Neurochemistry" (B.W. Agranoff and M.H. Aprison, eds.), Vol II, pp.277-341, Plenum Press, New York.
2. Van Der Werf, P., and Koshland, D.E., Jr. (1977). *J. Biol. Chem. 252*, 2793.
3. Kleene, S., Toews, M.L., and Adler, J. (1977). *J. Biol. Chem. 252*, 3214.
4. Armstrong, J.B. (1972). *Can. J. Microbiol. 18*, 591.
5. Aswad, D., and Koshland, D.E., Jr. (1974). *J. Bacteriol. 118*, 640.
6. Adler, J., and Dahl, M.M. (1967). *J. Gen. Microbiol. 46*, 161.
7. Kort, E.N., Goy, M.F., Larson, S.A., and Adler, J. (1975). *Proc. Nat. Acad. Sci. USA 72*, 3939.
8. Springer, W.R., and Koshland, D.E., Jr. (1977). *Proc. Nat. Acad. Sci. USA 74*, 533.
9. Goy, M.F., Springer, M.S., and Adler, J. (1977). *Proc. Nat. Acad. Sci. USA 74*, 4964.
10. Axelrod, S., and Daly, J. (1965). *Science 150*, 892.
11. Clarke, S., Sparrow, K., and Koshland, D.E., Jr., In preparation.
12. Kim, S., Lew, B., and Chang, F.N. (1977) *J. Bacteriol. 130*, 839.
13. Parkinson, J.S. (1977). *Ann. Rev. Genetics 11*, 397.
14. DeFranco, A., Parkinson, J.S., and Koshland, D.E., Jr. (1979) *J. Bacteriol.* (Submitted).
15. Stock, J.B., and Koshland, D.E., Jr. (1978) *Proc. Nat. Acad. Sci. USA 75*, 3659.
16. Parkinson, J.S., and Revello, P.T. (1978). *Cell 15*, 1220.
17. Macnab, R.M., and Koshland, D.E., Jr. (1972). *Proc. Nat. Acad. Sci. USA 69*, 2509.
18. Springer, M.S., Goy, M.F., and Adler, J. (1977). *Proc. Nat. Acad. Sci. USA 74*, 3312.
19. Kondoh, H., Ball, C.B., and Adler, J. (1979). *Proc. Nat. Acad. Sci. USA 76*, 260.
20. Kim, S., and Paik, W.K. (1970). *J. Biol. Chem. 245*, 1806.

21. Kim, S. (1974). *Arch. Biochem. Biophys. 161*, 652.
22. Diliberto, E.J., Jr., and Axelrod, J. (1974). *Proc. Nat. Acad. Sci. USA 71*, 1701.
23. Iqbal, M., and Steenson, T. (1976). *J. Neurochem. 27*, 605.

… Modulation of Protein Function

REQUIREMENT OF TRANSMETHYLATION REACTIONS FOR EUKARYOTIC CELL CHEMOTAXIS[1]

Ralph Snyderman and Marilyn C. Pike

Laboratory of Immune Effector Function, Howard Hughes Medical Institute; Division of Rheumatic and Genetic Diseases, Department of Medicine and Department of Microbiology and Immunology, Duke University Medical Center, Durham, North Carolina 27710

ABSTRACT Transmethylation reactions mediated by S-adenosyl methionine are required for the chemotactic response of at least two types of eukaryotic cells, human monocytes and guinea pig macrophages. While the transmethylation reaction(s) required for eukaryotic cell chemotaxis are as yet unknown, incubation of chemotactic agents with the cell types studied did not produce a measurable stimulation of carboxy-0-methylation. Phospholipid methylation was, however, depressed by chemotactic factors in intact chemotactically responsive cells. It is suggested that polarized depression of phospholipid methylation occurs when responsive cells sense chemotactic gradients. Local alterations of the composition of methylated phospholipid near sites of occupied chemotactic factor receptors may be necessary for the directed migratory response of eukaryotic cells.

INTRODUCTION

The rapid accumulation of phagocytic cells at sites of antigen is essential for immunologically mediated host defense. It has been clearly shown that phagocytes of the granulocytic and monocytic series are capable of sensing and migrating along certain chemical gradients in vitro(1-3). Moreover, human blood granulocytes, rabbit peritoneal neutrophils and guinea pig peritoneal macrophages have surface receptor sites capable of binding chemotactic factors with high affinity and specificity(4-5). Although many metabolic alterations occur following the binding of chemotactic factors to leukocytes, the actual mechanism by which receptor

[1]This work was supported in part by a grant from the National Institute of Dental Research, No. 5 R01 DE03738-06.

occupancy is translated into directed migration remains largely unknown. In bacteria, the occupancy of surface chemoreceptors with chemoattractants results in S-adenosyl-L-methionine mediated methylation of certain membrane proteins which has been shown to be required for the chemotactic response of these organisms(6-8). In earlier work, we had found that depressed levels of the enzyme, adenosine deaminase(ADA), were associated with abnormal macrophage chemotaxis(9). This enzyme appears to play an important regulatory role in S-adenosyl-L-methionine(AdoMet)-mediated methylation reactions in that inhibition of ADA leads to the accumulation of S-adenosyl-L-homocysteine(AdoHcy) in cells incubated with exogenous adenosine and L-homocysteine(10). AdoHcy, a product of AdoMet-mediated methylation reactions, is a potent competitive inhibitor of methyltransferase reactions(11). Appreciating the importance of methylation for bacterial chemotaxis and the requirement of ADA for monocyte chemotaxis, we sought to determine if AdoMet-mediated methylation was required for the chemotaxis of eukaryotic cells and, if so, which type of methylation reaction was involved in this response.

METHODS

Chemotaxis. Chemotaxis *in vitro* was performed in modified Boyden chambers as previously described(12,13) using the following types of leukocytes as responder cells: human blood monocytes and glycogen-induced guinea pig peritoneal macrophages. Polycarbonate filters with a 5.0µ pore size were used to separate the cells from the chemotactic stimuli.

Phagocytosis. Immune phagocytosis by human monocytes and guinea pig macrophages of opsonized sheep erythrocytes labelled with ^{51}Cr was performed as previously described(9).

Protein Carboxy-O-Methylation. Carboxy-O-methylation was measured by a modification of a previously published method(14,15). Briefly, leukocytes were incubated with L-[methyl-^{3}H] methionine for various periods of time, washed once and resuspended in media containing bovine serum albumin (20 mg/ml). Cellular and carrier proteins were precipitated with perchloric acid and the precipitates washed with ethanol. Supernatants were neutralized and saved for AdoHcy and AdoMet assays. The washed precipitates were dissolved in pH 11.0 sodium borate buffer and, after incubation, the solution was extracted with toluene/3-methyl-1-butanol. Portions of the extractable material were counted directly for radioactivity or were evaporated at 70°C. The difference in radioactivity before and after evaporation was used to measure carboxy-0-methylation.

AdoHcy and AdoMet Determination. AdoHcy and AdoMet were assayed as previously described using high pressure liquid chromatography on a Whatman Partisil-10 SCX column(15).

Phospholipid (PL) Methylation. The methylation of phosphatidylethanolamine was measured in intact cells or membrane preparations by determining the incorporation of L-[methyl-^{3}H] methionine into phospholipids as previously described(16). Briefly, cells were incubated with L-[methyl-^{3}H] methionine for various times, then pelleted, washed once and extracted with chloroform/methanol. The organic phase was applied to a cellulose TLC plate and separation of free labelled methionine derivatives from phospholipids was achieved with chloroform/methanol/water. Phospholipid methyltransferase activity in cell extracts or membrane preparations was assayed by incubating the preparations with L-[methyl-^{3}H] AdoMet for various times followed by extraction with chloroform/methanol. The organic phase was applied to a silica gel G plate and developed in chloroform/methanol/water to separate phospholipids from other contaminating radiolabelled compounds. Samples from the plates containing phospholipids were scraped into scintillation vials and the radioactivity determined. In experiments where individual reaction products were measured, the [^{3}H]-methylated phospholipids contained in the scrapings of the cellular TLC plates were eluted with chloroform/methanol and further fractionated on silica gel G plates as previously described(16,17).

Cellular Phospholipid Synthesis. Total PL synthesis by monocytes or macrophages was measured by determining the incorporation of ^{32}Pi into organic solvent extractable radioactivity as previously described(16).

RESULTS AND DISCUSSION

Requirement of AdoMet Mediated Methylation for Eukaryotic Cell Chemotaxis. Previous studies have shown that the incubation of lymphoblastoid cells with the ADA inhibitor erythro-9-(2-hydroxy-3-nonyl) adenine(EHNA) plus adenosine and L-homocysteine leads to the accumulation of AdoHcy which results in the concomitant inhibition of AdoMet mediated methylation reactions(10). To determine if AdoMet mediated methylation was required for the chemotactic responses of eukaryotic cells, human mononuclear leukocytes isolated from blood by ficoll-hypaque density sedimentation, were exposed to various concentrations and combinations of EHNA, adenosine, and L-homocysteine thiolactone and were then tested for chemotaxis to N-formyl-methionyl-leucyl-phenylalanine (fMet-Leu-Phe) (Fig. 1). Protein carboxy-O-methylation, used as a monitor of transmethylation reactions, and AdoHcy levels were also measured in cells treated with these agents. EHNA in combination with adenosine (Ado) produced slight but measurable increases in AdoHcy as well as inhibition of both carboxy-O-methylation and chemotactic responsiveness. When cells were incubated in the presence of 10^{-5}M EHNA, 10^{-4}M L-homocysteine thiolactone and with 0.05 mM or 0.1 mM Ado, more

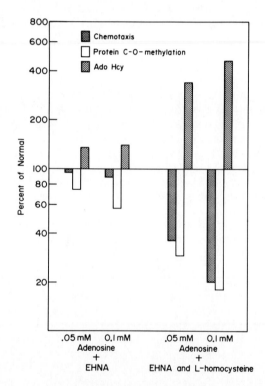

FIGURE 1. Effects of EHNA, adenosine and L-homocysteine thiolactone on the chemotaxis, protein carboxy-O-methylation and intracellular AdoHcy levels of human monocytes.

striking increases in intracellular AdoHcy levels were observed and were accompanied by marked inhibition of both Carboxy-O-methylation and chemotactic responsiveness. Substitution of D-homocystine for L-homocysteine thiolactone did not lead to any more inhibition of methylation or depression of chemotaxis than that seen with EHNA and Ado alone. The chemotaxis of guinea pig macrophages is likewise markedly depressed by EHNA, Ado and L-homocysteine thiolactone.

3-Deazaadenosine, a competitive inhibitor of transmethylation reactions(18) was also studied for its effects on human monocyte and guinea pig macrophage chemotaxis. Cells were treated with various doses of 3-deazaadenosine then tested for chemotactic responsiveness to fMet-Leu-Phe(monocytes) or zymosan-activated serum(macrophages). The chemotaxis of guinea pig macrophages was inhibited by as much as 83% by 10^{-4}M 3-deazaadenosine, and significant inhibition was noted at a dose as little as 10^{-8}M (Fig. 2). Human monocyte chemotaxis was unaffected by deazaadenosine alone, but marked inhibition of this response was noted when 10^{-4}M L-homocysteine thiolactone was added along with 3-deazaadenosine. Monocyte chemotaxis was inhibited by as much as 97% under these conditions. These results are not unexpected, since the ability of 3-deazaadenosine to inhibit methylation in cells depends at least in part on the formation of S-3-deazaadenosyl-L-homocysteine and on increases in AdoHcy(18). Human monocytes presumably have less available intracellular homocysteine than their more metabolically active counterparts, guinea pig macrophages, thus the formation of the aforementioned compounds would be limited in the absence of exogenous L-homocysteine. Experiments are underway to test this hypothesis directly.

These data indicate that inhibition of AdoMet mediated transmethylation reactions produces marked inhibition of the chemotaxis of human monocytes and guinea pig macrophages.

Immune phagocytosis has certain similarities to chemotaxis in that it requires leukocyte receptor occupancy followed by a cellular "metabolic burst" and membrane movement. There are, however, substantial differences between chemotaxis and phagocytosis in that the latter phenomenon does not require sustained polarized cellular movement. To determine if transmethylation reactions were also required for phagocytosis, human monocytes or guinea pig macrophages were incubated with various doses of EHNA, Ado and L-homocysteine thiolactone. Doses of these agents which produced more than 90% inhibition of chemotaxis had no depressive effect whatsoever on immune phagocytosis by human monocytes. Phagocytosis by guinea pig macrophages was, however, inhibited by doses of Ado exceeding 10^{-5}M in the presence of 10^{-5}M EHNA

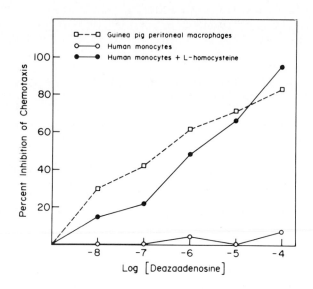

FIGURE 2. Effects of 3-deazaadenosine and L-homocysteine thiolactone on human monocyte and guinea pig macrophage chemotaxis. Cells were preincubated for 30 min. at 37°C with the indicated concentrations of 3-deazaadenosine in the presence and absence of 10^{-4}M L-homocysteine thiolactone and tested for in vitro chemotactic responsiveness.

and 10^{-4}M L-homocysteine thiolactone, but, at all doses of adenosine tested in the presence of EHNA and L-homocysteine thiolactone, chemotaxis was clearly inhibited more than phagocytosis.

The foregoing data demonstrate that transmethylation reactions mediated by AdoMet are required for the chemotactic responsiveness of at least two types of leukocytes. Since immune phagocytosis by human monocytes has either a strikingly lower or no requirement for methylation, it can be surmised that the metabolic requirement for the two seemingly similar biological phenomena are different in this cell type. It is interesting to speculate that AdoMet mediated methylation is required for the sustained polarized cellular movement that is associated with a chemotactic response.

Effects of Chemotactic Factors on Carboxy-O-Methylation and Phospholipid Methylation in Macrophages. While the foregoing data showed that inhibition of AdoMet mediated methylation produced inhibition of chemotaxis, they did not indicate which transmethylation reaction(s) is actually required for this phenomenon. Since cell division does not occur during the chemotactic response and since new protein synthesis is not a prerequisite for chemotaxis(15), we felt that the methylation of DNA or RNA was unlikely to be required for a chemotactic response. In bacteria, carboxy-O-methylation is indeed necessary for chemotaxis(6-8) and O'Dea et al. have reported a rapid stimulation of this type of methylation in rabbit peritoneal neutrophils incubated with chemotactic factors(19). We therefore studied the effects of chemotactic factors on carboxy-O-methylation in guinea pig macrophages. Macrophages were incubated with or without an optimal chemotactic dose of fMet-Met-Met(10^{-8}M), and the carboxy-O-methylation of acid precipitable proteins was tested at times ranging from one to sixty minutes later (Fig. 3). No significant alteration in protein carboxymethylation was observed at any time tested. Total [^{3}H]-methanol production, furthermore, was not altered in cells treated with fMet-Met-Met or C5a. In other experiments, macrophages were incubated with doses of fMet-Leu-Phe, fMet-Met-Met or C5a ranging from 10^{-7} to 10^{-10}M and then tested for protein carboxymethylation one hour later. No stimulation of this type of methylation reaction was noted under any of the conditions tested. Thus, in guinea pig macrophages, a burst of protein carboxy-O-methylation is not seen following exposure of the cells to a chemotactic stimulus.

In contrast to their effects on carboxy-O-methylation, chemotactic factors produced a striking alteration in phospholipid(PL) methylation by guinea pig macrophages (Fig. 4A). The chemotactic factors, fMet-Met-Met, fMet-Leu-Phe, fNor-Leu-Phe, fMet-Leu and C5a all substantially inhibited phospholipid methylation at doses which corresponded to their chemotactic activity in vitro (Fig. 4B). Neither nonformylated Met-Met-Met nor fPhe-Met is a competitive antagonist of fMet peptide receptor binding and chemotaxis(4,20).

To determine if chemotactic factors actually depressed PL methylation or stimulated the demethylation of methylated PL, macrophages were prelabelled with [^{3}H-methyl] methionine

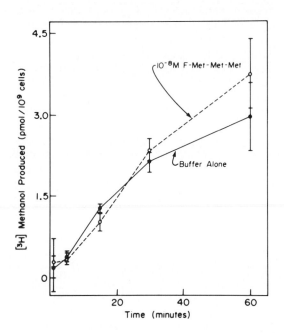

FIGURE 3. Kinetics of protein carboxy-O-methylation in guinea pig macrophages incubated in the presence and absence of the chemotactic peptide, fMet-Met-Met (10^{-8}M).

for one hour, then incubated with chemotactic factor for times ranging from ten minutes to ten hours (Fig. 5). No evidence could be found for increased turnover or degradation of methylated PL by chemotactic factor.

In studies of the kinetics of inhibition of PL methylation by chemotactic factors, significant inhibition of this reaction could be detected as early as five minutes after incubation of cells with fMet-Met-Met and persisted for the entire sixty minute incubation period studied.

The effects of immune phagocytosis on PL methylation by guinea pig macrophages was also studied. Cells were incubated with opsonized sheep erythrocytes at a ratio of 5:1 or 50:1 erythrocytes to macrophages. Neither dose of erythrocytes inhibited PL methylation, thereby indicating another dichotomy between phagocytosis and chemotaxis so far

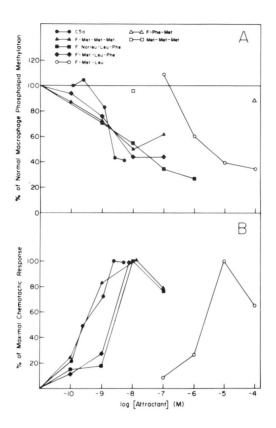

FIGURE 4 A. Effect of chemotactic agonists and antagonists on total phospholipid methylation in guinea pig macrophages.   B. Chemotactic activity (expressed as the % of maximal activity obtained with each substance) of chemotactic agonists.

as transmethylation reactions are concerned. In other experiments, guinea pig lymphocytes were incubated with doses of chemotactic factors which produced greater than 50% inhibition of PL methylation in guinea pig macrophages. No inhibition of methylation was seen in the lymphocytes. It should be noted that these cells do not respond chemotactically to the formyl-methionyl peptides nor to C5a. Thus, chemotactic factors do not appear to depress PL methylation in

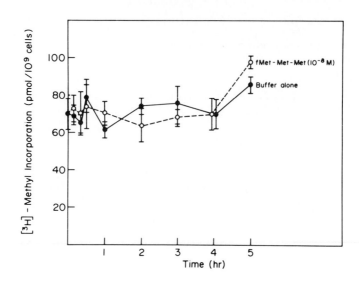

FIGURE 5. Effect of chemotactic factor on preformed methylated phospholipid derivatives in guinea pig macrophages. Macrophages prelabeled for 1 hour at 37°C with [^{3}H-methyl] methionine were further incubated in the presence or absence of fMet-Met-Met(10^{-8}M). Aliquots of cells were removed at the indicated times and total residual phospholipid associated radioactivity was determined.

nonchemotactically responsive cells.

It was also important to determine if chemotactic factors specifically inhibited PL methylation by macrophages or more generally depressed PL synthesis. We therefore measured total PL synthesis in macrophages in the presence or absence of chemotactic factors. No inhibition of total PL synthesis was induced by fMet-Leu-Phe, fMet-Met-Met or C5a at doses ranging from 10^{-10} to 10^{-8}M. Chemotactic factors thus selectively inhibit PL methylation rather than total PL synthesis. We then studied whether chemotactic factors selectively inhibited the formation of certain methylated derivatives of phosphatidylethanolamine. Macrophages were incubated with [^{3}H-methyl] methionine plus fMet-Met-Met, and the isolated phospholipids were separated on a TLC system which allows the quantification of newly formed phosphatidyl mono-, di- and tri- methylethanolamine. It was found that the synthesis of all three methylated derivatives of phosphatidylethanolamine were uniformly depressed by chemotactic factor in guinea pig macrophages.

Effects of Ionophore A23187 on PL Methylation by Macrophages. Incubation of chemotactic factors with leukocytes results in the rapid influx of Ca^{++} into the cells(21). We therefore sought to determine whether a calcium ionophore affected phospholipid methylation in macrophages. Cells were incubated with concentrations of the divalent cation specific ionophore, A23187 ranging from 10^{-7}M to 10^{-5}M, and tested for phospholipid methylation and chemotactic responsiveness. Phospholipid methylation was inhibited by as much as 60 percent by 10^{-5}M A23187(22). Similarly, as has been reported by others, macrophage chemotaxis was inhibited in a dose dependent manner by A23187(23). The inhibitory effects of A23187 on phospholipid methylation were augmented in the presence of exogenously added Ca^{++} and were partially reversed in the presence of Mg^{++}(Table I). The addition to the cells of 1.5mM $MgCl_2$ reduced the ionophore induced inhibition of phospholipid methylation from 60 percent to 30 percent. $MgCl_2$ alone (1.5mM) did not alter phospholipid methylation in macrophages. These findings indicate that the influx of Ca^{++} into macrophages is capable of inhibiting phospholipid methyltransferase reactions.

Effects of Chemotactic Factors and Ionophore A23187 on Phospholipid Methyltransferase Activity in Extracts of Guinea Pig Macrophages. We next determined whether chemotactic factors of A23187 exerted their phospholipid methylation inhibitory activity by direct action on the macrophage phosphatide methyltransferase(s). Guinea pig macrophage extracts were prepared by brief sonication of the cells followed by centrifugation at 12,000 X g for fifteen minutes. The cell-free extracts were incubated for one hour at 37°C in 0.1M Tris-acetate buffer, pH 8.5, in the presence or absence of doses of the chemotactic factors, fMet-Met-Met, fMet-Leu-Phe or C5a ranging from 10^{-10} to 10^{-7}M or with 10^{-5}M or 10^{-6}M A23187 and the transfer of the $[^3H]$-CH_3 from $[^3H$-methyl$]$ Ado-Met to phosphatidylethanolamine was measured. The chemotactic factors did not significantly alter the phospholipid

TABLE I
EFFECT OF Mg^{+2} AND Ca^{+2} ON IONOPHORE A23187
INDUCED INHIBITION OF MACROPHAGE
PHOSPHOLIPID METHYLATION

Cells incubated with[1]	Incorporation of $[^3H]$-CH_3 into phospholipids (pmol/10^9 cells)	% inhibition[2]
Buffer	37.5	–
10^{-5}M A23187	15.0	60
10^{-5}M A23187 +		
$CaCl_2$ 0.5 mM	13.6	64
1.5 mM	11.0	71
$MgCl_2$ 0.5 mM	27.1	28
1.5 mM	26.3	30

[1]Cells were preincubated for 15 minutes at 37°C under the indicated conditions before addition of 10 µCi [^{3}H-methyl] methionine. Total phospholipid methylation was measured after an additional hour of incubation at 37°C.

[2]% inhibition = $(1 - \frac{E}{C}) \times 100$ where E = methylation of macrophages under the indicated experimental conditions and C = methylation of macrophages in the presence of buffer alone.

methyltransferase activity in macrophage extracts. Similarly, A23187 in the presence or absence of 1.5mM $CaCl_2$ or $MgCl_2$ did not affect phospholipid methylation in such preparations. Doses of $CaCl_2$ or $MgCl_2$ alone ranging from 0.1mM to 10mM also did not alter the methyltransferase activity contained in macrophage extracts. These findings indicate that intact macrophages are required for both chemotactic factor and A23187 induced inhibition of phospholipid methylation. Moreover, the depression of methylation in intact macrophages by these substances cannot be attributed to a direct inhibitory effect of Ca^{++} or Mg^{++} on the phospholipid methyltransferase(s). Following from the aforementioned data, we are currently investigating whether the cytoskeletal system of intact macrophages plays a role in the inhibition of phospholipid methylation by chemotactic factors.

CODA

These studies demonstrate that transmethylation reactions mediated by AdoMet are required for the chemotactic responsiveness of several types of leukocytes. Unfortunately, the precise role such reactions play in the initiation or propagation of a chemotactic response by eukaryotic cells is as yet unknown. In bacteria, stimulation of protein carboxy-0-methylation is clearly demonstrable when cells are exposed to chemotactic stimuli(6-8,24). In the cell types studied here, such a stimulation of protein carboxy-0-methylation could not be found. O'Dea et al. have reported a rapid, albeit transient, stimulation of carboxymethylation by chemotactic factors in rabbit neutrophils, a cell type which we have not studied(19). If carboxymethylation is indeed a general requirement for eukaryotic cell chemotaxis, it is certainly more difficult to demonstrate in leukocytes than it is in bacteria. Our studies show that chemotactic factors depress phospholipid methylation in macrophages. The requirement for AdoMet mediated transmethylation for chemotaxis, coupled with the findings of inhibition of PL methylation by chemotactic factors, might seem paradoxical but can be reconciled by the following hypothesis. During a non-chemotactic state, leukocyte PL methylation is ongoing in a uniform manner throughout the cell. When the cell is exposed to a chemotactic gradient, PL methylation is inhibited locally at sites of chemotactic factor binding. Since inhibition of PL methylation by chemotactic factors is not associated with a depression of total PL synthesis, local changes in the relative concentration of phosphatidylethanolamine to phosphatidylcholine in newly synthesized membrane might well occur. Such changes would be expected to alter the biophysical properties of the membrane in the microenvironment where the chemotactic factor binds to its receptor. If these alterations occur, they might affect local membrane microviscosity and fluidity, thereby altering receptor movement or the ability of the membrane to adhere to external surfaces or to anchor internal cytoskeletal elements. Global inhibition of AdoMet-mediated methylation reactions by EHNA, Ado and L-homocysteine or by 3-deazaadenosine could prevent the local alterations in PL methylation induced by chemotactic factors and thereby prevent a directed migratory response. It is, of course, quite possible that other AdoMet mediated transmethylation reactions are also required for eukaryotic cell chemotaxis. In any case, the study of transmethylation reactions and leukocyte chemotaxis are now experimentally approachable and will certainly lead to a better understanding of the cellular physiology of eukaryotic cell movement.

REFERENCES

1. Boyden, S.V. (1962). J. Exp. Med. 115, 453
2. Gallin, J.I., and Quie, P.G. (1978). "Leukocyte Chemotaxis." Raven Press, New York.
3. Wilkinson. P.C. (1974). "Chemotaxis and Inflammation." Churchill Livingstone, Edinburgh.
4. Williams, L.T., Snyderman, R., Pike, M.C., and Lefkowitz, R.J. (1977). Proc. Natl. Acad. Sci. USA 74, 1204.
5. Aswanikumar, S., Corcoran, B., Schiffmann, E., Day, A.R., Freer, R.J., Showell, H.J., and Pert, C.B. (1977). Biochem. Biophys. Res. Commun. 74, 810.
6. Kort, E.N., Goy, M.F., Larsen, S.H., and Adler, J. (1975). Proc. Natl. Acad. Sci. USA 64, 1300.
7. Springer, W.R., and Koshland, D.E. Jr. (1977). Proc. Natl. Acad. Sci. USA 74, 533.
8. Springer, M.S., Goy, M.F., and Adler, J. (1977). Proc. Natl. Acad. Sci. USA 74, 3312.
9. Snyderman, R., Pike, M.C., Fischer, D.G., and Koren, H.S. (1977). J. Immunol. 119, 2060.
10. Kredich, N.M., and Martin, D.W. Jr. (1977). Cell 12, 931.
11. Hurwitz, J., Gold, M., and Anders, M. (1964). J. Biol. Chem. 239, 3474.
12. Snyderman, R., Gewurz, H., and Mergenhagen, S.E. (1968). J. Exp. Med. 128, 259.
13. Snyderman, R., Altman, L.C., Hausman, M.S., and Mergenhagen, S.E. (1972). J. Immunol. 108, 857.
14. Diliberto, E.J. Jr., Viveros, O.H., and Axelrod, J. (1976). Proc. Natl. Acad. Sci. USA 73, 4050.
15. Pike, M.C., Kredich, N.M., and Snyderman, R. (1978). Proc. Natl. Acad. Sci. USA 75, 3928.
16. Pike, M.C., Kredich, N.M., and Snyderman, R. (1979). Submitted.
17. Hirata, F., Viveros, O.H., Diliberto, E.M. Jr., and Axelrod, J. (1978). Proc. Natl. Acad. Sci. USA 75, 1718.
18. Chiang, P.K., Richards, H.H., and Cantoni, G.L. (1977). Mol. Pharmacol. 13, 939.
19. O'Dea, R.J., Viveros, O.H., Axelrod, J., Aswanikumar, S., Schiffmann, E., and Corcoran, B.A. (1978). Nature 272, 462.
20. Aswanikumar, S., Schiffmann, E., Corcoran, B.A., and Wahl, S.M. (1976). Proc. Natl. Acad. Sci. USA 73, 2439.
21. Boucek, M.M., and Snyderman, R. (1976). Science 193, 905.
22. Pike, M.C., Kredich, N.M., and Snyderman, R. (1979). J. Supramol. Structure (in press).
23. Gallin, J.I., Sandler, J.A., Clyman, R.I., Manganiello, V.C., and Vaughan, M. (1978). J. Immunol. 120, 492.
24. Silverman, M., and Simon, M. (1977). Proc. Natl. Acad. Sci. USA 74, 3317.

ROLE FOR METHYLATION IN LEUKOCYTE CHEMOTAXIS

E. Schiffmann,[1] R.F. O'Dea,[3] P.K. Chiang,[2]
K. Venkatasubramanian,[1] B. Corcoran,[1]
F. Hirata,[2] and J. Axelrod[2]

[1]National Institute of Dental Research and
[2]National Institute of Mental Health
National Institutes of Health, Bethesda, Md. 20014
and
[3]University of Minnesota Medical School
Minneapolis, Minnesota 55455

Key words: Chemotaxis, leukocyte, methylation, protein carboxyl groups, membrane phospholipids

Evidence is presented for the role of methylating reactions in leukotaxis. A protein carboxylmethylation reaction in neutrophils is stimulated in a dose dependent manner by formylmethionyl attractants. Antagonists of the fmet peptides blocked the increase in methylation. Peptide attractants were also found to stimulate degradation of methyl-labelled phospholipids in these cells and this too was inhibited by specific peptide antagonists. These degradative events were also inhibited by a phospholipase A_2 inhibitor. The methylating enzyme and its substrate were found in the 30,000 g pellet from homogenized neutrophils. De novo protein synthesis was not required for methylation to occur. Also, transmethylation reactions were implicated in chemotaxis. 3-Deazaadenosine, an inhibitor of adenosylhomocysteine hydrolase, caused inhibition of chemotaxis and an increase in adenosylhomocysteine, itself an inhibitor of transmethylation. Methylation of protein carboxyl groups and phospholipids was also depressed in the presence of this inhibitor. The results suggest a requirement for protein carboxymethylation and a role for phospholipids in leukotaxis.

A Role for Methylation in Leukocyte Chemotaxis

Leukocytes respond chemotactically to a variety of naturally occurring substances including factors derived from complement (1), certain lipids (2), denatured proteins (3), and products of bacterial metabolism (4). Many of these are

not well characterized and, therefore, difficult to study.
We have found that simple formylmethionyl peptides are potent
leukoattractants and have properties similar to attractants
from bacteria (5,6). Rabbit and human neutrophils have a
specific receptor for these peptides (7,8). It has also been
found that the cell hydrolyzes the peptides, probably by a
peptidase associated with the receptor (9). Hydrolysis may
serve to free the receptor and allow the cell to detect additional molecules of the attractant. More recently the molecular events in the cell triggered by the interaction between
receptor and attractant have been studied and indicate that
methylation reactions play a role. These studies were based
on the role of methylation reactions in bacterial chemotaxis,
where the reversible carboxymethylation of specific proteins
transmits the signal (10,11). Moreover membrane perturbations produced by other agonists in different cells has been
shown to involve alterations in levels of methylated phospholipids in the cell membrane (12). The present study describes
the changes observed in the levels of methylation of both
protein carboxyl groups and membrane phospholipids in response to chemoattractants. These reactions appear to transmit the signal from the activated receptor to the neutrophil's
motility apparatus. Inhibitors of these reactions blocks
chemotaxis.

METHODS

Chemotactic assays on rabbit peritoneal exudate neutrophils (PMN's) were performed with the modified Boyden chamber
method using micropore filters (4). Binding of labelled
peptide attractant to cells was measured by previously
described procedures (7). Determination of protein carboxylmethylation (^{3}H methyl groups donated by ^{3}H-methionine) was
accomplished by alkaline hydrolysis of labelled ester and by
the isolation and counting of ^{3}H-methanol (12). The extent
of methylation of phospholipids was determined by $CHCl_3$-CH_3OH
extraction of ^{3}H-methionine-labelled cells, evaporation of
solvents and counting the lipid residue (13). Separation and
measurement of intermediates in methionine metabolism including S-adenosylmethionine (AdoMet) and S-adenosylhomocysteine (AdoHcy), was achieved with the aid of high performance liquid chromatography.

RESULTS

Protein carboxylmethylation

F-met peptides elicited a rapid, but transient stimulated
transfer of methyl groups from ^{3}H-methionine to endogenous

protein (12) (Fig. 1). The formation of protein methyl esters in neutrophils reached a maximum thirty seconds after the addition of attractant and subsided within one to two minutes. In the absence of attractant the methylesterification reaction increased linearly with time.

The specificity of the reaction was indicated in a variety of observations. The time courses of both uptake of ^{3}H-methionine by cells and incorporation of label into protein were measured (Fig. 1). During the period of stimulated methyl ester formation the uptake of methionine was not stimulated by attractant, and incorporation of methionine into protein was stimulated less than carboxylmethylation. The stimulation of carboxylmethylation was found to be independent of de novo protein synthesis (Table I). Additional evidence for specificity was suggested by results obtained with antagonists of chemotactic peptides (Table II). Antagonists prevented the stimulated carboxylmethylation in the presence of N-formyl peptides, while the antagonists alone did not appreciably affect incorporation. In addition, the shape of the dose response curve for stimulated carboxylmethylation was found (Fig. 2) to be similar to that for chemotaxis (6). The optimal concentration of agonist-stimulated methylation was 10 nM while it was 1 nM for chemotaxis. This difference is probably due to a need for a gradient to obtain a chemotactic response. The methylation reaction was measured in the absence of a gradient and at a

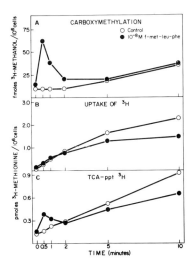

Fig. 1. Stimulation of leukocyte protein carboxymethylation uptake of ^{3}H-methionine and incorporation of ^{3}H-methionine into protein by fmet-leu-phe. Determination of reactions was performed in accordance with previously described procedures (12).

five-fold higher concentration of cells. At this cell density, hydrolysis of the attractant may reduce its concentration significantly (14).

Correlation between inhibition of carboxylmethylation and inhibition of chemotaxis

Agents that inhibit the methylation reaction also inhibited chemotaxis (Table II). Carbobenzoxy-Phenylalanyl-Methionine (Z-Phe-Met) an antagonist of chemotactic peptides at the receptor level, eliminated both stimulated methylation as well as chemotaxis. Additionally, 5 mM EGTA, which does not affect binding to the receptor, inhibits the stimulation of methylation by about 70 percent and also eliminated chemotaxis. Ca^{++} appears to be required for optimal methylation reaction and has been shown previously to be required for chemotaxis (15).

Subcellular localization of methylase activity and substrate

The presence of the protein carboxylmethylase (PCM) and its substrate, methyl acceptor protein (MAP), were identified in subfractions of rabbit neutrophils (Table III). In the presence of added gelatin the 30,000 g supernatant fraction contained higher enzyme activity than the pellet. However, in the absence of added substrate, the specific activities of both subfractions were about equal with the particulate

TABLE I

STIMULATION OF NEUTROPHIL PROTEIN CARBOXYMETHYLATION

BUT NOT PROTEIN SYNTHESIS BY A LEUKOTACTIC PEPTIDE

Effect	% of control[a]	
	Cycloheximide absent	Cycloheximide added
Protein carboxymethylation	217 ± 23	232
^{3}H-incorporation into protein	124 ± 7	22

[a] Control value (minus fMet-Leu-Phe) for ^{3}H-methionine incorporation into protein was approximately 20 pmol per 10^6 and for protein carboxylation approximately 20 fmol per 10^6 cells of ^{3}H-methanol liberated.

fraction containing the greater total activity. It appears that the particulate fraction contains an effective substrate. In the presence of added PCM from beef adrenal glands the MAP content of the supernatant fraction was shown to be almost three-fold that of the pellet. This indicates that the endogenous PCM in the leukocyte is specific for certain substrates. Addition of attractant to an homogenate of leukocytes did not stimulate protein carboxymethylation activity.

Alterations in phospholipid metabolism induced by chemoattractants

We also studied the effects of an attractant upon alterations in the methylation of phospholipids (Fig. 3). When ^{3}H-methionine and attractant were added simultaneously to cells and allowed to incubate for 30 min (Fig. 3a), an initial decrease in the rate of methylation (25 percent at 10 min) was observed. The reaction then accelerated slightly suggesting a small stimulation by the attractant, and then continued at a rate parallel to that in the unstimulated cells. To determine whether the attractant stimulated the degradation of the phospholipids or inhibited their synthesis, we labelled cells with ^{3}H-methionine and subjected them to an additional incubation with unlabelled methionine in both the presence and absence of attractant (Fig. 3b). The attractant

TABLE II

INHIBITION OF STIMULATED PROTEIN CARBOXYMETHYLATION

BY NON-CHEMOTACTIC ANTAGONISTS

Additions	% Control Carboxymethylation[a]	Inhibition of CTX[b]
Control	100	-
FMet-Leu-Phe 10^{-8}M	142	-
ZPhe-Met 10^{-4}M	92	+
FMET-Leu-Phe 10^{-8}M + ZPhe-Met 10^{-4}M	89	+
EGTA 5 mM	113	+

[a] PCM control was 23 fmol. ^{3}H Methanol liberated per 10^6 cells.

[b] Chemotaxis determinations done with modified Boyden assay ( ).

stimulated the release of labelled methyl groups by 35 percent after 10 min. The results indicated that the chemoattractant increased the degradation of phospholipids, rather than inhibiting their synthesis. We have found that under these conditions phospholipid synthesis, as measured by incorporation of choline, is not stimulated by attractants (not shown). Therefore, the incorporation of ^{3}H-methionine (Fig. 3a) is undoubtedly the result of methylation of the preformed phospholipids themselves.

The stimulated release appears to be specific in that the dose response curve for this reaction shows a close correlation with that observed for chemotaxis (not shown). In addition mepacrine, a specific inhibitor of phospholipase A_2, inhibited both the disappearance of methyl groups and chemotaxis (not shown). The results indicate that chemoattractants increase the catabolism of phospholipid components and that these reactions are required for chemotaxis.

Role of transymethylation in chemotaxis

It is known that the transfer of methyl groups from AdoMet, the probable intermediate in methylation, to methyl acceptors is inhibited by S-adenosylhomocysteine (AdoHcy) (16) (Fig. 4). This compound is cleaved to Ado and Hcy by the enzyme AdoHcy hydrolase (16) which maintains equilibrium conditions markedly in the direction of synthesis ($K_e \sim 10^{-6}$).

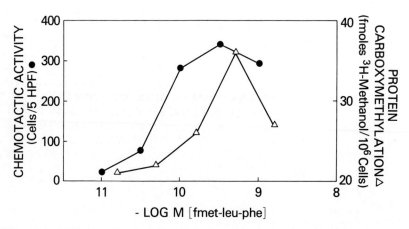

Fig. 2. Effect of varying concentrations of fmet-leu-phe on leukocyte carboxymethylation in the case of protein carboxymethylation reactions were terminated sixty seconds after the addition of peptide and ^{3}H-methionine. Chemotactic response and carboxymethylation were normalized to the same number of cells (10^7 cells/ml) (12).

Where this enzyme is inhibited, a rise in AdoHcy and a depression of both methylation and chemotaxis would be expected. Accordingly, we determined the effect of inhibition of this enzyme upon the accumulation of metabolities of methionine, and upon both methylation and chemotaxis.

1. Alterations in chemotaxis and the metabolism of methionine in cells treated with 3-deazaadenosine (DZA)

Incubation of cells with DZA, a specific inhibitor of AdoHcy hydrolase (17) resulted in a three-fold increase of AdoHcy over control cells within 1.5 min (Fig. 5). The levels of AdoMet were not changed. The increased levels of AdoHcy in the presence of DZA were not maintained long (Table IV), but with added Hcy a slow accumulation of 3-deazaadenosylhomocysteine (DZAHcy) occurred. The formation of this compound may result from the lack of full inhibition of AdoHcy hydrolase, with added DZA acting as a substrate. In this respect it has been found that AdoHcy hydrolase, isolated from neutrophils, was not irreversibly inhibited by DZA (18). These results indicate that certain inhibitors of transmethylation, AdoHcy and DZAHcy, are formed in the presence of DZA and that the total level of purine riboside-homocysteine conjugates (AdoHcy + DZAHcy) is increased in the presence of the inhibitor, under these conditions. AdoMet, however, appears to be turned over rapidly.

2. Inhibition of methylation

It might be expected that increases in intracellular levels of inhibitors of transmethylation would cause the

TABLE III

PROTEIN CARBOXYMETHYLASE (PCM) AND METHYL-ACCEPTOR PROTEIN (MAP)[a]

IN RABBIT PERITONEAL NEUTROPHILS

Fraction	PCM				MAP	
	-gelatin		+gelatin			
	activity[b]	% total	activity[b]	% total	activity[b]	% total
SUPERNATANT (30,000 x g)	3.5	27.3	21.0	67.4	12.4	71.6
PELLET (30,000 x g)	3.2	72.7	3.5	32.6	1.0	28.4

[a] PCM activity was determinent in presence of an express of exogenous MAP (gelatin)
MAP activity was determinent in presence of an express of exogenous PCM

[b] activity is expressed as number of fmoles/10 min/μg protein

decreased methylation of cellular substrates. In Table V, the effects of pretreating cells with DZA, Hcy, and DZA + HCY upon both protein carboxyl and lipid methylation in the absence of attractant are shown. The methylation in both proteins and lipids was reduced 25, 35 and almost 50 percent by pretreating cells with DZA, Hcy and DZA + Hcy respectively. The largest degree of inhibition observed, produced by the combination of DZA and Hcy, is consistent with the formation of intracellular DZAHcy as described above. In addition, treatment of cells with mepacrine, a phospholipase A_2 inhibitor, caused an 80 percent inhibition of lipid methylation.

3. Inhibition of chemotaxis

3-Deazaadenosine inhibits chemotaxis completely at a level of 15 µM (Table VI), but to an extent of 60 percent at 10 µM. The addition of 1 mM Hcy to 10 µM DZA increased the inhibition of chemotaxis (75 percent). This effect is consistent with the previously demonstrated inhibition by DZA of AdoHcy hydrolase isolated from neutrophils (18). 2-chloroDZA, found to be inactive against the hydrolase, was also not found to inhibit chemotaxis. The concentration of DZA producing half-maximal inhibition was estimated to be 0.5 µM (Fig. 6). The inhibition of methylation by DZA (Table V) was clearly demonstrated at 100 µM. This, however, represents the inhibition of all methylation reactions occurring in the cell. The stimulated methylation represent only a small proportion of this, especially during the chemotactic assay (2 hrs).

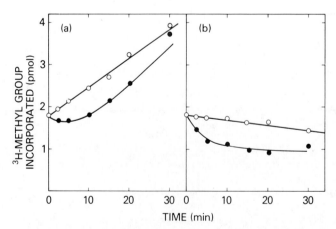

Fig. 3. Stimulated degradation of methylated phospholipids by fmet-leu-phe. (a) Attractant and ^{3}H-methionine added simultaneously to leukocytes. (b) Attractant added to leukocytes prelabeled with ^{3}H-methionine.

● - with attractant
o - without attractant

Incorporation of ^{3}H was determined as previously described (23).

It can also be seen (Fig. 6) that inhibition of chemotaxis by DZA was not due to an effect on the binding of attractant by a receptor. We have also found (not shown) that chemotaxis in cells exposed to DZA or DZA + HCY was irreversibly inhibited and could not be restored by washing the cells even though the extracted neutrophil AdoHcy hydrolase was not irreversibly inhibited (18). In other studies (Venkatasubramanian et al., unpublished data), it has been found that chemotaxis in neutrophils was inhibited by the combination of an adenosine deaminase inhibitor, adenosine and Hcy. This would be expected to raise intracellular levels of AdoHcy (19,20).

In agreement with the inhibitory effects on methylation of mepacrine, the phospholipase A_2 inhibitor, we have found that this compound also inhibits chemotaxis (half-maximal effect at 8×10^{-5} M not shown). This suggests that the formation of lysophospholipids may have a role in chemotaxis.

DISCUSSION

It appears that there is an important role in leukotaxis for both protein carboxylmethylation and the methylation of phospholipids. Subcellular particulate fractions have been shown to contain both enzyme and substrate for the carboxymethylation. Changes in the levels of these protein and lipid components in neutrophils are specifically stimulated

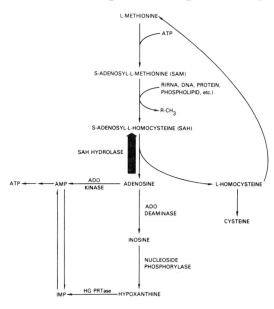

Fig. 4. Schematic diagram of metabolism of S-adenosylmethionine (AdoMet) (20).

by formylmethionyl peptide attractants. The specificity of this effect is indicated by the close correlation between the dose-response data of peptide-induced chemotaxis and methylation and by the inhibition of both these stimulated responses by specific peptide antagonists of the formylated attractants. Neither the uptake of methionine nor its incorporation into protein is altered by the attractant. The methylation reactions appear to be stimulated subsequent to the initial interaction between agonist and cell binding site as judged by the ability of specific inhibitors of methylation to block the effect.

Our results suggest that AdoHcy hydrolase has an important role in the regulation of chemotaxis. If the level of AdoHcy rises with respect to that of AdoMet, inhibition of transmethylation is favored. Therefore, the inhibition of the hydrolase by an agent such as DZA would result in an increase in AdoHcy as a consequence of its generation during transmethylation by AdoMet. The formation of DZAHcy in the presence of DZA would provide an additional inhibitor of transmethylation. The studies of Pike et al (20) have shown that conditions which cause increases in the products of the hydrolase reaction, adenosine and Hcy (as in adenosine deaminase inhibition) can result in concomitant increases in AdoHcy because the equilibrium constant is markedly in the direction of synthesis. There is evidence, therefore, for two modes of regulation of methylation reaction (and, as a result,

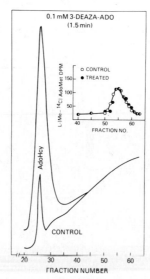

Fig. 5. The effect of 3-Deazaadenosine (DZA) upon the formation of adenosylhomocysteine and adenosylmethionine in leukocytes. The metabolites of methionine were determined by high performance liquid chromatography (18).

chemotaxis), by AdoHcy hydrolase and Ado deaminase, respectively: increase in AdoHcy by direct inhibition of the enzyme and increase in AdoHcy by failure to remove products of hydrolysis of AdoHcy during Ado deaminase inhibition.

The manner in which these methylation reactions participate in the chemotactic response is not clear. In the case of carboxylmethylation, it has been suggested in other systems that this process may play a role in stimulating secretion, as of catechol amines in the adrenal medulla (21). Here it was found that chromaffin vesicles membranes contained a high proportion of methyl acceptor protein. It was postulated that catecholamine stimulation of the medulla led to methylation of the vesicle membrane, reducing its negative surface potential and allowing the vesicle to fuse with the plasma membrane. This could facilitate exocytosis of vesicular contents. Gallin et al (22) have suggested that in the neutrophil an analogous process might occur in which new chemotaxis receptors could be brought to the cell surface by chemotactically stimulated fusion of specific granules with the plasma membrane. These granules have been shown to contain a high proportion of attractant-binding sites compared to other organelles. Another possible role of carboxymethylation could be reversible esterification of cell surface membrane protein carboxyl groups as in the case in bacterial chemotaxis (10,11). It is conceivable that such a reversible reaction may provide rapid conformational changes in proteins linked to contractile elements.

TABLE IV

LEVELS OF ADOHCY AND 3-DEAZA-ADOHCY IN RABBIT NEUTROPHILS INCUBATED WITH 0.2 mM 3-DEAZAADENOSINE AND 0.2 mM HOMOCYSTEINE THIOLACTONE[a]

Time (min)	AdoHcy[b]	3-Deaza-AdoHcy[b]	AdoHcy + 3-Deaza-AdoHcy[b]
		(nmol/33 x 10^6 cells)	
0	185	0	185
15	175	14	189
30	182	49	232
60	142	81	233

[a] Neutrophils were incubated in Gey's balanced salt solution containing 2% bovine serum albumin, 37°C. AdoMet concentration was undetectable under these conditions.

[b] Determinations were made with the aid of high performance liquid chromatography.

In bacteria, protein carboxymethylation has been demonstrated as a requirement for chemotaxis using mutants lacking either a methy-accepting protein or a protein carboxymethylase. Lipids have not yet been shown to play a role in bacterial systems.

It has been demonstrated that other agonists caused alterations in the levels of methylation of the lipid components in their respective target cells (23,24). This was correlated with increases in membrane fluidity. These changes might be expected to cause profound effects upon the cell. Chemoattractants stimulate decreased cell surface charges, rapid changes in membrane polarization, increased cationie fluxes (25) and marked generation and utilization of energy. The production of lysophospholipids, demonstrated in the present work, might be crucial in these events. Such studies may lead to an understanding of how localized membrane perturbations lead to profound changes in the whole cell's behavior.

Whatever the specific roles carboxylmethylation and lipid methylation may have in leukotaxis, it seems evident that a fundamental biochemical process, transfer of the methyl group of methionine to a variety of substrates is involved in cellular motility over a broad spectrum of the phylogenetic scale.

TABLE V

INHIBITION OF METHYLATION

Additions	Per Cent Control	
	Lipid Methylation[a]	Carboxylmethylation[b]
None	100	100
1 mM Hcy	65.3	30.1
0.1 mM DZA	75.4	41.3
1 mM Hcy + 0.1 mM DZA	47.5	21.3

[a] Control ^{3}H CPM in Lipid, $27284/10^7$ cells.
[b] Control ^{3}H CPM in Protein, $4201/10^7$ cells.

TABLE VI

EFFECT OF 3-DEAZAADENOSINE ON RABBIT NEUTROPHIL

CHEMOTAXIS WITHOUT PRE-INCUBATION[a]

Additions	% Inhibition of chemotaxis
A. 3-Deazaadenosine	
1 μM	0
1 μM + 1 mM Hcy-thiolactone	0
10 μM	60
10 μM + 1 mM Hcy-thiolactone	75
15 μM	100
15 μM + 1 mM Hcy-thiolactone	100
B. 2-Chloro-3-deazaadenosine	
1 mM	0
1 mM + 1 mM Hcy-thiolactone	0

[a]Chemotactic activity was assayed in the presence of 10^{-9} M fMet-Leu-Phe for 2 hours; 3-deazaadenosine and L-Hcy-thiolactone were added simultaneously.

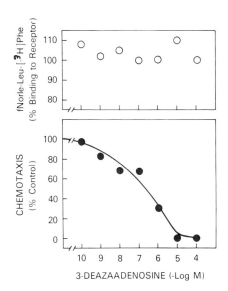

Fig. 6. The effect of 3-Deazaadenosine upon chemotaxis and binding of fNorle-Leu-^{3}H-Phe to leukocytes. Chemotaxis and binding were performed as previously described (4) (7).

REFERENCES

1. Snyderman, R., Phillips, J.K., and Mergenhagen, S.E. (1970). Infec. Immunity 1, 521-525.
2. Turner, S.R. and Lynn, W.S. (1978). In "Leukocyte Chemotaxis" (J.I. Gallin and P.G. Quie, eds.), pp. 289-298. Raven Press, New York.
3. Wilkinson, P.C. (1974). "Chemotaxis and Inflammation" p. 54. Churchill, Livingstone. Edinburgh and London.
4. Schiffmann, E., Showell, H.U., Corcoran, B.A., Ward, P.A., Smith, E. and Becker, E.L. (1975). J. Immunol. 114, 1831-1837.
5. Schiffmann, E., Corcoran, M.A. and Wahl, S.M. (1975). Proc. Natl. Acad. Sci. U.S.A. 72, 1059-1062.
6. Showell, H.J., Freer, R.J., Zigmond, S.H., Schiffmann, E., Aswanikumar, S., Corcoran, B.A. and Becker, E.L. (1976). J. Exp. Med. 143, 1154-1169.
7. Aswanikumar, S., Corcoran, B.A., Schiffmann, E., Day, A.L., Freer, R.J., Showell, H.J., Becker, E.L. and Pert, C.B. (1977). Biochem. Biophys. Res. Commun. 74, 210-217.
8. Williams, L.T., Snyderman, R., Pike, M.C. and Lefkowitz, R.J. (1977). Proc. Natl. Acad. Sci. U.S.A. 74, 1204-1208.
9. Aswanikumar, S., Schiffmann, E., Corcoran, B.A. and Wahl, S.M. (1976). Proc. Natl. Acad. Sci. U.S.A. 73, 2439-2442.
10. Kleene, S.J., Toews, M.L. and Adler, J. (1977). J. Biol. Chem. 252, 3214-3218.
11. Werf, P. Van der and Koshland, D.E., Jr. (1977). J. Biol. Chem. 252, 2793-2795.
12. O'Dea, R.F., Viveros, O.H., Axelrod, J., Aswanikumar, S., Schiffmann, E. and Corcoran, B.A. (1978) Nature 272, 462-464.
13. Hirata, F. and Axelrod, J. (1978). Nature 275, 219-220.
14. Aswanikumar, S., Schiffmann, E., Corcoran, B.A. and Wahl, S.M. (1976). Proc. Natl. Acad. Sci. U.S.A. 73, 2439-2442.
15. Gallin, J.G. and Rosenthal, A.S. (1974). J. Cell Biol. 62, 594-609.
16. De La Haba, G. and Cantoni, G.L. (1959). J. Biol. Chem. 234, 603-608.
17. Chiang, P.K., Richards, H.H. and Cantoni, G.L. (1977). Molec. Pharmacol. 13, 939-947.
18. Chiang, P.K., Venkattasubramanian, K., Richards, H.H., Cantoni, G.L. and Schiffmann, E. (1979). "Conference on Transymethylation: (E. Usdin, C.R. Creveling, and R. Borchardt, eds.). Elsevier North Holland, New York.

19. Kredich, N.M. and Martin, D.V., Jr. (1978). Cell 12, 931-938.
20. Pike, M.C., Kredich, N.M., Snyderman, R. (1978). Proc. Natl. Acad. Sci. U.S.A. 75, 3928-3932.
21. Diliberto, E.J., Jr., Viveros, O.H. and Axelrod, J. (1976). Proc. Natl. Acad. Sci. U.S.A. 73, 4050-4054.
22. Gallin, J.I., Wright, D.G. and Schiffmann, E. (1978) J. Clin. Invest. 62, 1364-1374.
23. Hirata, F., Strittmatter, W.J. and Axelrod, J. (1979). Proc. Natl. Acad. Sci. U.S.A. 76, 368-372.
24. Hirata, F., Axelrod, J. and Crews, F.T. (1979). J. Biol. Chem. (in press).
25. Becker, E.L., Showell, H.J., Naccache, P.H. and Sha'afi, R. (1978) In "Leukocyte Chemotaxis" (J.I. Gallin and P.G. Quie, eds.), pp. 113-121. Raven Press, New York.

ENZYMATIC FORMATION OF CYCLIC CMP BY MAMMALIAN TISSUES[1]

Louis J. Ignarro and Stella Y. Cech

Department of Pharmacology, Tulane University School of
Medicine, New Orleans, Louisiana 70112

ABSTRACT Cyclic CMP (cytidine 3',5'-monophosphate) was identified as a product of an enzymatic reaction involving mammalian tissue, CTP and a divalent cation ($Fe^{2+} > Mn^{2+} > Mg^{2+} > Ca^{2+}$). Cytidylate cyclase activity was localized primarily to the total sedimentable fraction, and less than 15% of homogenate activity was present in the soluble fraction. The sedimentable fraction lost activity upon repeated washing but regained full activity upon addition of soluble fraction. Cyclic CMP was identified as a reaction product by comparison with authentic compound in several analytical systems: column chromatography on Dowex 1-formate and on PEI cellulose; thin layer chromatography; crystallization to constant specific activity. The reaction was enzymatic in that product formation was pH-, temperature-, time- and substrate-dependent, and was inhibited by proteolytic digestion, acid or alkaline pH, boiling, high ionic strength and organic solvents. Using 0.3 mM Mn^{2+} as cation, the K_m and V_{max} were 0.19 mM and 0.98 nmole/min, respectively. Cyclic CMP formation was stimulated by homogenates of: mouse liver, spleen, kidney, heart and brain; livers from rat, guinea pig and rabbit; mouse myeloid leukemic tumors. Mouse liver cytidylate cyclase activity was inhibited by certain reductants and stimulated by certain oxidants.

INTRODUCTION

The natural occurrence of cyclic CMP in mammalian cells was first reported by Bloch (1), who also demonstrated that

---

[1] This work was supported by grants from: The Cancer Assoc. of Greater New Orleans; the Edward G. Schlicder Foundation; Smith Kline & French; Merck Sharp & Dohme; and Lilly.

cyclic CMP abolishes the temperature-dependent lag phase and stimulates resumption of growth of leukemia L-1210 cells. The occurrence of cyclic CMP in mammalian tissues was recently confirmed by specific radioimmunoassay (2). Last year, Kuo and co-workers (3) and Cheng and Bloch (4) identified phosphodiesterase activity with specificity for cyclic CMP in mammalian tissues. Studies from the above laboratories revealed that cellular cyclic CMP levels were higher, and cyclic CMP phosphodiesterase activities were lower, in more rapidly proliferating than in quiescent cells. Recent studies from this laboratory demonstrated that mammalian tissues contain cytidylate cyclase activity and can, therefore, synthesize cyclic CMP from CTP (5,6). The capacity of mammalian tissues to synthesize cyclic CMP could not be confirmed by Gaion and Krishna (7), who claimed that 5'-CMP and 5'-CDP, rather than cyclic CMP, were the reaction products formed employing our assay procedure. Data illustrated in the present report indicate that: cyclic CMP can be formed by mammalian tissues; cyclic CMP formation is enzymatic in nature; the cyclic CMP product formed in the reaction with homogenate appears to be bound to protein and this property markedly alters the elution profiles of cyclic CMP on neutral alumina and Dowex-50(H^+) column chromatography.

METHODS

Preparation of tissue homogenates and determination of cytidylate cyclase activity were described previously (6). Briefly, cytidylate cyclase activity was determined by incubation (37°C for 10 min) of homogenate (2-3 mg protein) in 1.0 ml of reaction medium containing 40 mM Tris HCl (pH 7.4), 1 mM CTP, 3×10^5 cpm of α-[32-P] CTP, 0.3 mM Me^{2+} (usually Mn^{2+} or Fe^{2+}) and 3×10^4 cpm of [3-H] cyclic CMP (to monitor product recoveries). Reactions were terminated by addition of 0.1 ml of ice-cold 60 mM EDTA and cooling the samples on ice. Reaction mixtures were then chromatographed on properly aged and humidified neutral alumina columns (6). Alumina chromatography resulted in the retention of over 99.99% of added CTP, and blank values were less than 0.005% of total added CTP when the α-[32-P] CTP was purified on columns (6 cm x 0.7 cm) of Dowex-50(H^+) prior to use. Chromatography of authentic CTP, CDP, CMP, cytidine 2',3'-monophosphate and cyclic CMP, in the presence of tissue, on columns of neutral alumina revealed that only cyclic CMP was recovered in the first 3 ml of eluate after elution with 50 mM Tris HCl, pH 7.6. Recoveries of [3-H] cyclic CMP were 60-75%. Specific methods employed in this study are described in the section on RESULTS, as they are applied in certain experiments.

RESULTS

Presence and Distribution of Cytidylate Cyclase Activity in Mouse Liver.

Cytidylate cyclase activity was first identified in mouse liver homogenates when the freshly prepared homogenate fraction was incubated at 37°C (pH 7.4) in the presence of 0.1 mM or 1 mM CTP and 0.3 mM Mn^{2+} or 0.3-1 mM Fe^{2+} (5,6). A series of experiments revealed that (a) cyclic CMP formation was linear for 10 min at 37°C, (b) cyclic CMP formation was linear throughout a protein concentration range of 0.25-3.6 mg per ml of reaction volume incubated at 37°C for 10 min, (c) the optimal pH for the reaction was 7.4-7.6, (d) the optimal temperature for the reaction was 37°C, (e) nonionic detergents such as 0.1-1% Triton X-100 or Tween-20 markedly inhibited cyclic CMP formation, and (f) less than 15% of total homogenate activity was found in the soluble fraction (supernatant from 105,000g for 60 min). The concentration of CTP (0.1-1 mM) remained relatively constant over 5-10 min of incubation at 37°C with up to 4 mg of homogenate protein. This was estimated by determining the amount of α-[32-P] CTP eluting with authentic CTP on columns of Dowex-50(H^+) and PEI cellulose. Moreover, the inclusion of a powerful CTP regenerating system (10 mM phosphocreatine and 200 Units/ml of phosphocreatine kinase) did not affect product formation under any condition tested. Analysis of the subcellular distribution of cytidylate cyclase activity in mouse liver (Fig. 1) indicated that approximately 68% of the total homogenate activity was associated with the total sedimentable or particulate fraction and about 12% was found in the soluble fraction. The particulate enzyme activity was distributed throughout the three sedimentable fractions tested, which were the nuclear-plasma membrane fraction (900g x 10 min once washed sediment), mitochondrial-lysosomal-peroxisomal fraction (subsequent 9,000g x 20 min once washed sediment) and the microsomal fraction (subsequent 105,000g x 60 min once washed sediment).

Identification of Cyclic CMP as the Reaction Product.

Cyclic CMP was identified as the product of the reaction by comparison with authentic compound in several different analytical systems. In each type of product verification experiment, reactions of mouse liver homogenate, 1 mM CTP and 0.3 mM Mn^{2+} were conducted in a 1 ml volume at 37°C (pH 7.4) for 10 min. Product formation was terminated by addition of 0.1 ml of ice-cold 60 mM EDTA and samples were chromatographed on

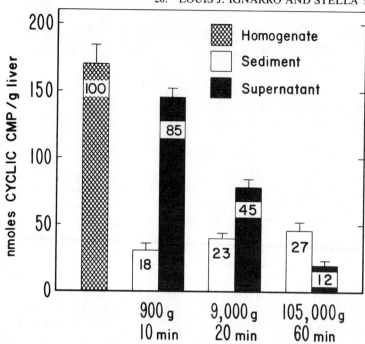

FIGURE 1. Subcellular distribution of cytidylate cyclase activity in mouse liver. Liver was homogenized in 50 mM Tris HCl (pH 7.4) - 0.25 M sucrose and filtered through 50 μm pore nylon. Each sediment was washed once and the resulting supernatant combined with original supernatant. Reactions were conducted with 0.3 mM Mn^{2+}. Numbers within bars signify % of total homogenate activity. Data represent the mean ± S.E.M. of 9 determinations from 3 separate experiments.

neutral alumina (0.5 g in a 0.7 cm diam. column). Alumina eluates (3 ml) were lyophilized, reconstituted in 0.5-1 ml of H_2O and used in the product verification procedures described below. The data illustrated in Figs. 2 and 3 indicate that the [32-P] reaction product displayed similar elution profiles to [3-H] cyclic CMP in the presence of excess unlabeled cyclic CMP during column chromatography on Dowex 1-formate and PEI cellulose. In other experiments using Dowex 1-formate column chromatography, elution of columns with 10 ml of 0.03 N formic acid prior to elution with 0.5 N formic acid failed to release any [3-H] or [32-P] from the columns. Thus, our data are not in agreement with those of Gaion and Krishna (7), where [32-P] eluted in the fractions containing 0.01-0.04 N formic acid. In the PEI cellulose column chromatographic procedure, [3-H] cyclic CMP and [32-P] reaction product eluted with 0.05 M LiCl (Fig. 3). Experiments with

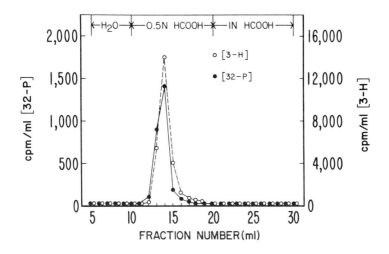

FIGURE 2. Dowex 1-formate column chromatography of eluates from neutral alumina columns. One ml of reconstituted, lyophilized eluate from alumina chromatography of standard reaction mixtures was chromatographed on columns (4 cm x 0.8 cm) of Dowex 1-formate as indicated. Data represent mean values from 5 separate experiments.

unlabeled compounds, using spectrophotometric methods of analysis, indicated that CMP, CDP and CTP eluted with 0.3 M, 1.0 M and 1.5 m LiCl, respectively. Thus, it is unlikely that the [32-P] reaction product was CMP, CDP or CTP. Thin layer chromatography of neutral alumina eluates (after centrifugation and lyophilization) on PEI cellulose plates revealed that authentic cyclic CMP (20 μmoles), [3-H] cyclic CMP and [32-P] reaction product migrated to exactly the same spot (Rf 0.34), after ascending chromatography in 0.1 M LiCl for a distance of 15 cm. Authentic CMP, CDP and CTP (20 μmoles each) yielded Rf values of 0.08, 0.03 and zero, respectively. Thus, again, the data indicate that the [32-P] reaction product was cyclic CMP.

The data in Table 1 indicate that the [32-P] reaction product and [3-H] cyclic CMP crystallized in the presence of excess unlabeled cyclic CMP to constant specific activity, thus, revealing the very similar chemical properties between

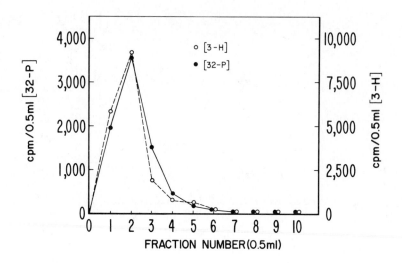

FIGURE 3. PEI cellulose column chromatography of eluates from neutral alumina columns. 0.5 ml of reconstituted, lyophilized eluate from alumina chromatography of standard reaction mixtures was chromatographed on columns (2 cm x 0.6 cm) of PEI cellulose, using H_2O as eluant. Data represent mean values from 5 separate experiments.

the [32-P] reaction product and authentic cyclic CMP. Similar experiments using excess unlabeled CMP, CDP, CTP, cytidine 2',3'-monophosphate or uridine 3',5'-monophosphate failed to demonstrate crystallization of either [3-H] cyclic CMP or [32-P] product to constant specific activity. Indeed, crystals resulting from the third crystallization with the above pyrimidines contained less than 50 cpm per 10 mg of crystals. These data indicate that the [32-P] reaction product was cyclic CMP.

Incubation of lyophilized alumina eluates of reaction mixtures (EDTA was omitted; reactions were terminated by cooling on ice), reconstituted in 10 mM Tris HCl (pH 8.5) containing 5 mM Mg^{2+}, with 1 mg/ml of snake venom 5'-nucleotidase for 60 min at 37°C failed to metabolize either [3-H] cyclic CMP or [32-P] product, as determined by subsequent column or thin layer chromatography on PEI cellulose. Since snake venom 5'-nucleotidase hydrolyzes CMP, but not

TABLE I
CRYSTALLIZATION OF PRODUCT TO CONSTANT SPECIFIC ACTIVITY

Sample[a]	Specific activity (cpm/mg authentic cCMP)	
	[3-H]	[32-P]
First crystallization	782	135
Second crystallization	709	121
Third crystallization	763	130

[a] Reaction mixtures were chromatographed on neutral alumina columns, the eluates were boiled and centrifuged, authentic cCMP was added, and successive crystallizations conducted in cold H_2O-isopropanol mixtures.

cyclic CMP, to cytidine, these data indicate that the [32-P] reaction product was not CMP. Incubation of lyophilized alumina eluates of reaction mixtures (as described above), reconstituted in 10 mM Tris HCl (pH 7.1), with 1 mg/ml of crystalline bovine pancreatic ribonuclease for 60 min at 37°C failed to metabolize either [3-H] cyclic CMP or [32-P] product, as determined by subsequent PEI cellulose thin layer chromatography. Since ribonuclease opens the phosphodiester ring of 2',3'-cyclic CMP, but not that of 3',5'-cyclic CMP, these data indicate that the [32-P] product was not 2',3'-cyclic CMP. Further, addition of beef heart phosphodiesterase to cytidylate cyclase reaction mixtures fortified with 5 mM Mg^{2+}, after the standard 10 min incubation period at 37°C to generate [32-P] product, failed to metabolize either [3-H] cyclic CMP or [32-P] product to CMP even after 6 hr at 37°C. Under these conditions cyclic GMP and cyclic AMP were hydrolyzed to GMP and AMP, respectively, within 5 min, as determined by either Dowex 1-formate or PEI cellulose column chromatography.

Effects of Cations, Anions, Ionic Strength, Solvents, pH and Temperature on Cytidylate Cyclase Activity. The data in Table II indicate the effects of various divalent and two trivalent cations on mouse liver cytidylate cyclase activity. Mn^{2+} and Fe^{2+} were the two most active divalent cations tested and Fe^{3+}, a trivalent cation, was very active. Mg^{2+},

TABLE II
EFFECTS OF CATIONS ON CYTIDYLATE CYCLASE ACTIVITY

Cation added[b]	Cytidylate cyclase activity[a] (pmoles cCMP/min/mg protein)				
	Concentration of cation (mM)				
	0	0.1	0.3	1	3
No addition	51				
Mn^{2+}		82	221	206	11
Mg^{2+}		62	105	92	26
Ca^{2+}		50	82	80	22
Fe^{2+}		79	225	576	246
Fe^{3+}		90	362	883	738
La^{3+}		52	50	112	73
Co^{2+}		48	49	65	12
Ni^{2+}		53	44	37	18
Cu^{2+}		54	41	33	22
Zn^{2+}		50	43	26	15
Ba^{2+}		52	48	44	36

[a]Reactions using mouse liver homogenate were conducted as described in the text. Data represent mean values of triplicate determinations from 3 separate experiments.

[b]The salts used were Mn^{2+}, Mg^{2+} and Ca^{2+} acetate; Fe^{2+} and Fe^{3+} ammonium sulfate; La^{3+}, Co^{2+}, Ni^{2+}, Cu^{2+}, Zn^{2+} and Ba^{2+} chloride.

Ca^{2+} and La^{3+} showed some activity whereas Co^{2+}, Ni^{2+}, Cu^{2+}, Zn^{2+} and Ba^{2+} were essentially without effect. Activities of Mn^{2+} and Ca^{2+} at 0.3 mM were similar to those at 1 mM, but inhibition occurred at 3 mM. Fe^{2+}, on the other hand, showed maximal activity at 1 mM, as did La^{3+}. Most of the experiments to be described were conducted with Mn^{2+} and Fe^{2+}, but not with Fe^{3+} because the latter yielded variable effects from one experiment to another.

Although the data are not illustrated, various anions were tested and some were observed to inhibit cytidylate cyclase activity at concentrations above 1 mM. One to 10 mM concentrations of NaF, KBr, KI, Na_2SO_4 and KNO_3 elicited no observable effects. $NaNO_2$, $Na_2S_2O_4$ and NaCN, respectively,

TABLE III
EFFECTS OF PROTEASES AND pH ON CYTIDYLATE CYCLASE ACTIVITY

Treatment[b]	Cytidylate cyclase activity[a] (pmoles cCMP/min/mg protein)	
	0.3 mM Mn^{2+}	0.3 mM Fe^{2+}
None	246	262
Trypsin 10 mg/ml	61	68
Chymotrypsin 10 mg/ml	43	51
Pronase 10 mg/ml	11	9
None	237	243
pH 2	21	23
pH 4	56	64
pH 6	107	121
pH 7.4	201	210
pH 8	193	198
pH 10	60	63
pH 12	25	20

[a] Reactions using mouse liver homogenate were conducted as described in the text. Data represent averaged values of triplicate determinations from a single experiment for the protease and for the pH experiments.

[b] Proteases (at pH 7.4) and 50 mM buffers of varying pH were preincubated with homogenate at 37°C for 60 min prior to standard assay.

produced the following % inhibition of cytidylate cyclase activity at the concentrations indicated: 1 mM zero, 5 mM 42, 10 mM 71; 1 mM zero, 5 mM 65, 10 mM 93; 1 mM zero, 5 mM 14, 10 mM 35. The effect of ionic strength on cytidylate cyclase activity was determined with NaCl and KCl. NaCl and KCl, at 10 mM and 100 mM, did not affect cyclic CMP formation. However, at 1 M, both salts inhibited enzyme activity 95-100%. The following effects on cytidylate cyclase activity were obtained with various solvents at a concentration of 10% v/v: methanol, 62% decrease; ethanol, 90% decrease; acetone, 74% decrease; DMSO, 11% decrease; acetonitrile, 85% decrease; glycerol, 32% increase. At 1% v/v, none of the solvents affected enzyme activity. It is of interest to note

TABLE IV
STABILITY OF CYTIDYLATE CYCLASE UNDER VARIOUS CONDITIONS

Condition[b]	Cytidylate cyclase activity[a] (pmoles cCMP/min/mg protein)	
	0.3 mM Mn^{2+}	0.3 mM Fe^{2+}
None	243	251
Boiled homogenate	0	2
After 1 hr at 0°C	264	278
After 6 hr at 0°C	256	269
After 24 hr at 0°C	219	221
After 1 hr at 37°C	211	220
After 6 hr at 37°C	48	124
After 24 hr at -70°C	136	197
After 72 hr at -70°C	87	126
One freeze-thaw cycle	198	253
Five freeze-thaw cycles	140	219
Ten freeze-thaw cycles	92	194

[a]Reactions using mouse liver homogenate were conducted as described in the text. Data represent mean values of triplicate determinations from 2 separate experiments.

[b]The various conditions signify manipulations of the liver homogenate as prepared, prior to standard assay.

that glycerol was the only solvent that stimulated cytidylate cyclase activity.

Several types of experiments were conducted to ascertain whether the conversion of CTP to cyclic CMP was in fact enzymatic in nature. Some of the data presented above suggest that the reaction was enzymatic. Other data also substantiate this view. For example, treatment of mouse liver homogenates at 37°C (pH 7.4) for 60 min with pronase, trypsin or chymotrypsin destroyed the capacity of homogenate to form product (Table III). Likewise, preincubation of homogenate at low and high pH values greatly reduced homogenate activity (Table III). Boiled homogenate had no activity, and homogenate activity was markedly reduced by incubation at 37°C for 6 hr or by freeze-thawing (Table IV). Homogenate activity was stable for up to at least 6 hr at 0°C, and retained 88-90%

of its original activity after 24 hr at 0°C. Consistently, homogenate activity was reduced after storage at -70°C for 24 hr or longer. Double-reciprocal plots of velocity of product formation vs. substrate concentration (0.01 to 1.0 mM CTP) yielded straight lines (r=0.99) for reactions conducted in the presence of 0.3 mM Mn^{2+} or 1 mM Fe^{2+}. The K_m and V_{max} values in the presence of Mn^{2+} were 0.19 mM and 0.98 nmole/min, respectively. The K_m and V_{max} values in the presence of 1 mM Fe^{2+} were 0.11 mM and 2.62 nmoles/min, respectively. Thus, velocity increased with substrate concentration in the manner required by the Michaelis-Menten principle. The linearity of the plot indicates that a simple interaction occurred between enzyme and substrate and that the enzyme-substrate complex was at a relatively constant concentration and in equilibrium with respect to free substrate and enzyme during the time of incubation.

Effects of Chemical Compounds on Cytidylate Cyclase Activity. A variety of chemical compounds were tested for their effects on cytidylate cyclase activity in mouse liver homogenate (Table V). Generally, reducing agents inhibited whereas oxidizing agents stimulated cytidylate cyclase activity. Agents which activate guanylate cyclase, such as NaN_3 and nitroprusside, inhibited cytidylate cyclase activity. It is of interest that pyrimidine compounds were more potent inhibitors than were purine compounds. Many other compounds were tested at various concentrations and were found to have no appreciable effect on cytidylate cyclase activity. These included a variety of pyrimidine and purine nucleoside mono- and di-phosphates and 3',5'-cyclic monophosphates, catecholamines, muscarinic agonists, prostaglandins of the A, E and F type, other 20-carbon polyunsaturated fatty acids, pyridine nucleotides and glucagon.

Cyclic CMP Formation in Other Tissues. Cytidylate cyclase activity was found also in liver homogenates from the rat, guinea pig and rabbit. Activities in the presence of 0.3 mM Mn^{2+}, expressed as nmole cyclic CMP/min/mg protein (mean ± S.E.M.), were: 0.19 ± 0.013 (rat); 0.23 ± 0.016 (guinea pig); 0.23 ± 0.020 (rabbit). Figure 4 illustrates cytidylate cyclase activities in the liver, kidney, spleen, heart and lung of the rat, using 0.3 mM Mn^{2+} or 1 mM Fe^{2+} as the added divalent cation. Activities were 3 to 10-fold higher with 1 mM Fe^{2+} than with 0.3 mM Mn^{2+}. Figure 5 depicts cytidylate cyclase activity, using 0.3 mM Mn^{2+}, Mg^{2+} or Fe^{2+}, in rapidly proliferating murine myeloid leukemic tumors. Myeloid leukemic tumors, which originated spontaneously in

TABLE V
EFFECTS OF CHEMICALS ON CYTIDYLATE CYCLASE ACTIVITY

Chemical test agent[b]	Cytidylate cyclase activity[a] (pmoles cCMP/min/mg protein)	
	0.3 mM Mn^{2+}	0.3 mM Fe^{2+}
None	227	245
Dithiothreitol 5 mM	205	208
Mercaptoethanol 20 mM	206	202
N-ethylmaleimide 2 mM	181	247
$Na_2S_2O_5$ 4 mM	156	178
$Na_2S_2O_4$ 4 mM	95	92
Ascorbate 2 mM	186	167
Dehydroascorbate 2 mM	287	293
Ferricyanide 1 mM	319	304
Ferrocyanide 1 mM	224	249
H_2O_2 5 mM	236	304
Catalase 5 mg	293	371
H_2O_2 + catalase	342	448
NaN_3 1 mM	118	137
Nitroprusside 1 mM	157	-
UTP 1 mM	54	55
TTP 1 mM	75	162
GTP 1 mM	132	204
ATP 1 mM	143	211
Cytosine arabinoside 1 mM	158	150
β-D-O^2,2'-cyclocytidine 1 mM	120	139

[a]Reactions using mouse liver homogenate were conducted as described in the text. Data represent mean values of triplicate determinations from 2 separate experiments.

[b]Chemical test agents were added at start of assay incubation.

liver and spleen, were obtained as subcutaneous transplants in C57BL mice (Jackson Laboratory), and were transplanted in C57BL/6J or DBA/1J male mice. The data in Fig. 5 indicate that cyclic CMP formation increased markedly during the first 72 hr of proliferation, decreased sharply at 96 hr and oscillated in a consistent manner thereafter.

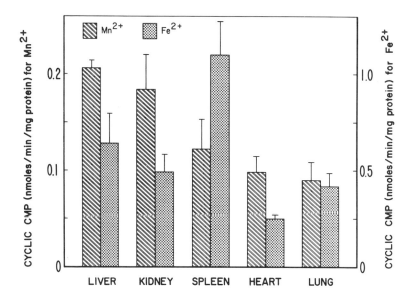

FIGURE 4. Cytidylate cyclase activities in homogenates of various mouse tissues. Reactions were conducted with either 0.3 mM Mn^{2+} or 1 mM Fe^{2+}, as indicated. Data represent the mean ± S.E.M. of 4 separate experiments.

Problems Associated with Cytidylate Cyclase Assay. During the reaction between mouse liver homogenate, divalent cation and α-[32-P] CTP at 37°C, a [32-P] product was formed which was precipitated to the extent of about 75% with TCA or boiling. Chromatography of the incubated reaction mixture on a column of neutral alumina (0.5 g) resulted in the concomitant elution of [32-P] product and added [3-H] cyclic CMP, where the % recovery of [3-H] was 70. However, passage of the alumina eluate through a second neutral alumina column resulted in only a 21% recovery of [32-P], whereas [3-H] recovery was 69% (Fig. 6). A significant amount of protein passed through the columns as well. About 20% of the protein in the first alumina eluate was recovered in the second alumina eluate. The finding that the recoveries of [32-P] and protein were practically identical after the second passage through alumina columns suggested that the [32-P] product was protein bound. Treatment of the first alumina eluate with 1 mg/ml of pronase at 37°C for 30 min resulted in a much higher recovery (53%) of [32-P] product and removal of 99% of the protein (Fig. 6). The [3-H] cyclic CMP remained

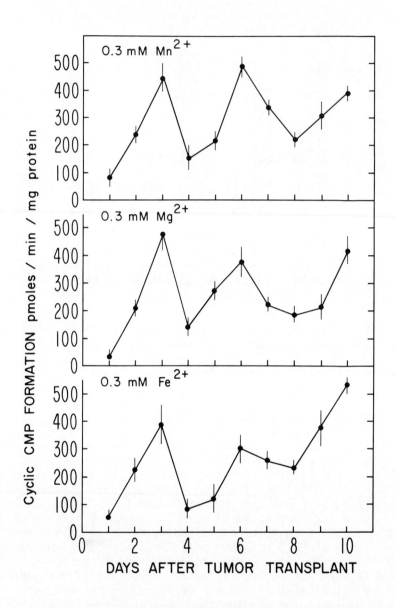

FIGURE 5. Cytidylate cyclase activity in growing transplanted myeloid leukemic tumors in mice. Reactions, using homogenate fractions of excised tumor tissue, were conducted with 0.3 mM Mn^{2+}, Mg^{2+} or Fe^{2+}, as indicated. Data represent the mean ± S.E.M. of 9 determinations from 3 separate experiments.

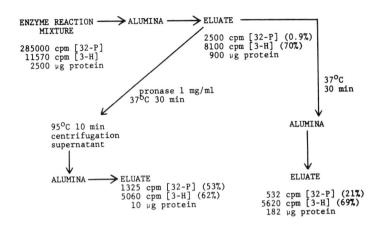

FIGURE 6. Effect of pronase on recovery of reaction product from neutral alumina columns. The enzyme reaction mixture contained the indicated amounts of α-[32-P] CTP, [3-H] cyclic CMP and protein. Eluates were 3 ml and contained the indicated amounts of [32-P] reaction product, [3-H] cyclic CMP and protein. Numbers in parentheses signify: (a) for the first eluate, the % cpm recovered from the enzyme reaction mixtures; and (b) for the second eluate, either with or without pronase treatment, the % cpm recovered from the first eluate. Data represent one of four separate experiments.

unaffected by the various procedures, thus, indicating that only the enzymatically formed [32-P] product was protein bound. Another problem with the assay as described is the failure of the [32-P] product, isolated by alumina chromatography, to elute with authentic cyclic CMP on columns of Dowex-50(H^+), unless the alumina eluate is incubated with pronase at 37°C for 30 min prior to Dowex chromatography (Fig. 7). This problem did not occur when undigested samples were chromatographed on Columns of Dowex 1-formate or PEI cellulose, although more consistent data and slightly higher values for cyclic CMP formation were obtained with pronase-treated samples. Finally, treatment of reaction mixtures (after EDTA) with 10 mg/ml of pronase (37°C for 60 min)

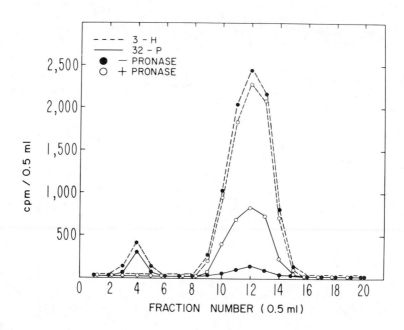

FIGURE 7. Effect of pronase on recovery of reaction product from Dowex-50(H^+) columns. One ml of reconstituted, lyophilized eluate from alumina chromatography of standard reaction mixtures was further incubated at 37°C for 30 min in the presence or absence of 1 mg/ml pronase and then chromatographed on columns (4 cm x 0.8 cm) of Dowex-50(H^+), using H_2O as the eluant. Data represent mean values from 4 separate experiments.

followed by boiling (10 min) and centrifugation (1,000g for 5 min) yielded a clear supernatant which contained unlabeled, [32-P] and [3-H] cyclic CMP, as identified by thin layer chromatography on PEI cellulose.

DISCUSSION

Using several different analytical techniques, cyclic CMP was identified as the product of a reaction at 37°C involving mouse liver homogenate, CTP and Mn^{2+}. The reaction appeared to be enzymatic in that double reciprocal plots of reaction velocity vs. substrate concentration were linear, catalytic activity was destroyed by boiling, proteolytic digestion, freeze-thawing, exposure to acid or alkali and organic solvents, and product formation was dependent on pH, temperature, time and substrate concentration. Thus, it appeared that an enzyme, i.e., cytidylate cyclase, catalyzed the conversion of CTP to cyclic CMP. Cytidylate cyclase activity was distributed to the total particulate or sedimentable fraction, and less than 15% of the total homogenate activity was located in the soluble fraction. The data on the subcellular distribution of cyclic CMP forming activity are difficult to interpret because apparently both the sedimentable and nonsedimentable fractions must be together for cytidylate cyclase activity to be detected. Thus, repeated washing of the total particulate fraction resulted in loss of enzymatic activity, which was restored by adding back the soluble fraction.

Although cyclic CMP formation proceeded to a limited extent in the absence of added divalent cation, addition of Mn^{2+}, Fe^{2+}, Mg^{2+} or Ca^{2+} stimulated cyclic CMP formation. The trivalent cations, Fe^{3+} and La^{3+}, also stimulated cyclic CMP formation. Within a concentration range of 0.1-3 mM, optimal effects of the cations tested were obtained with 0.3-1 mM cation. The actual substrate for cytidylate cyclase may be a substrate-metal complex. More definitive experiments must be conducted before any meaningful conclusions can be drawn. Preliminary data suggest that the substrate is a $CTP-Me^{2+}$ or $CTP-Me^{3+}$ complex and that (a) excess Me is not required for CTP hydrolysis, (b) excess Mn^{2+}, Mg^{2+} or Ca^{2+} is inhibitory and (c) excess Fe^{2+}, Fe^{3+} or La^{3+} is at first stimulatory, then inhibitory at higher concentrations.

Studies on the effects of various chemical agents on cytidylate cyclase activity indicated that cyclic CMP formation was inhibited with reducing agents, stimulated with oxidants, inhibited with nitric oxide forming substances, inhibited with certain pyrimidine and purine nucleotides and unaffected by NaF, glucagon, catecholamines, muscarinic agents and prostaglandins. These data, as well as those on subcellular distribution, indicate that the properties of cytidylate cyclase are much different from those of either adenylate or guanylate cyclase.

Recently, the existence of a phosphodiesterase that hydrolyzes cyclic CMP preferentially was demonstrated by two independent laboratories (3,4). The enzyme was partially purified from rat liver but was present also in many different mammalian tissues. Cyclic CMP phosphodiesterase appears to be a high K_m enzyme and requires mM concentrations of cyclic CMP and a divalent cation for activity. It is of interest that both cyclic CMP phosphodiesterase and cytidylate cyclase are activated markedly with Fe^{2+} and to a lesser extent with both

Mn^{2+} and Mg^{2+} (3,5). The findings of Kuo et al. provide an explanation of why we observed no cyclic CMP product hydrolysis under the experimental conditions employed to determine cytidylate cyclase activity, namely, that the phosphodiesterase requires the presence of a much higher cyclic CMP concentration than was present in our reaction mixtures. Preliminary data from this laboratory indicated that addition to reaction mixtures of 1-5 mM cyclic CMP plus either Fe^{2+} (1-3 mM) or Mn^{2+} (3-5 mM) resulted in loss of linearity of cytidylate cyclase activity with time, i.e., the reaction was not linear, even for 2 min. Incubations in excess of 10 min, under these conditions, resulted in product destruction, which could not be reversed with methylxanthines, thus, supporting the findings of Kuo et al. (3) that cyclic CMP phosphodiesterase is not inhibited with methylxanthines. It is also of interest that, quantitatively, the organ distribution of rat cyclic CMP phosphodiesterase paralleled that of mouse cytidylate cyclase, i.e., liver > kidney > spleen > heart > lung.

The biologic significance of mammalian cyclic CMP is not known presently. Several findings suggest that cyclic CMP is a positive modulator of cell proliferation: (a) cyclic CMP is found in rapidly proliferating tissues (1,2), (b) cyclic CMP stimulates leukemia L-1210 tumor growth (8), (c) cyclic CMP phosphodiesterase activity is lower in rapidly proliferating tissues than in more slowly proliferating tissues (3,9). Cell cyclic CMP formation is more rapid in rapidly proliferating tissues (5, this report), and (e) CTP synthetase and UDP kinase activities are higher in rat hepatomas than in normal liver (10,11).

The inherent technical problems associated with the cytidylate cyclase assay developed in this laboratory has been emphasized by Gaion and Krishna (7). Indeed, the latter investigators claim that our assay procedure does not measure cyclic CMP formation and that they could not detect any cyclic CMP formation using their assay procedures. Experiments from this laboratory, as well as from those of Dr. J. Stone and Dr. A. Bloch (personal communications), indicate that the [32-P] product formed is indeed cyclic CMP, but that it is bound to at least one protein or peptide immediately after formation in homogenate fractions. Apparently, addition of [3-H] cyclic CMP to such fractions does not bind to protein. Protein bound cyclic CMP product does not behave like authentic cyclic CMP on column chromatography using Dowex-50(H^+) or neutral alumina (second passage through alumina). Pretreatment of reaction mixtures or alumina eluates with pronase or mixtures of other neutral proteases results in a [32-P] product which is partially protein-free and partially peptide-bound. Samples treated in this manner show the presence of cyclic CMP product on column chromatography using Dowex-50(H^+) and/or neutral alumina. Preliminary data indicate that the reported (5,6, this report) absolute values for cytidylate cyclase activity (pmoles cyclic CMP/min/mg protein) may be slightly lower than what they should be, because of the problem of protein-bound product formation. Because of the problems associated with determinations of cytidylate cyclase activity by our method, and the report (7) that cyclic CMP formation could not be detected by other methods, the suggestion is made that before any investigator studies cytidylate cyclase activity or cyclic CMP formation by either of the two published procedures (6,7), the technical problems associated with what appears to be protein-bound cyclic CMP should be clearly elucidated. Further, two different procedures should be employed to measure cytidylate cyclase activity, one procedure being a specific radioimmunoassay for cyclic CMP, as the one reported recently by Cailla et al. (2).

ACKNOWLEDGMENTS

The majority of this work was conducted by Stella Y. Cech, in partial fulfillment of the requirements for the Ph.D. degree in Pharmacology.

REFERENCES

1. Bloch, A. (1974). Biochem. Biophys. Res. Commun. 58, 652.
2. Cailla, H. L., Roux, D., Delaage, M., and Goridis, C. (1978). Biochem. Biophys. Res. Commun. 85, 1503.
3. Kuo, J. F., Brackett, N. L., Shoji, M., and Tse, J. (1978). J. Biol. Chem. 253, 2518.
4. Cheng, Y., and Bloch, A. (1978). J. Biol. Chem. 253, 2522.
5. Cech, S. Y., and Ignarro, L. J. (1977). Science 198, 1063.
6. Cech, S. Y., and Ignarro, L. J. (1978). Biochem. Biophys. Res. Commun. 80, 119.
7. Gaion, R. N., and Krishna, G. (1979). Biochem. Biophys. Res. Commun. 86, 105.
8. Bloch, A., Dutschman, G., and Maue, R. (1974). Biochem. Biophys. Res. Commun. 59, 955.
9. Helfman, D. M., Brackett, N. L., and Kuo, J. R. (1978). Proc. Natl. Acad. Sci. USA 75, 4422.
10. Williams, J. C., Weber, G., and Morris, H. P. (1975). Nature 253, 567.
11. Williams, J. C., Kizaki, H., Weber, G., and Morris, H. P. (1978). Nature 271, 71.

CYCLIC CMP PHOSPHODIESTERASE: BIOLOGICAL INVOLVEMENT
AND ITS REGULATION BY AGENTS[1]

J. F. Kuo, Mamoru Shoji, David M. Helfman and
Nancy L. Brackett

Departments of Pharmacology and Medicine,
Emory University School of Medicine,
Atlanta, Georgia 30322

ABSTRACT Occurrence of cyclic CMP phosphodiesterase (C-PDE) was shown recently in a number of mammalian tissues. Levels of C-PDE are generally 2-3 orders of magnitude lower than those of cyclic AMP phosphodiesterase (A-PDE) and cyclic GMP phosphodiesterase (G-PDE). We found that decreased activity levels of C-PDE are invariably associated with tissues undergoing rapid cell proliferation, either physiologic or pathologic, as exemplified by the regenerating liver, the ontogenesis of guinea pig tissues, and the fast-growing Morris hepatoma 3924A. Changes in the activity levels of A-PDE and G-PDE, on the other hand, were variable and tissue-specific, suggesting that catabolism of cyclic CMP may be more critically related to cell proliferation than that of cyclic AMP or cyclic GMP, or both. C-PDE, which requires Fe^{2+} or Mg^{2+} for its maximal activity and has a Km for cyclic CMP of about 2-6 mM, was partially purified from pig or rat livers to remove most of the contaminating A-PDE and G-PDE. Classical phosphodiesterase inhibitors (such as papaverine) are without effects on C-PDE at concentrations that inhibit A-PDE and G-PDE 70-90%. Imidazole stimulates both A-PDE and G-PDE, but not C-PDE. 2'-Deoxy cyclic AMP (a specific inhibitor of A-PDE) and 2'-deoxy cyclic GMP (a specific inhibitor of G-PDE) are poor and non-specific inhibitors of C-PDE. Potassium phosphate and adenosine phosphates (mono-, di-, tri-, and tetra-), while greatly inhibiting C-PDE and slightly inhibiting G-PDE, conversely stimulate A-PDE. The partially purified C-PDE is activated by mercaptoethanol, and this activated (reduced) enzyme requires Mg^{2+} as a metal activator while its activity is now no longer dependent on Fe^{2+}. It is

[1]This work was supported by USPHS Grants HL-15696, CA-23391 and T32-GM-07594.

suggested that the newly identified C-PDE is a potential site of bioregulation particularly in proliferative processes, and of pharmacological interventions as well.

INTRODUCTION

The occurrence of cyclic CMP (cytidine 3':5'-monophosphate) was first reported in leukemia L-1210 cells by Bloch (1-3). He demonstrated that addition of cyclic CMP causes an initiation of growth of leukemic cells in culture, that cyclic CMP is greatly elevated (about 100-fold) in the regenerating rat liver, and that cyclic CMP is detectable in the urine of patients with acute myelocytic leukemia, but not in the normal subjects (2,3). The occurrence of cyclic CMP in biological materials has been confirmed subsequently by Cailla et al. (4) and Murphy and Stone (5), both using radioimmunoassay methods, and by us (6) using the binding assay method. Levels of cyclic CMP (pmol/g fresh tissue) obtained for the normal rat liver by various laboratories are compared as follows: Bloch (3), 1,100-1,500; Cailla et al. (4), 1.0 ± 0.4; Murphy and Stone (5), 100-150; Kuo et al. (6), 200-400. The reasons for this variability of the cyclic CMP values are not clear, probably due to differences in the isolation and measurement procedures employed, and partly due to the demonstrated presence of the "bound" cyclic CMP (5,7) which renders detection of the "total" cyclic CMP impossible by certain procedures. The existence of cytidylate cyclase, an enzyme that catalyzes the formation of cyclic CMP from CTP, has been reported by Cech and Ignarro (8,9). The validity of its assay procedures, however, has been questioned by Gaion and Krishna (10,11), largely based upon a discrepancy in the chromatographic behavior of the reaction product on Dowex 50. This discrepancy appears to result either from an excessive binding of cyclic CMP to protein(s) (5,7) or from the bound cyclic CMP behaving differently from the free cyclic CMP on Dowex 50.

By analogy to the more conventional and better studied cyclic AMP and cyclic GMP systems, the intracellular concentration of cyclic CMP may be regulated by, in addition to the presumed cytidylate cyclase, a phosphodiesterase specific for, or with a high affinity toward, this pyrimidine cyclic nucleotide. This appears indeed to be the case. In the present paper, we describe certain biological involvements of cyclic CMP phosphodiesterase (C-PDE) and the regulation of its activity by certain agents.

METHODS

The C-PDE activity was assayed (12-16) in a reaction mixture containing, in 0.1 ml, Tris/Cl buffer (pH 7.5), 5 µmol; $FeSO_4 \cdot 7H_2O$ or $MgSO_4$, 1.0 µmol; 5'-nucleotidase (snake venom), 20 µg; cyclic [5-^{3}H]CMP (Amersham), 0.1 nmol or 0.1 µmol, containing 80,000 cpm; and appropriate amounts of the enzyme preparations, as indicated. The reaction product, [5-^{3}H]cytidine, was purified by AG1 X-8 as described elsewhere (12-16). The assay methods for cyclic AMP phosphodiesterase (A-PDE) and cyclic GMP phosphodiesterase (G-PDE) were modifications (17) of that originally described by Thompson and Appleman (18), using 10-20 mM Mg^{2+} to activate the enzymes. C-PDE from the rat liver or the pig liver extract was purified through the step of pH 4.5 precipitation (12). Its further purification was carried out by either BioGel HTP treatment (16), Sephadex G-200 gel filtration (19) or DEAE-cellulose chromatography (19).

RESULTS

The rat liver is the first mammalian tissue from which we detected the C-PDE activity (12). The majority (over 90%) of its activity was found in the soluble fraction. When assayed using 1 µM substrate concentration (for the low Km enzyme form), its activity levels (expressed as pmol hydrolyzed/min/g fresh tissue) are about 125, compared to 11,200 for A-PDE and 3,590 for G-PDE. Their activity levels are more comparable when assayed using 1 mM substrate concentration (for the high Km enzyme form). Their values are C-PDE, 113,400; A-PDE, 157,100; G-PDE, 417,900. We found that C-PDE is present in every one of the many rat tissues examined (12), with the highest levels found in the liver, kidney and intestine, and the lowest levels seen in the skeletal muscle, cerebellum, aorta and blood cells (Table I). This can be compared with the highest levels (7,000-12,000 pmol/min/g fresh tissue) of the low K_m form of A-PDE seen in the brain tissues, kidney, lung and intestine, and the highest levels (6,000-10,000 pmol/min/g fresh tissue) of the low K_m form of G-PDE seen in the lung, spleen, intestine and brain tissues.

It appears that C-PDE in the liver extract is of a high K_m species, with an apparent K_m for cyclic CMP of 2.4 mM (Fig. 1D). The absence of a low K_m species of the enzyme is obvious, because its activity is linear at least up to a cyclic CMP concentration of 200 µM (Fig. 1, A and B), and its activity begins to plateau only when the substrate concentration reaches the mM range (Fig. 1C). This is intriguing,

TABLE I
DISTRIBUTION OF C-PDE IN RAT TISSUES[a]

Rat tissue	cCMP (1 μM)	cCMP (1 mM)
Liver	165	106,100
Cerebral cortex	20	9,300
Cerebellum	8	3,100
Lung	17	4,500
Heart	25	17,800
Aorta (thoracic)	8	6,000
Kidney	225	75,100
Spleen	41	19,600
Skeletal muscle	5	4,900
Intestine	122	49,400
Testis	34	17,200
Blood cells (unfractionated)	8	7,500

[a]The 30,000 x g supernatants were used as the enzyme source, assayed using either 1 μM or 1 mM cyclic CMP as substrate. Values given are expressed as pmol hydrolyzed/min/g wet wt. Taken from ref. 12 with permission.

since the low K_m species of A-PDE and G-PDE has been shown to exist in all tissues examined, including the liver (20). Quantitatively similar results are also obtained for the partially purified C-PDE preparations.

Cyclic CMP has been postulated by Bloch (1-3) to function as a positive effector of cell proliferation. Provided that the cytidylate cyclase activity remains unaltered, decreased C-PDE levels in theory would lead to elevated cyclic CMP and thus the accelerated cell proliferation. It is conceivable, therefore, that decreased C-PDE activity levels are likely to be seen in tissues undergoing rapid growth. This appears to be the case. We (15) found that in the cerebral cortex of the developing guinea pig, tissue levels of C-PDE, as those of A-PDE and G-PDE, assayed with both 1 μM and 1 mM substrate concentrations, are the lowest in the fetus (Fig. 2). Their levels, with a possible exception of the low K_m form of G-PDE, increase rapidly as the tissue develops, and reach the adult values at the neonatal stage. Results similar to those seen for the cerebral cortex are also noted for the developing heart. In the kidney, the lowest C-PDE levels are also found in the fetus (Fig. 3). Interestingly, the

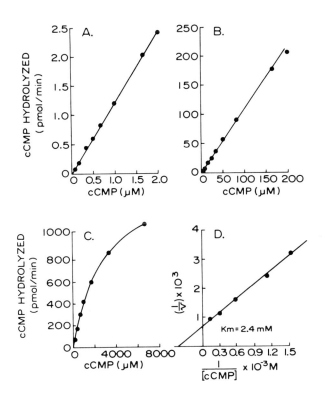

FIGURE 1. C-PDE activity from the rat liver extract as a function of cyclic CMP concentration. The crude extract (385 µg protein) was incubated under the standard assay conditions in the presence of 10 mM Fe^{2+} and varying concentrations of cyclic [5-^{3}H]CMP, as indicated. Reaction velocity (v), pmol hydrolyzed/min. Taken from Ref. 12, with permission.

high K_m form of G-PDE in the kidney from the neonate and the pup increases to levels that are even higher than those of the adult, whereas C-PDE activity (assayed with both substrate concentrations) and those of the low K_m form of A-PDE and G-PDE increase only to intermediate values in the neonate. In the intestine, in contrast to the cerebral cortex, heart and kidney mentioned above, C-PDE is the only enzyme that is depressed in the fetus, which remains fully depressed in the neonate, and increases subsequently as the animals mature (Fig. 4). No changes in the intestinal levels of

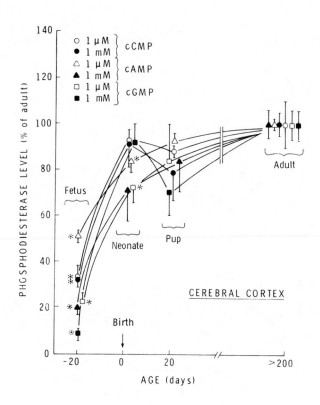

FIGURE 2. Changes in activity level of PDE's in homogenates of the cerebral cortex of developing guinea pigs. The data are presented as % of the enzyme levels seen in the adult tissue, which are taken as 100%. Substrates used were either 1 µM of 1 mM of cyclic CMP, cyclic AMP or cyclic GMP. The individual enzyme level (pmol hydrolyzed/min/g wet wt) in the adult tissue for 1 µM cyclic CMP (○), cyclic AMP (△), and cyclic GMP (□) were 45, 23,160, and 31,300, respectively; the corresponding velues with 1 mM cyclic CMP (●), cyclic AMP (▲), and cyclic GMP (■) were 31,360, 2,426,300, and 2,444,170 respectively. The data presented are the means ± SE from three to six animals. Asterisks, significantly different from the adult (P<0.05 to P<0.0005). Taken from Ref. 15, with permission.

A-PDE and G-PDE were observed during the entire course of development. In addition to the above tissues, we noted that only C-PDE and the high K_m form of G-PDE are depressed in the fetal liver, and that both C-PDE and A-PDE are depressed

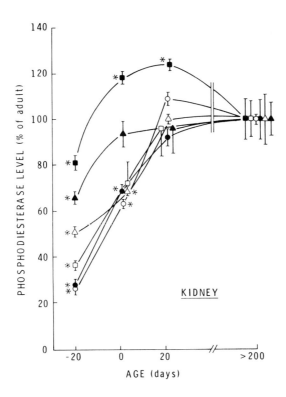

FIGURE 3. Changes in the activity levels of phosphodiesterases in the kidney of developing guinea pigs. Assay conditions, presentation of data, and symbols used in the figure were the same as in Fig. 2. The individual mean enzyme activity levels (pmol hydrolyzed/min/g wet wt) seen in the adult kidney for 1 μM cyclic CMP, cyclic AMP, and cyclic GMP were 109, 17,830, and 22,130, respectively; the corresponding values with 1 mM substrates were 62,950, 682,600, and 524,600, respectively. The data presented are from three to six animals. Asterisks, significantly different from the adult ($P<0.05$ to $P<0.005$). Taken from Ref. 15, with permission.

whereas G-PDE is elevated in the fetal lung (15). It seems that the depressed C-PDE activity in the guinea pig fetal tissues is due to decreased enzyme levels, rather than the presence of a separate species of the enzyme having a lower catalytic activity. This tentative conclusion is based upon the double-reciprocal kinetic analysis, which indicates that,

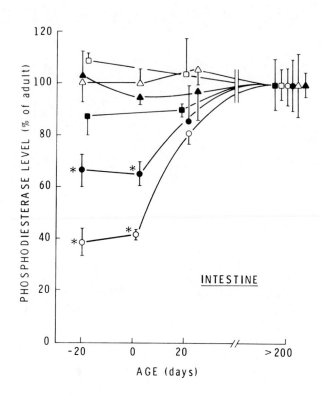

FIGURE 4. Changes in the activity levels of phosphodiesterases in the intestine of developing guinea pigs. Assay conditions, presentation of data, and symbols used in the figure were the same as in Fig. 2. The individual mean enzyme activity levels (pmol hydrolyzed/min/g wet wt) seen in adult intestine for 1 μM cyclic CMP, cyclic AMP, and cyclic GMP were 108, 17,270, and 8,580, respectively; the corresponding values with 1 mM substrates were 38,060, 1,600,300, and 415,430, respectively. The data presented are from three to six animals. Asterisks, significantly different from the adult ($P<0.05$ to $P<0.0005$). Taken from Ref. 15, with permission.

using the same amount of the tissue extract for assays, the fetal enzymes exhibit lower V_{max} values whereas their K_m values (3-6 mM) are identical to those of the adult enzyme (15).

Next, we examined changes in the enzyme activities in the regenerating liver (13). We found in young rats (120-150 g) that the activity of C-PDE (assayed using both 1 μM and 1 mM cyclic CMP) decreases as early as 8 hrs after

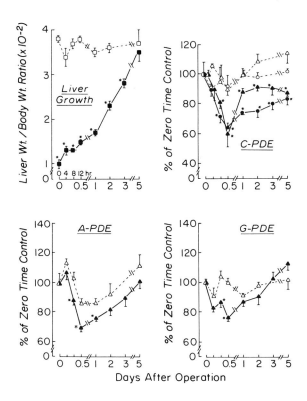

FIGURE 5. Liver growth (□,■) and changes in phosphodiesterase activities after partial hepatectomy on young rats. Open symbols indicate the control animals (with sham operations) and filled symbols indicate the hepatectomized animals. The enzyme activities in homogenates were assayed with either 1 μM (Δ,▲) or 1 mM (O,●) concentrations of the respective cyclic nucleotides as substrates. The data shown are expressed as percentages of the respective enzyme activities (pmol/min/g wet wt) seen at day zero, which were taken as 100%. The individual enzyme activities for 1 μM substrate at day zero were: cyclic CMP phosphodiesterase, 152; cyclic AMP phosphodiesterase, 11,300; and cyclic GMP phosphodiesterase, 6,600. The cyclic CMP phosphodiesterase activity for 1 mM cyclic CMP at day zero was 99,900. The data presented are the means ± S.E. from 3 to 4 rats. Asterisks, significantly different from the control animals (with sham operations) of the same or the closest time points ($P<.05$ to $P<.0005$). Taken from Ref. 13, with permission.

partial hepatectomy (Fig. 5) in which about 75% of the total liver mass was removed. The C-PDE activity is lowest 12 hrs after the surgery, then gradually recovers thereafter, but remains depressed during the entire experimental period; at day 5, when the liver had grown back to nearly its original weight, the C-PDE activity is still about 20% lower than in the control rats that received sham operations. Although the changes in the low and high K_m forms of A-PDE and G-PDE are less pronounced than the changes that occur in C-PDE, their activity levels are similarly depressed during the earlier phase of the liver regeneration (Fig. 5). However, their activities return to the control values during days 2 and 3 after partial hepatectomy, at which time the liver is still undergoing rapid growth. We also noted that the patterns of changes on the high K_m forms of A-PDE and G-PDE are similar to the low K_m enzymes (13). In old rats (350-400 g) the rate of liver regeneration after hepatectomy is slower than the young rats. We observed that in old rats the maximum decrease in C-PDE is smaller and occurs at a much later time (2 days as opposed to 12 hrs) after the surgery (13), suggesting that there is an inverse relationship between C-PDE activity and the rate of liver regeneration.

Depression of C-PDE also occurs in neoplastic growth (14). We noted that the C-PDE activity (assayed using 1 μM and 1 mM cyclic CMP) from the homogenates of the fast-growing Morris hepatoma 3924A is much lower than the normal or the host liver (Table II). The depression of C-PDE in the hepatoma prevails regardless of whether the C-PDE level is expressed on the basis of tissue weight, total tissue protein or DNA (14). The decrease in the C-PDE level, however, is less pronounced when the data are expressed on the protein basis; this is due to the fact that the protein content of the hepatoma is only 63% of that of the control liver (normal or host). The low K_m form of A-PDE activity is elevated, whereas the low K_m form of G-PDE is conversely depressed, in the hepatoma (Table II). The double-reciprocal plots of the kinetic data (Fig. 6) reveal that C-PDE from the normal liver and the hepatoma has an identical K_m of 4.2 mM for cyclic CMP, with a lower V_{max} for the hepatoma, suggesting that the hepatoma has a decreased level of C-PDE, and that the hepatoma enzyme may be indistinguishable from the liver enzyme. Compared to hepatoma 3924A shown above, the C-PDE level in the extremely slow-growing Morris hepatoma 9618A (a minimal deviation tumor) remains unaltered, whereas both A-PDE and G-PDE levels are lower (14). These observations obtained with hepatomas (14), coupled with findings made for the developing guinea pig tissues (15) and the regenerating rat liver (13), clearly suggest that catabolism of cyclic CMP may

TABLE II
C-PDE, A-PDE and G-PDE in MORRIS HEPATOMA 3924A[a]

Sub-strate	Tissue	Activities of PDE's	
		pmol/min/g wet wt	pmol/min/mg DNA
cCMP (1 µM)	Normal liver	162 ± 6	61 ± 4
	Host liver	150 ± 5	52 ± 1
	Hepatoma 3924A	24 ± 2*	8 ± 1*
cCMP (1 mM)	Normal liver	87,400 ± 3,517	32,564 ± 1,588
	Host liver	78,267 ± 3,254	26,978 ± 294
	Hepatoma 3924A	14,633 ± 1,578*	4,750 ± 502*
cAMP (1 µM)	Normal liver	8,966 ± 359	3,344 ± 187
	Host liver	8,505 ± 256	2,939 ± 127
	Hepatoma 3924A	12,248 ± 212	4,005 ± 158*
cGMP (1 µM)	Normal liver	4,759 ± 216	1,768 ± 62
	Host liver	5,805 ± 289*	2,000 ± 50
	Hepatoma 3924A	2,784 ± 130*	914 ± 67*

[a]The enzyme activities in homogenates of the hepatoma and the liver of ACI/c rats, with or without the tumor, were assayed using 1 µM or 1 mM of cyclic nucleotides as substrates, as indicated. Data, presented as means ± S.E., are from 3-6 rats. Asterisks, significantly different from the normal liver ($P<0.025 - 0.0005$). Taken from Ref. 14, with permission.

be more crucially related to the proliferative processes or states of cells than that of cyclic AMP and cyclic GMP, and that depression of C-PDE may be inversely related to the rate of tissue growth.

Another line of evidence suggesting C-PDE is a separate enzyme comes from the findings that the effects of many agents on this pyrimidine cyclic nucleotide phosphodiesterase are distinguishable from those on A-PDE and G-PDE (16). We found that papaverine effectively inhibits A-PDE and G-PDE when assayed using 1 µM substrates (Fig. 7). The inhibition, however, is largely diminished when a high concentration (1 mM) of substrates was used in the assays, agreeing with the findings (for example, see Ref. 21) that papaverine is a competitive inhibitor of both enzymes. Interestingly, papa-

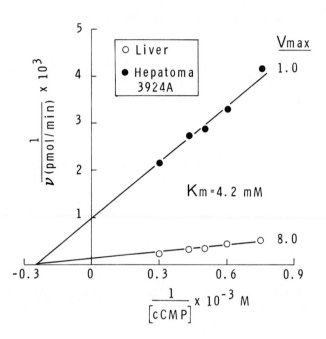

FIGURE 6. Double-reciprocal plots of the C-PDE activity in homogenates of normal liver from ACI/c rats and Morris hepatoma 3924A. The amount of homogenates used contained 10 mg fresh tissue. Taken from Ref. 14, with permission

verine at a concentration as high as 100 µM is without effect on C-PDE, assayed with either 1 µM or 1 mM cyclic CMP (Fig. 7). The effects of other classical phosphodiesterase inhibitors, such as 1-methyl-3-isobutylxanthine (MIX) and theophylline on these three enzyme activities are qualitatively similar to those shown for papverine (16), i.e. they are much less inhibitory to C-PDE than to either A-PDE or G-PDE.

2'-Deoxy cyclic AMP is more specific and potent in inhibiting A-PDE than G-PDE, whereas 2'-deoxy cyclic GMP is more specific and potent in inhibiting G-PDE than A-PDE in the liver extract (Fig. 8), in agreement with our findings made earlier for the purified enzymes from the guinea pig lung (21). These two deoxy purine cyclic nucleotides were found to be less specific and potent in inhibiting the C-PDE activity (Fig. 8). Based upon the demonstrated specific effects of the deoxy compounds, it is speculated that 2'-deoxy cyclic CMP would be a specific and potent inhibitor for C-PDE.

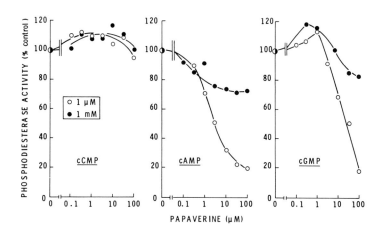

FIGURE 7. Comparative effects of papaverine on the C-PDE, A-PDE and G-PDE activities in the rat liver extract. The data are presented as % of the enzyme activities observed in the absence of papaverine (control activities), which were taken at 100%. The control activities for C-PDE, A-PDE, and G-PDE, assayed using 1 µM of the respective cyclic nucleotides were 2.4, 197 and 261 pmol/min/mg protein, respectively; the corresponding values using 1 mM concentrations of substrates were 2,160, 17,666 and 21,696 pmol/min/mg protein, respectively. The amounts of enzyme proteins in extracts used for assaying C-PDE, A-PDE and G-PDE (both substrate concentrations) were 205, 33 and 33 µg, respectively. Taken from Ref. 16, with permission.

Imidazole has been reported to stimulate A-PDE and G-PDE when they are assayed with a mM range of substrates (22,23). The effect of imidazole is confirmed for the two enzyme activities from the rat liver extract (Fig. 9). Interestingly, imidazole could not stimulate the liver C-PDE at either 1 µM or 1 mM substrate concentration, and moreover, it is slightly inhibitory to C-PDE at its highest concentration (20 mM) tested (Fig. 9). The effects of papaverine, MIX, deoxy cyclic nucleotides and imidazole on the C-PDE activity in the liver extract (Figs. 7-9) are similar to those seen for the partially purified C-PDE, which is shown to be largely devoid of the contaminating A-PDE and G-PDE activities (16).

It was observed, in contrast to some agents described above, that potassium (or sodium) phosphate is rather specific and potent in inhibiting C-PDE, compared to its inhibi-

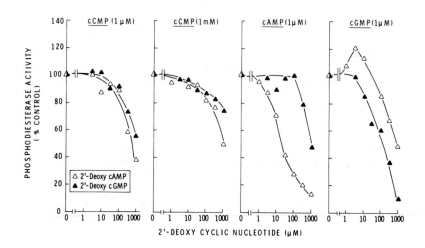

FIGURE 8. Comparative effects of 2'-deoxy cyclic AMP and 2'-deoxy cyclic GMP on the C-PDE, A-PDE, and G-PDE activities in the rat liver extract. The presentation of the data, the assay conditions and the amounts of enzyme activities used were as in Fig. 7 except that A-PDE and G-PDE were assayed only in the presence of 1 μM substrates. Taken from Ref. 16, with permission.

tion on A-PDE and G-PDE (Table III). For example, phosphate at 5 mM inhibits the partially purified C-PDE from the rat liver about 70% without affecting A-PDE or G-PDE. The phosphate, at 20 mM, while inhibiting C-PDE about 85%, also inhibits G-PDE about 70% without affecting A-PDE (Table III). The order of inhibition by the phosphate is therefore: C-PDE >> G-PDE > A-PDE. This is true whether C-PDE is activated by 10 mM Fe^{2+} or 10 mM Mg^{2+}.

ATetP (adenosine tetraphosphate), ATP, ADP and AMP, at 2 mM, like potassium phosphate, all inhibit C-PDE to greater extents than G-PDE. They, unlike the inorganic phosphate, conversely stimulate A-PDE (Table IV). Interestingly, adenosine and adenine (which do not contain phosphate), while having little effect on C-PDE and G-PDE, also stimulate A-PDE. GMP, IMP, CMP, and UMP, like AMP mentioned above, all stimulate A-PDE while inhibiting both C-PDE and G-PDE. Cyclic AMP, cyclic GMP, cyclic IMP, cyclic CMP and cyclic UMP, and their 2':3'-cyclic analogs, are again in general more inhi-

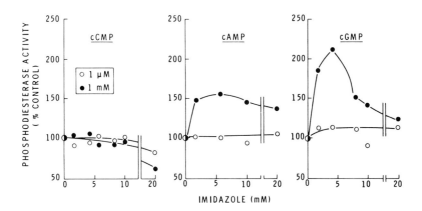

FIGURE 9. Comparative effects of imidazole on the C-PDE, A-PDE and G-PDE activities in the rat liver extract. The presentation of the data, the assay conditions and the amounts of enzyme activities used were as in Fig. 7. Taken from Ref. 16, with permission.

bitory to C-PDE and G-PDE than to A-PDE; this is true even when the substrate inhibitions of the respective phosphodiesterases are taken into acount (Table IV).

C-PDE is activated in a dose-dependent manner by mercaptoethanol (Table V). Other thiols, such as dithiothreitol and reduced glutathione, are also effective. One interesting feature in this reduction-related activation of C-PDE is the apparent loss of its Fe^{2+} dependency, suggesting that stimulation of the C-PDE activity by Fe^{2+} in the absence of the thiol reagent may be due to the possibility that Fe^{2+} itself may be acting like a reducing agent. In the thiol-activated C-PDE, Mg^{2+} now becomes the most effective metal activator, further stimulating the enzyme activity (Table V). A partially purified C-PDE preparation was used in the experiments described above.

DISCUSSION

Several lines of evidence indicate that C-PDE is a distinct, new class of phosphodiesterase; its cyclic CMP-hydrolyzing activity is not likely intrinsic to the conventional A-PDE or G-PDE, or both. The evidence includes non-parallel changes in levels of the three enzyme activities in tissues

TABLE III
EFFECTS OF PHOSPHATE ON C-PDE, A-PDE AND G-PDE[a]

K- or Na-Phosphate (mM)	cCMP (1 μM)		cCMP (1 mM)		cAMP (1 μM)	cGMP (1 μM)
	Fe^{2+}	Mg^{2+}	Fe^{2+}	Mg^{2+}	Mg^{2+}	Mg^{2+}
0	100%	100%	100%	100%	100%	100%
	(7.2)	(5.6)	(4200)	(3150)	(232)	(96)
5	26	23	31	25	106	103
10	19	20	17	20	101	75
20	13	12	10	14	99	30

[a]The enzyme activities were assayed in the presence of 1 μM or 1 mM substrates, Fe^{2+} (5 mM) or Mg^{2+} (10 mM), and varying concentrations of phosphate (pH 7.5), as indicated. C-PDE (59 μg) was from the Bio Gel HTP step (16). The rat liver extract, containing 25 μg protein, was used directly as the source of A-PDE and G-PDE. Activities (pmol/min/mg protein) for the individual activities are given in parentheses, which are normalized as 100%. Taken from Ref. 16, with permission.

of certain physio-pathologic states; purification of C-PDE removing most of the contaminating A-PDE and G-PDE activities; and differential effects of many agents on the three phosphodiesterase activities. Biological involvements of C-PDE are probably best suggested by our observations that its levels, as opposed to those of A-PDE and G-PDE, are invariably lower in the rapidly growing tissues (fetal and regenerating) and hepatoma 3924A. Depression of C-PDE activity in theory would lead to elevated cyclic CMP concentrations in cells, and thus their accelerated proliferation. The findings that cyclic CMP is greatly elevated (about 100-fold) in regenerating liver (3,5) lend a strong support to the contention that the C-PDE activity maybe inversely related to the cellular cyclic CMP concentration (12-16) and that cyclic CMP may act as a positive effector to promote cell division (1-3). It remains to be seen, however, whether depressed C-PDE levels in the fetal tissues and hepatoma 3924A would translate into elevated cyclic CMP concentrations in them.

The persistent decrease in C-PDE levels in the growing tissues, compared to variable and tissue-specific changes in A-PDE and G-PDE levels, also suggest a possibility that cyc-

TABLE IV
EFFECTS OF NUCLEOTIDES ON C-PDE, A-PDE AND G-PDE[a]

Addition (2 mM)	cCMP (1 μM)	cCMP (1 mM)	cAMP (1 μM)	cGMP (1 μM)
None (control)	100% (7.5)	100% (5500)	100% (11.4)	100% (9.9)
ATetP	41	24	196	87
ATP	29	51	129	100
ADP	44	55	173	109
AMP	18	34	250	44
Adenosine	106	77	223	88
Adenine	92	-	191	99
3':5'-cyclic AMP	8	22	28	0
2':3'-cyclic AMP	13	17	77	58
K- or Na- Phosphate	36	57	99	47
GMP	37	44	220	16
3':5'-cyclic GMP	31	59	69	0
2':3'-cyclic GMP	18	29	109	25
IMP	31	38	188	11
3':5'-cyclic IMP	40	55	50	0
CMP	45	47	127	59
3':5'-cyclic CMP	81	87	124	102
2':3'-cyclic CMP	36	33	104	102
UMP	25	40	154	36
3':5'-cyclic UMP	64	82	104	82
2':3'-cyclic UMP	38	66	115	77

[a]C-PDE (110 μg) of pig livers was from the Sephadex G-200 step; A-PDE (31 μg) was from the crude extract of pig livers; G-PDE (0.04 μg) of guinea pig lungs was from the preparative polyacrylamide gel electrophoresis (17). The activity (pmol hydrolyzed/min), assayed in the absence of additions with the indicated concentrations of cyclic nucleotides as substrates, given in parentheses, is normalized at 100%.

lic CMP is most likely involved in the basic proliferative process of all cell types, rather than the specific functions or metabolisms of the individual tissues. The clue regarding

TABLE V
EFFECTS OF MERCAPTOETHANOL ON C-PDE[a]

Mercapto-ethanol (mM)	Cyclic CMP (1 μM)			Cyclic CMP (1 mM)		
	None	Mg^{2+}	Fe^{2+}	None	Mg^{2+}	Fe^{2+}
0	0.7	0.6	1.3	466	477	812
3	1.3	1.8	1.6	809	817	913
30	2.3	3.3	1.9	1054	1254	1152

[a]C-PDE (260 μg), from the DEAE-cellulose step, was assayed in the presence of varying concentrations of mercaptoethanol, with or without 10 mM Mg^{2+} or Fe^{2+}, using 1 μM or 1 mM of cyclic CMP as substrate. Activity: pmol hydrolyzed/min.

the relationship between the C-PDE level and the tissue growth rate is provided by the following observations. a) The depression of the C-PDE level is smaller in the regenerating liver of slower growth rate seen in old rats, as compared to the greater depression of C-PDE and faster rate of liver regeneration seen in young rats (13). b) C-PDE is not depressed in the slowest growing Morris hepatoma 9618A, (14), a minimal deviation tumor. We also noted that C-PDE is not lower in the isoproterenol-induced cardiac hypertrophy in which A-PDE is elevated (12,24). Since the cardiomegaly (heart weight and protein increase about 80%) involves primarily the enlargement of the existing cells (i.e., hypertrophy), rather than the increase in number of new cells (i.e., hyperplasia), it is further suggested that cyclic CMP may play a role in tissue growth in which cell proliferation, but not cell growth, is the prominent event.

Vast amounts of experimental evidence obtained by many investigators, although not all conclusive and sometimes controversial or even contradictory, suggest that cyclic AMP may act as a negative effector in cell proliferation (for example see Ref. 25 for a review). Subsequent studies have implicated cyclic GMP as a positive effector in cell division (for example see Ref. 26 for a review). It appears that the proposed roles of cyclic nucleotides in cell proliferation have evolved from the simple hypothesis of cyclic AMP inhibition (Model 1, Fig. 10) to a more complicated dualistic mechanism consisting of the inhibitory effect of cyclic AMP and the stimulatory effect of cyclic GMP (Model 2). In light of the recent findings concerning the cyclic CMP system, we wish to propose that the stimulatory effect of cyclic CMP may also be involved in the intricate balance of bioregulation, thus forming a "three-ring circus" type of interaction with cyclic AMP and cyclic GMP. Model 3 depicts the hypothesis that cyclic CMP and cyclic GMP both counteract the effect of cyclic AMP without themselves mutually interacting. Model 4 shows interactions of all three cyclic nucleotides on an equal basis. In view of the findings that C-PDE, but not A-PDE and G-PDE, appears to be more critically involved in the normal and malignant growth, a more prominent role of cyclic CMP compared to the others may be suggested, as depicted in Model 5 (Fig. 10).

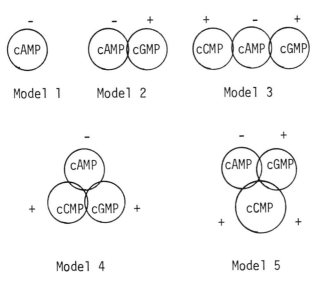

FIGURE 10. Hypothetical models of cyclic nucleotides in cell proliferation. - indicates inhibition; + indicates stimulation.

More recently, we have identified a cyclic CMP-binding protein from extracts of the liver and other tissues (27). It is not clear whether it is related to the presumed cyclic CMP-dependent protein kinase. Nevertheless, the finding of the binding protein provides additional evidence to support the existence of a cyclic CMP system in tissues, and to hint at the mechanism of action of this pyrimidine cyclic nucleotide.

One interesting feature of C-PDE is the effects of various agents on this enzyme that are different from those on A-PDE and G-PDE, suggesting its unique catalytic and molecular properties. It appears, therefore, that regulation, or lack of regulation, of C-PDE can be achieved independently of the others. This coupled with unilateral alterations in C-PDE levels found in many growing tissues and in the hepatoma, strongly implicate C-PDE as a potential site not only for bioregulation in various physio-pathologic processes but also for pharmacological interventions. The exact mechanism by which cyclic CMP exerts its presumed effects, and the precise functional relationship that may exist among these three cyclic nucleotides remains yet to be fully explored.

Recent kinetic and physical studies of A-PDE and G-PDE activities reveal the presence of multiple catalytic sites exhibiting different kinetic properties, and the interconvertibility of various forms of A-PDE and G-PDE (28,29). Furthermore, thiol agents have been shown to prevent the conversion of the amoebal A-PDE from the low K_m to the high K_m form (30), and to increase the V_{max} of the liver A-PDE (29). We have now found that mercaptoethanol activates C-PDE and causes a change in the metal activator specificity for the enzyme (Table V). Whether the reduction-related activation of C-PDE is also involved in changes in its physical, molecular and catalytic properties is not clear. A complete purification and thorough characterization of this new pyrimidine cyclic nucleotide phosphodiesterase is clearly in order.

REFERENCES

1. Bloch, A. (1974). Biochem. Biophys. Res. Commun. 58, 652.
2. Bloch, A., Dutschman, G., and Maue, R. (1974). Biochem. Biophys. Res. Commun. 59, 995.
3. Bloch, A. (1975). Adv. Cyclic Nucleotide Res. 5, 331.
4. Cailla, H.L., Roux, D., DeLaage, M., and Goridis, C. (1978). Biochem. Biophys. Res. Commun. 85, 1503.
5. Murphy, B.E., and Stone, J.E. (1978). The Pharmacologist. 20, 232 (abstract); personal communications.
6. Kuo, J.F., Brackett, N.L., Schatzman, R., and Shoji, M. (1979), In preparation.
7. Ignarro, L.J. (1979). Science, 203, 673.
8. Cech, S.Y., and Ignarro, L.J. (1977). Science, 198, 1063.
9. Cech, S.Y., and Ignarro, L.J. (1978). Biochem. Biophys. Res. Commun. 80, 119.
10. Gaion, R.M., and Krishna, G. (1979). Biochem. Biophys. Res. Commun. 86, 105.

11. Gaion, R.M., and Krishna, G. (1979). Science. 203, 672.
12. Kuo, J.F., Brackett, N.L., Shoji, M., and Tse, J. (1978) J. Biol. Chem. 253, 2518.
13. Shoji, M., Brackett, N.L., and Kuo, J.F. (1978). Science. 202, 826.
14. Shoji, M., Brackett, N.L., Helfman, D.M., Morris, H.P., and Kuo, J.F. (1978). Biochem. Biophys. Res. Commun. 83, 1140.
15. Helfman, D.M., Brackett, N.L., and Kuo, J.F. (1978). Proc. Nat. Acad. Sci. USA. 75, 4422.
16. Kuo, J.F., Shoji, M., Brackett, N.L., and Helfman, D.M. (1978). J. Cyclic Nucleotide Res. 4, 463.
17. Davis, C.W., and Kuo, J.F. (1977). J. Biol. Chem. 252, 4078.
18. Thompson, W.J., and Appleman, M.M. (1971). Biochemistry. 10, 311.
19. Helfman, D.M., Shoji, M., and Kuo, J.F. (1979). In preparation.
20. Appleman, M.M., Thompson, W.J., and Russell, T.R. (1973). Adv. Cyclic Nucleotide Res. 3, 65.
21. Davis, C.W., and Kuo, J.F. (1978). Biochem. Pharmacol. 27, 89.
22. Butcher, R.W., and Sutherland, E.W. (1962). J. Biol. Chem. 237, 1244.
23. Donnelly, T.E., Jr. (1976). Arch. Biochem. Biophys. 173, 375.
24. Tse, J., Brackett, N.L., and Kuo, J.F. (1978). Biochim. Biophys. Acta. 542, 399.
25. Ryan, W.L., and Heidrick, M.L. (1974). Adv. Cyclic Nucleotide Res. 4, 81.
26. Goldberg, N.D., Haddox, M.K., Dunham, E., Lopez, C., and Hadden, J.W. (1974). In "The Cold Spring Harbor Symposium on the Regulation of Proliferation in Animal Cells" (B. Clarkson and R. Baserga, eds.), pp. 609-625. Cold Spring Harbor Laboratory, New York.
27. Schatzman, R., Brackett, N.L., Shoji, M., and Kuo, J.F. (1979). In preparation.
28. Pichards, A.-L., and Cheung, W.Y. (1976). J. Biol. Chem. 251, 5726.
29. Van Inwegen, R.G., Plesger, W.J., Strada, S.J., and Thompson, W.J. (1976). Arch. Biochem. Biophys. 175, 700.
30. Chassy, B.M. (1976). Science. 175, 1016.

REGULATION OF PHOSPHOENOLPYRUVATE CARBOXYKINASE (GTP) SYNTHESIS[1]

Richard W. Hanson, Michele A. Cimbala,[3] J. Garcia-Ruiz[2], Kathelyn Nelson, and Dimitris Kioussis

Department of Biochemistry, Case Western Reserve University School of Medicine, Cleveland, Ohio 44106 and The Fels Research Institute, Temple University School of Medicine, Philadelphia, Pennsylvania 19140

ABSTRACT The synthesis of phosphoenolpyruvate carboxykinase from the cytosol of rat liver is regulated by cyclic AMP and insulin as well as by glucocorticoids. Cyclic AMP causes a rapid induction of enzyme synthesis (a 10-fold increase by 90 min. after Bt_2cAMP administration) whereas the glucocorticoids alter the rate of synthesis of the enzyme more slowly. In the livers of diabetic animals the synthesis of phosphoenolpyruvate carboxykinase is 10-20 fold higher than the synthesis rate of the enzyme in normal fed rats and the injection of insulin causes a rapid de-induction of enzyme synthesis ($t_{\frac{1}{2}}$ 40 min.). A similar decrease in the rate of phosphoenolpyruvate carboxykinase synthesis occurs when starved animals are refed a diet high enough in carbohydrate. These rapid changes in the synthesis of hepatic phosphoenolpyruvate carboxykinase are accompanied by equally rapid changes in translatable mRNA coding for the enzyme. There is a direct correlation between the levels of translatable mRNA and the synthesis rate of the enzyme as measured _in vivo_ in all conditions studied to date and the kinetics of both parameters is remarkably similar. The turnover of hepatic phosphoenolpyruvate carboxykinase mRNA can be estimated to be approximately 40 min. by using cordycepin, an inhibitor of mRNA processing. Enzyme mRNA can be stabilized if rats are injected with the protein synthesis inhibitor cycloheximide, suggesting that if specific mRNA is being translated it is protected from degradation. It is thus

[1] Supported in part by grant AM 18034 from the National Institutes of Health.
[2] Present address: Department of Biochemistry and Molecular Biology, Autonoma University of Madrid.
[3] Recipient of a postdoctoral fellowship from the Juvenile Diabetes Foundation.

possible that the levels of phosphoenolpyruvate carboxykinase mRNA can be controlled at transcription (or at processing of mature mRNA) or at translation which stabilizes the mRNA against degradation.

INTRODUCTION

A wide variety of hormones have been demonstrated to alter the synthesis rate of phosphoenolpyruvate carboxykinase (GTP) (EC 4.1.1.32) (PEPCK) in rat tissues. As shown in Table I, hepatic PEPCK synthesis responds rapidly to induction by cAMP and to de-induction by insulin. These two compounds, however, have no effect on renal PEPCK. Glucocorticoids stimulate PEPCK synthesis in liver and kidney cortex but decrease the synthesis rate of the enzyme in adipose tissue. Changes in hepatic PEPCK synthesis caused by cAMP and insulin are of considerable interest as a model for studying the mechanism of hormone action because of the rapidity of their action and the magnitude of their effect. If a rat is starved for 24 hours and refed a normal diet, the synthesis rate of PEPCK in the liver is decreased in 90 min. to 10% of the initial value noted before refeeding (1,2) (Fig. 1). Conversely the administration of Bt_2cAMP to animals causes a 10-fold induction of hepatic PEPCK synthesis within 90 min. (3). In this review we will summarize our studies on the

TABLE I

FACTORS KNOWN TO REGULATE PEPCK SYNTHESIS

Hormone or Metabolic Status	Tissue	Effect on Enzyme Synthesis	Reference
glucagon	liver, adipose tissue	rapid[a] increase	12,13,31
Bt_2cAMP	kidney cortex	no effect	15
glucocorticoids	liver, kidney cortex	slow[b] increase	32,15
	adipose tissue	slow decrease	31
insulin	liver, adipose tissue	rapid decrease	2,31
	kidney cortex	no direct effect	15
acidosis	liver, adipose tissue	no effect	15
	kidney cortex	slow increase	15
alkalosis	liver, adipose tissue	no effect	15
	kidney cortex	slow decrease	15

[a] Rapid synthesis induction-(Bt_2cAMP increases hepatic PEPCK synthesis 8-fold in 90 min.); de-induction (insulin decreases hepatic PEPCK synthesis 50% in 30 min.).

[b] Slow synthesis-(triamcinolone increases renal PEPCK synthesis 2-fold in 8 hrs.).

MODULATION OF PROTEIN FUNCTION 359

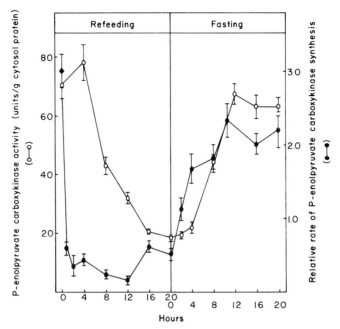

FIGURE 1. Effect of starvation and refeeding on the activity and synthesis rate of hepatic PEPCK. Experimental details are from Hopgood et al. (1) and the figure is taken from a review by Tilghman, Hanson and Ballard (30).

hormonal regulation of PEPCK synthesis focusing on the mechanisms by which cAMP and insulin alter the rate of enzyme synthesis.

REGULATION OF ENZYME LEVELS

The earliest studies on the induction of hepatic PEPCK were by Lardy and his associates (4,5,6) who clearly demonstrated that the enzyme was increased in activity by starvation and decreased in activity by refeeding a diet containing carbohydrate. Shrago et al. (4) also noted the relationship between the decrease in PEPCK activity seen after insulin administration and the decrease in PEPCK activity normally noted during carbohydrate refeeding. Multiple injections of glucocorticoids also were shown to induce PEPCK activity in rat liver (4) as was the oral administration of large doses of tryptophan (7). The use of inhibitors of RNA and of protein synthesis in this work provided the initial evidence that the inductive effects of starvation and glucocorticoids were the result of a stimulation in enzyme synthesis (4,7).

Studies of the development of PEPCK in rat liver cytosol by Ballard and Hanson (8) and by Yeung and Oliver (9,10)

demonstrated that the activity of the enzyme was absent in
fetal rat liver and appeared initially at birth. Premature
delivery (9) or the injection of glucagon or Bt_2cAMP in utero
(10) induced the development of PEPCK activity. The administration of actinomycin D could block the induction process.
Furthermore, the injection of glucose or of insulin to fetuses before delivery prevented the normal appearance of the
enzyme. The results of early studies of the development of
hepatic PEPCK were consistent with work on the adult animal.
They indicated that the activity of the enzyme was induced
(or induced to develop initially) by agents such as glucagon
and Bt_2cAMP which increased cAMP levels in the liver (11).
The activity of the enzyme was decreased by the administration of glucose, which elicited insulin secretion, or by
direct injection of insulin itself. Since the induction
process could be blocked by inhibitors of RNA synthesis (10)
or of protein synthesis (7) it was inferred that the induction involved de novo enzyme synthesis.

Direct Studies of Enzyme Synthesis. Immunochemical
techniques were used to determine the rate of hepatic PEPCK
synthesis and degradation. The first of these studies
involved the development of PEPCK at birth and demonstrated
that the initial appearance of the enzyme involved its rapid
synthesis, in the absence of observable rates of enzyme
degradation (12). Hopgood et al. (1) also noted that starvation increased enzyme synthesis in the adult rat and that
refeeding caused a rapid "de-induction" of PEPCK synthesis in
the liver (Fig. 1), but not in adipose tissue. In a subsequent report, Tilghman et al. (2) more thoroughly studied the
de-induction of PEPCK synthesis caused by refeeding starved
animals and noted that the synthesis rate of the enzyme
decreased with a half-life of about 30 min. This de-induction required glucose and insulin and could be blocked by the
injection of Bt_2cAMP. These rapid changes in the rate of
PEPCK synthesis after refeeding glucose to a starved rat were unexpected since the alteration in enzyme activity during the
first hour after refeeding is very small because of the relatively long half-time of degradation of the enzyme (8-10 hrs.)
(1,2).

An interesting aspect of the regulation of PEPCK synthesis is the marked differences in tissue response to the same
hormone. These differences are outlined in Table I. Hepatic
PEPCK synthesis is stimulated by glucagon and Bt_2cAMP (13,14,
15) and decreased by insulin, whereas renal synthesis of the
enzyme is not directly responsive to either of these compounds
(16). Glucocorticoids, which were shown by Reshef et al. (17)
to decrease the activity of hepatic PEPCK in normal rats,

actually increased the synthesis rate of the enzyme in diabetic animals (18). This difference was due, in part, to the glucocorticoid-induced release of insulin which subsequently decreased the synthesis of hepatic PEPCK (18). The enzyme in the kidney cortex is stimulated by glucocorticoids (16, 19), whereas adipose tissue PEPCK synthesis is decreased by glucocorticoid administration (31). Finally, the induction of metabolic acidosis by ammonium chloride gavage causes an increase in synthesis of only the renal form of PEPCK, an effect which can be reversed by the administration of bicarbonate (16). The induction of PEPCK synthesis by glucocorticoids and by acidosis is relatively slow compared to Bt_2cAMP induction of the enzyme and the subsequent rate of de-induction of the renal enzyme after bicarbonate administration is also slow (16).

These differences in the hormonal response of PEPCK in rat tissues are consistant with the role which the tissue plays in whole body metabolism. The liver is responsive to alterations in blood glucose and maintains glucose homeostasis by taking up glucose from the portal blood during periods when glucose is being actively absorbed by the intestine. The liver then converts this glucose to glycogen and fatty acids. During starvation the glycogen is mobilized and the rate of hepatic gluconeogenesis increased. PEPCK is synthesized to accomodate an increased flux over the gluconeogenic pathway. The synthesis rate of the enzyme increases as the ratio of insulin to glucagon decreases. These effects are rapid and constantly going on during the course of a day as the animal alternates periods of eating and fasting. The kidney cortex, on the other hand maintains acid-base balance, excretes weak acids during metabolic acidosis and generally regulates the acidity of the tubular urine. This process requires NH_4^+ generated within the kidney by the deamination of glutamine by successive steps to α-ketoglutarate (19). This α-ketoglutarate is further metabolized in the citric acid cycle to oxalacetate which is then converted to PEP by PEPCK. The PEP formed in this sequence can be used to synthesize glucose. Alternately, it can be oxidized completely to CO_2 and water via pyruvate kinase, pyruvate dehydrogenase and the TCA cycle. The kidney can synthesis glucose from glutamine, particularly during starvation and in diabetes and renal PEPCK is induced under these conditions to accomodate flux over the pathway. However, the rate of response of the kidney enzyme is slower, which reflects the longer time course of onset of metabolic acidosis.

<u>Mechanism of the Hormonal Regulation of PEPCK Synthesis.</u> Earlier studies on the mechanisms of cAMP and glucocorticoid

action on the levels of PEPCK were carried out by Wicks (20) using fetal rat liver explants or Reuber H35 cells. Using actinomycin D to block RNA synthesis, Wicks and McGibben (21) showed that although glucocorticoid stimulation of PEPCK activity in Reuber H35 cells was inhibited, cAMP still caused a rapid but transient (3 hr.) induction of the enzyme. From these studies and a variety of others reviewed in detail by Wicks (22), he suggested that glucocorticoids act at transcription and increase the amount of specific mRNA for PEPCK, whereas cAMP acts post-transcriptionally (perhaps at translation of the mRNA) to stimulate the synthesis of new enzyme protein. A number of more recent studies using Reuber H35 cells (23) and isolated perfused rat liver (24) have noted responses of PEPCK to glucocorticoids and cAMP consistent with this proposed mechanism.

Measurements of translatable mRNA coding for hepatic PEPCK have not provided direct support for this mechanism (15,24). In general, all of our studies have shown a close correlation between the levels of PEPCK mRNA in the liver and rate of enzyme synthesis. An example of this relationship is shown in Fig. 2., in which the de-induction of PEPCK synthesis caused by refeeding a starved animal (insert) is closely paralleled by a decrease in PEPCK mRNA. The half-time of decay for both processes is approximately 40 min. If these starved-refed animals are injected with Bt_2cAMP the level of translatable mRNA for the enzyme increases rapidly; so that by 90 min. after administration of the cyclic nucleotide there is an eight-fold induction of PEPCK mRNA (15). Furthermore, this induction in PEPCK mRNA caused by Bt_2cAMP can be blocked by prior injection of cordycepin (15). The fact that Bt_2cAMP increases the amount of PEPCK mRNA and that this induction can be blocked by inhibitors of mRNA processing strongly suggests that the effect of Bt_2cAMP is not exerted at the level of mRNA translation but rather at some step involved in message synthesis or processing.

Studies of the development of PEPCK mRNA in fetal rat liver also suggest that cAMP acts to stimulate the appearance of mature enzyme template. We could detect no PEPCK mRNA in the livers of fetal rats before birth (Fig. 3). The injection of Bt_2cAMP causes the initial appearance of enzyme template and also stimulates enzyme synthesis, with both events occurring in parallel (26). The administration of actinomycin D blocks the induction of PEPCK mRNA by Bt_2cAMP suggesting that de novo RNA synthesis is involved in the rapid response of enzyme mRNA to the administration of the cyclic nucleotide. Again, both the appearance and decay of enzyme mRNA in the liver of fetal rats is rapid (Fig. 3).

A rapid decay of PEPCK mRNA in the liver of adult rats

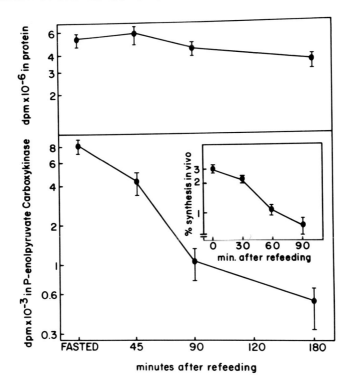

FIGURE 2. <u>Changes in the levels of hepatic PEPCK mRNA after refeeding.</u> Rats fasted for 24 hr. were refed glucose by stomach tube. The poly(A)+ RNA was isolated from the livers of animals at the times indicated in the figure. The mRNA (10 μg) was translated in the wheat germ translation system and total released protein (upper panel) and PEPCK synthesized were measured as outlined by Kioussis <u>et al.</u> (25) The rate of PEPCK synthesis <u>in vivo</u> is shown in the insert and is redrawn from Tilghman <u>et al.</u> (2). This figure is from Kioussis <u>et al.</u> (25).

was noted after the administration of cordycepin. The halftime of this decay is approximately 40 min. (Fig. 4) and directly follows the decay kinetics of enzyme synthesis noted in an earlier study by Tilghman <u>et al.</u> (2). Furthermore, Iynedjian and Hanson (15) reported on inhibition of the induction of PEPCK mRNA in the livers of starved-refed rats injected with Bt_2cAMP if cordycepin was administered before the cyclic nucleotide. These effects of cordycepin on PEPCK mRNA were not due to a general inhibition of total mRNA since the half-life of most hepatic messages is longer than that of

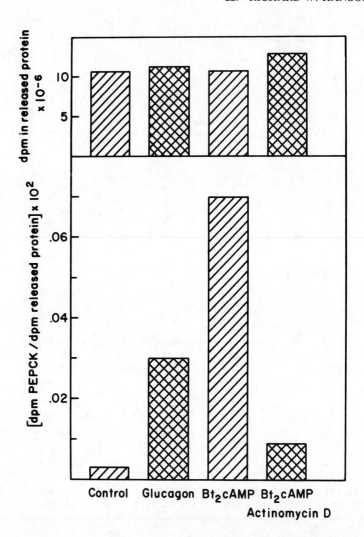

FIGURE 3. <u>Hormonal induction of PEPCK mRNA in the livers of 21-day fetal rats in utero</u>. Total liver RNA was extracted and the poly(A)$^+$ RNA was isolated and assayed in a wheat germ translation system. Values are expressed as the percentage of [^{3}H]-leucine incorporated into PEPCK relative to newly synthesized proteins released by the wheat germ system after the addition of 10 μg of mRNA. Actinomycin D (1 mg/kg body wt.) or saline was injected into the animals 30 min. before Bt$_2$cAMP (1 μmol/fetus). Glucagon was administered at a dose of 200 μg/fetus. Animals were killed 3 hrs later. Further details are given in Ruiz <u>et al</u>. (26).

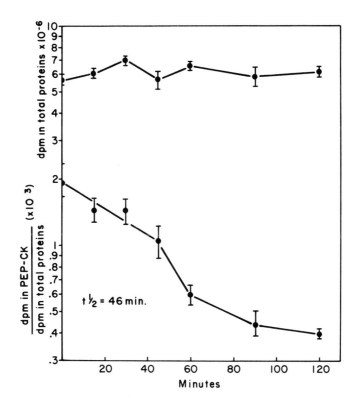

FIGURE 4. Effect of cordycepin on the levels of PEPCK mRNA in the livers of starved rats. Cordycepin (2 mg/100 g body wt.) was injected into 24 hr. starved rats and the total hepatic mRNA isolated and translated in the wheat germ system. Values are expressed as described in Fig. 3.

PEPCK mRNA. These studies also demonstrate the dependence of PEPCK mRNA on continued mRNA processing for the maintenance of template levels in the liver. Since decay of PEPCK mRNA is rapid in the absence of mRNA processing and since cordycepin can block the stimulation of mRNA levels caused by Bt_2cAMP administration, it is possible that the rapid alterations in enzyme mRNA noted during fasting-refeeding (25) and after Bt_2cAMP (15) are due to some effect on enzyme processing, or to a direct effect on gene transcription. More definitive information on this point will require hybridization analysis of the hepatic mRNA to determine the existence of unique sequences for PEPCK mRNA. The demonstration of such sequences in fetal rat liver for example would strongly implicate RNA processing as a regulatory event in the mechanism of cAMP induction of PEPCK synthesis.

A recent report by Roper and Wicks (27) presents evidence that cAMP, when added to Reuber H35 cells, stimulates the rate at which ribosomes translate pre-existing mRNA for tyrosine aminotransferase by acting specifically on chain elongation. In contrast, Ernst and Feigelson (28) and Noguchi et al. (29) noted that cAMP increased the levels of translatable mRNA for tyrosine aminotransferase in adult rat liver. They also suggested as outlined above for PEPCK, that cAMP stimulates the synthesis of tyrosine aminotransferase by acting to increase transcription or template processing. There may be, however, alternative explanations of the results of our studies and those of Ernst and Feigelson (28) and Noguchi et al. (29). It is possible, for example, that mRNA for PEPCK or tyrosine aminotransferase is very labile when not being actively translated, so that an increase in the rate of initiation of protein synthesis would effectively increase the concentration of translatable mRNA for these enzymes by protecting it from degradation. For this mechanism to be tenable, however, it would be necessary to explain why cordycepin causes a rapid decrease in PEPCK mRNA in the absence of any known effects of this compound on the initiation of protein synthesis.

FUTURE STUDIES

Our understanding of the regulation of enzyme synthesis in mammalian tissues is benefiting greatly from the application of new techniques for the study of gene expression. In the past it was necessary to speculate on the mechanism by which a hormone regulated the level of an enzyme based on the effects of protein synthesis inhibitors on enzyme activity. A refinement in available techniques has reduced the extent of this speculation but definative statements concerning the mechanisms by which hormones act remains elusive. The availability of a specific cDNA to PEPCK will allow a considerable reduction in speculation and permit us to rule out several possible mechanisms for the hormonal regulation of PEPCK synthesis.

REFERENCES

1. Hopgood, M. F., Ballard, F. J., Reshef, L. and Hanson, R. W. (1973). Biochem. J. 134, 445-453.
2. Tilghman, S. M., Hanson, R. W., Reshef, L., Hopgood, M. F. and Ballard, F. J. (1974). Proc. Nat. Acad. Sci., U.S.A. 71, 1304-1308.
3. Srago, E., Lardy, H. A., Nordlie, R. C. and Foster, D. O. (1963). J. Biol. Chem. 238, 3188.

4. Foster, D. O., Ray, P. D. and Lardy, H. A. (1964). Biochem. 5, 555.
5. Shrago, E., Young, J. W. and Lardy, H. A. (1967). Science 158, 1572.
6. Foster, D. O., Lardy, H. A., Ray, P. D. and Johnston, J. B. (1967). Biochem. 6, 2120.
7. Ballard, F. J. and Hanson, R. W. (1967). Biochem. J. 104, 866.
8. Yeung, D. and Oliver, I. T. (1968). Biochem. J. 108, 325.
9. Yeung, D. and Oliver, I. T. (1968). Biochem. 7, 3231.
10. Girard, J. R., Cuendet, G. S., Morliss, E. B., Kervran, A., Rieutort, M. and Assan, R. (1973). J. Clin. Invest. 52, 3190.
11. Philippidis, H., Hanson, R. W., Reshef, L., Hopgood, M. F. and Ballard, F. J. (1972), Biochem. J. 126, 1127.
12. Wicks, W. D., Lewis, W. and McKibbin, J. B. (1972). Biochem. Biophys. Acta 264, 177.
13. Hanson, R. W., Fisher, L., Ballard, F. J. and Reshef, L. (1973). Enzyme 15, 97.
14. Iynedjian, P. B. and Hanson, R. W. (1977). J. Biol. Chem. 252, 655.
15. Iynedjian, P. B., Ballard, F. J. and Hanson, R. W. (1975). J. Biol. Chem. 250, 5596.
16. Reshef, L., Ballard, F. J. and Hanson, R. W. (1970). J. Biol. Chem. 245, 5979.
17. Gunn, J. M., Hanson, R. W., Meyuhas, O., Reshef, L. and Ballard, F. J. (1975). Biochem. J. 150, 195.
18. Longshaw, I. D., Alleyne, G. A. O. and Pogson, C. I. (1972). J. Clin. Invest. 51, 2284.
19. Pitts, R. F. (1974). "Physiology of the Kidney and Body Fluids". 3rd Ed., pp. 198-241. Year Book Medical Publishers, Chicago.
20. Wicks, W. D. (1971). J. Biol. Chem. 246, 217.
21. Wicks, W. D. and McKibbin (1972). Biochem. Biophys. Res. Commun. 48, 205.
22. Wicks, W. D. (1971). Annal. N. Y. Acad. Sci. 185, 152.
23. Tilghman, S. M., Gunn, J. M., Hanson, R. W., Reshef, L. and Ballard, F. J. (1975). J. Biol. Chem. 250, 3322.
24. Krone, W., Marquardt, W., Seitz, H. J. and Tarnowski, W. (1975). FEBS Letters 57, 64.
25. Kioussis, D., Reshef, L., Cohen, H., Tilghman, S., Iynedjian, P. B., Ballard, F. J. and Hanson, R. W. (1978). J. Biol. Chem. 253, 4327.
26. Garcia-Ruiz, J. P., Ingram, R. and Hanson, R. W. (1978). Proc. Nat. Acad. Sci., U.S.A. 75, 4189.
27. Roper, M. D. and Wicks, W. D. (1978). Proc. Nat. Acad. Sci. U.S.A. 75, 140.

28. Ernst, M. J. and Feigelson, P. (1978). J. Biol. Chem. 253, 319.
29. Noguchi, T., Diesterhaft, M. and Granner, D. (1978). J. Biol. Chem. 253, 1332.
30. Tilghman, S. M., Hanson, R. W. and Ballard, F. J. (1976). In "Gluconeogenesis; Its Regulation in Mammalian Species" (R. W. Hanson and M. A. Mehlman, eds.), pp. 47-92. John Wiley, New York.
31. Meyuhas, O., Gunn, J. M., Hanson, R. W., Ballard, F. J. and Reshef, L. (1976). Biochem. J. 158, 9.
32. Gunn, M. J., Meyuhas, O., Reshef, L., Ballard, F. J. and Hanson, R. W. (1975). Biochem. J. 150, 195.

Modulation of Protein Function

PEPTIDE-CHAIN INITIATION IN HEART AND SKELETAL MUSCLE[1]

Leonard S. Jefferson, Kathryn E. Flaim, and Howard E. Morgan

Department of Physiology, College of Medicine, The
Pennsylvania State University, Hershey, PA   17033

ABSTRACT   Rates of peptide-chain initiation were estimated in heart and skeletal muscle by measurements of the rate of protein synthesis and levels of ribosomal subunits. During perfusion of isolated rat heart or hemicorpus with buffer containing glucose and amino acids, polysomes decreased, levels of ribosomal subunits rose, and protein synthesis declined. These findings indicated development of a restraint on peptide-chain initiation in either tissue. In heart muscle, the restraint was relieved by addition of insulin, fatty acids or non-carbohydrate substrates. In skeletal muscle of mixed fiber type, only insulin was effective in relieving the restraint on initiation. In either tissue, activity of eIF-2, assayed by the formation of the ternary complex of met-tRNA$_f^{Met}$, GTP and eIF-2, was unchanged by insulin.
 Differences in regulation of peptide-chain initiation in heart as compared to skeletal muscle were expressed in vivo. Starvation or induction of diabetes decreased synthesis and increased subunit levels in mixed skeletal muscle. In soleus, a red skeletal muscle, and in heart, levels of ribosomal subunits did not increase and synthesis was reduced to a lesser extent. Reduction in synthesis in mixed skeletal muscle was due to two factors: decreased number of ribosomes and efficiency of synthesis (synthesis/ribosome). Efficiency in heart was sustained due to maintenance of rates of initiation by fatty acids and other non-carbohydrate substrates. Provision of these substrates to heart, but not mixed skeletal muscle, increased tissue levels of glucose-6-phosphate, an intermediate that facilitates peptide-chain initiation in reticulocyte lysates. Glucose-6-phosphate levels were also increased in heart, but not in mixed skeletal muscle, of diabetic animals. Although glucose-6-phosphate may mediate the substrate effect on peptide-chain initiation, insulin acted independently of this mechanism. These

[1]This work was supported by grants HL18258, HL20388, and AM15658 from the National Institutes of Health.

studies emphasize the importance of peptide-chain initiation as a rate-controlling step that is affected by availability of hormones and nutrients to heart and skeletal muscle.

INTRODUCTION

During the past few years, a number of factors have contributed to an advance in our understanding of both the chemical reactions and mechanisms of regulation of protein synthesis in heart and skeletal muscle. These include 1) a vast improvement in knowledge of the pathway of protein synthesis, 2) development of methods for measurement of rates of protein synthesis, 3) development and modification of procedures for in vitro perfusion of heart and skeletal muscle, and 4) development of methods for measurement of intermediates in the pathway of protein synthesis. At the present time, identification of the hormonal and non-hormonal factors affecting protein synthesis in heart and skeletal muscle is emerging, and, in some instances, the site of action of these factors on specific reactions or groups of reactions has been defined. The purpose of this chapter is to describe recent findings on the regulation of peptide-chain initiation in heart and skeletal muscle.

METHODS

Perfusion Methodology. Studies on the regulation of protein synthesis have been facilitated greatly by the development of procedures for in vitro perfusion of heart and skeletal muscle. Techniques used for the perfusion of the isolated rat heart and of the skeletal muscle contained in the isolated rat hemicorpus have been described in detail elsewhere (1,2).

Measurements of Rates of Protein Synthesis. Since the average half-time of turnover of heart and skeletal muscle protein is approximately 4-6 days and 9-15 days, respectively, net protein synthesis cannot be determined in most in vitro experiments. As a result, rates of synthesis have been estimated by tracing the incorporation of radioactive amino acids into muscle proteins. Estimates of this rate depend on knowledge of the specific activity of the pool of amino acids serving as the immediate precursor for formation of peptide bonds. While the most direct approach to measurement of the specific activity of the precursor pool is to determine this value for a selected amino acid in peptidyl-tRNA, this approach has a number of limitations (3). Therefore, the specific activity of amino acids acylated to tRNA, which is more easily determined, has been used by a number of investigators as an estimate of the specific activity of the precursor pool (3).

The amino acid phenylalanine is a suitable marker for determining rates of protein synthesis in the perfused heart and hemicorpus, since it is neither synthesized nor degraded by these preparations. Measurement of the specific activity of phenylalanyl-tRNA has been conducted in the isolated perfused rat heart and used to calculate rates of protein synthesis (4). When the perfusate concentration of phenylalanine was 0.01 mM, one-eighth the normal plasma concentration, specific activity of intracellular phenylalanine was 36% of that in the perfusate, whereas specific activity of phenylalanyl-tRNA was 59% of perfusate phenylalanine. An increase of phenylalanine concentration in the perfusate to 0.40 mM, five times the normal plasma level, expanded the intracellular pool of phenylalanine proportionally and equalized the specific activities of intracellular and perfusate phenylalanine. Under these conditions, the specific activity of phenylalanyl-tRNA reached approximately 95% of that of phenylalanine in the perfusate. When rates of protein synthesis were calculated by using specific activities of phenylalanyl-tRNA, the rate was essentially the same at phenylalanine concentrations of 0.01 and 0.4 mM. This finding was confirmed by measuring incorporation of [^{14}C]histidine over a range of phenylalanine concentrations varying from 0.01 to 3.6 mM. Increasing phenylalanine concentration had no effect on histidine incorporation. Therefore, when perfusions are carried out in the presence of phenylalanine at concentrations of 0.4 mM or higher, rates of protein synthesis can be calculated using the specific activity of phenylalanine in the perfusate or intracellular water. This approach has been employed successfully in determining the rates of protein synthesis in the perfused rat heart (4) and in the skeletal muscle contained in the isolated rat hemicorpus (5).

<u>Identification of Peptide-Chain Initiation as a Rate-Limiting Reaction in the Pathway of Protein Synthesis.</u> Factors modifying the rate of protein synthesis in heart and skeletal muscle could act at the following steps: 1) amino acid transport and availability of intracellular amino acids and aminoacyl-tRNA, 2) initiation of peptide chains, and 3) peptide-chain elongation and termination. Alterations in peptide-chain initiation, elongation, and termination have been evaluated by making concomitant estimates of rates of protein synthesis and levels of ribosomal subunits and polysomes. Data obtained from these determinations have been interpreted in a manner analogous to that applied to crossover

plots of intermediates in a metabolic pathway (6). When a reaction provides a significant limitation to flux through the pathway, its substrates accumulate and its products are depleted. For example, if peptide-chain initiation becomes rate-limiting for protein synthesis, ribosomal subunits accumulate and polysomes are depleted. When initiation is accelerated, protein synthesis increases, subunits are depleted, and polysomes accumulate. These estimates require that 1) rates of protein synthesis are estimated accurately, 2) ribosomal aggregation is maintained during tissue preparation, 3) ribosomal subunits are well resolved from polysomes on sucrose-density gradients and 4) ribosomal subunits are recovered quantitatively.

Isolation of polysomes and ribosomal subunits from muscle tissue has required the establishment of optimal Mg^{2+} and monovalent cation concentrations. In cell-free protein synthesizing systems, the optimal Mg^{2+} concentration for maintenance of the synthetic rate is approximately 5 mM (7). Under these conditions, monomeric ribosomes appear to be the products of peptide-chain termination. However, when homogenization buffers contain high Mg^{2+} concentrations (8,9) or are of low ionic strength (10,11), ribosomal subunits aggregate to form larger particles. When heart or skeletal muscle ribosomes are prepared in 0.25 M KCl, Mg^{2+} concentrations greater than 2 mM lead to aggregation of subunits into particles sedimenting as monomers on sucrose gradients. The presence of nucleotides, other ions, and a variety of potential binding sites for divalent cations complicates calculation of the intracellular Mg^{2+} concentration under physiological conditions (12). Indirect evidence suggests that the intracellular concentration of ionized Mg^{2+} ranges below 1 mM (13,14). Thus, in intact muscle cells, the concentration of monovalent cations and ionized Mg^{2+} would favor dissociation of monomeric ribosomes formed by peptide-chain termination into ribosomal subunits. Homogenization conditions for muscle have been chosen such that free ribosomes are present as subunits rather than monomers (15).

Estimates of the rates of protein synthesis and levels of polysomes and ribosomal subunits supply information that localize the effects of hormones and other factors to groups of reactions within the ribosome cycle; however, they provide no indication of the mechanism of these effects. Both initiation and elongation, as defined in these studies, combine many steps that require specific protein factors, tRNA, ribosomes, messenger RNA, and the binding and hydrolysis of GTP.

The partial reactions of peptide-chain initiation have been investigated in only a preliminary way in heart and skeletal muscle. Since the formation of a ternary complex of met-tRNA$_f^{Met}$, GTP, and the initiation factor eIF-2 is an early step in the pathway (16,17), possible changes at this step in association with altered rates of protein synthesis have been evaluated. Formation of the ternary complex requires GTP, is inhibited by GDP, Mg^{2+}, spermine, spermidine, and aurintricarboxylic acid, and is specific for initiator tRNA (18). The properties of the eIF-2 factor from heart and skeletal muscle are similar to those of the factor that has been implicated in the physiological control of initiation in cell-free systems derived from reticulocytes.

RESULTS AND DISCUSSION

In vivo protein synthesis in heart and skeletal muscle of fed, normal rats appeared to be limited by the rate of peptide-chain elongation, since peptide-chain initiation was sufficiently rapid to result in conversion of the bulk of ribosomal subunits into polysomes (15). It is not clear which reaction of chain elongation limits peptide bond formation in vivo. The supply of aminoacyl-tRNA, ribosomes, elongation factors, or GTP could determine how rapidly amino acids are added to growing peptide chains.

When the isolated rat heart was perfused in vitro with buffer containing normal plasma levels of amino acids and glucose, the rate of protein synthesis was linear during the first hour, but then declined sharply during the second and third hours (Figure 1, left panel). Addition of insulin to the perfusate had little, if any, effect during the first hour, but nearly maintained the initial rate of protein synthesis during the last 2 hours of perfusion. Early studies attributed the effects of insulin on protein synthesis to an accelerated rate of amino acid transport (19,20) and increased formation of aminoacyl-tRNA (21,22). Further investigation indicated that the hormone response illustrated in Figure 1 was not accounted for by an increased intracellular supply of amino acids (23). In rat hearts perfused for 3 hours with buffer containing normal plasma levels of amino acids and 15 mM glucose, insulin reduced the intracellular levels of 9 amino acids even though protein synthesis was accelerated.

Insulin could have accelerated protein synthesis by maintaining higher levels of high-energy phosphates. However, in both the presence and absence of the hormone, heart rate and ventricular pressure development were stable for 3 hours of perfusion (24). Furthermore, ATP levels did not decline significantly from unperfused tissue levels during 4.5 hours of perfusion (23). In hearts perfused for 1.5-3 hours, creatine phosphate was elevated in the presence of insulin com-

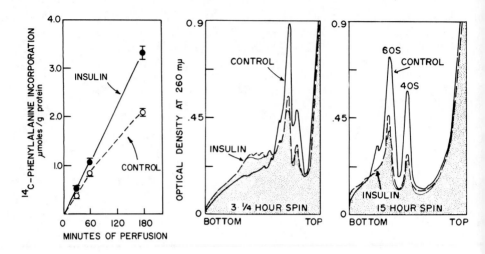

FIGURE 1. Effects of perfusion and insulin on protein synthesis and ribosomal aggregation in perfused rat heart. Hearts were perfused for 30-180 min with buffer containing 0.4 mM [^{14}C]phenylalanine and normal plasma levels of the other amino acids, 15 mM glucose, 3% albumin and, where indicated, 1 µg/ml insulin. Estimates of protein synthesis (left panel), as µmoles of phenylalanine incorporated per g protein, were calculated by dividing the dpm incorporated per g of muscle protein by the specific activity of the intracellular phenylalanine precursor pool. Sucrose gradient analysis of polysomes and ribosomal subunits (center and right panels) was performed using unperfused hearts from fed rats and hearts perfused for 180 min. Muscles were homogenized in 3 volumes of 10 mM Tris-HCl buffer, pH 7.4, containing 250 mM KCl and 2 mM $MgCl_2$. The postmitochondrial supernatants were layered onto sucrose gradients (0.4-2.0 M) and the gradients were centrifuged at 200,000 x g. Centrifugation was continued 3.25 or 15 h to better resolve polysomes or ribosomal subunits, respectively. Polysome and subunit profiles were resolved by monitoring continuously in a flow cell the UV absorbance at 260 mµ of the sucrose gradients. Shaded areas indicate the levels of polysomes and ribosomal subunits found in unperfused hearts. Positions of the 60S and 40S ribosomal subunits are indicated.

pared with control tissue. The significance of this observation with respect to the effect of the hormone on protein synthesis is not presently understood.

The alterations in the rate of protein synthesis produced by perfusion and insulin were accompanied by changes in the levels of polysomes and ribosomal subunits (Figure 1, center and right panels). An accumulation of ribosomal subunits and a fall in polysomes in association with the decline in the rate of protein synthesis indicated that a restraint on peptide-chain initiation had developed during perfusion and that this step had become rate-limiting for protein synthesis. Addition of insulin prevented this inhibition as evidenced by maintenance of the initial rate of protein synthesis and <u>in vivo</u> levels of polysomes and ribosomal subunits.

Virtually identical results were obtained in skeletal muscle perfused in the hemicorpus preparation. In skeletal muscle perfused for up to 1 hour (Figure 2, panel A and panel D), ribosomal subunit levels were maintained at the low levels of the unperfused condition. Polysomes were also maintained at the pre-perfusion level (5). For perfusions within this time period, the presence of insulin in the medium was not effective in increasing the rate of protein synthesis or in altering levels of ribosomal subunits or polysomes. With extension of perfusion time to 2 (panels B and E) or 3 (panels C and F) hours, the rate of protein synthesis declined from the rate that was measured during the first hour of perfusion and levels of ribosomal subunits were markedly elevated from the unperfused level (polysomes decreased from the pre-perfusion level). The decline in the rate of protein synthesis, therefore, resulted from a restraint on peptide-chain initiation which developed after 1 hour of perfusion. The presence of insulin in the initial buffer or addition of insulin after 2 hours of perfusion (panels C and F) effectively stimulated protein synthesis within 1 hour. This stimulation was brought about by the complete reversal of the restraint on peptide-chain initiation as evidenced by the return of ribosomal subunit and polysome levels to those observed in the unperfused condition. The ability of insulin to prevent or reverse the development of a restraint on peptide-chain initiation was not blocked by actinomycin D (unpublished data), suggesting that RNA synthesis is not essential for the hormone effect on initiation.

This effect of insulin on protein synthesis was established in muscle cells by measuring the rate of synthesis of myosin. During 3 hours of perfusion, myosin synthesis in either heart or skeletal muscle occurred at about 60% of the rate of total muscle protein and was increased 60-100% in both tissues when insulin was added to the perfusates.

From these studies, it is evident that protein synthesis in heart and skeletal muscle is limited in perfusion-induced insulin deficiency at the level of peptide-chain initiation.

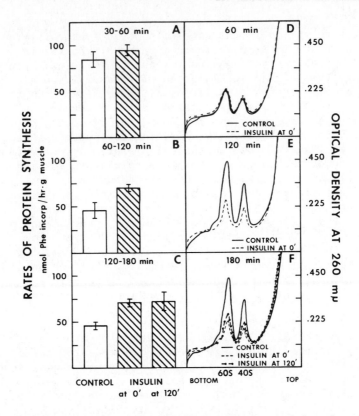

FIGURE 2. Rates of protein synthesis and levels of ribosomal subunits in perfused skeletal muscle of the rat. Rates of protein synthesis during the indicated time intervals were determined in the gastrocnemius muscle of the perfused hemicorpus as described in Figure 1. Levels of 60S and 40S ribosomal subunits were determined on samples of postmitochondrial supernatant from homogenized psoas muscle taken at the indicated times of perfusion. Preparations were perfused with buffer containing 3% bovine serum albumin (Fraction V), 15 mM glucose, 0.4 mM [^{14}C]phenylalanine, normal plasma levels of the other amino acids, and 30% washed bovine erythrocytes (Control), or with the above medium containing 1 µg/ml insulin (Insulin, shaded bars) added at the start of perfusion (0') or after 120 minutes of perfusion (120'). Protein synthesis rates are expressed as nmoles of phenylalanine incorporated per h per g muscle. Values are means ± SEM. The levels of ribosomal subunits seen in Panel D are equal to the levels found in unperfused muscle.

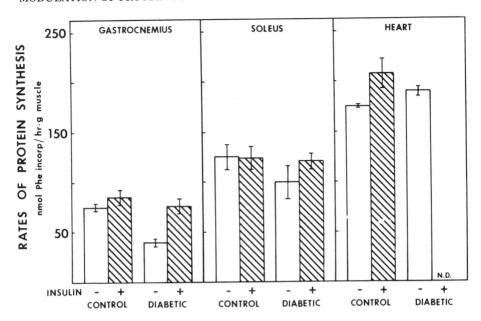

FIGURE 3. Effects of diabetes and insulin on rates of protein synthesis in perfused muscles of the rat. Rates of protein synthesis, as nmoles phenylalanine incorporated per h per g muscle, were determined by measuring the incorporation of [^{14}C]phenylalanine into muscles perfused for 60 min. Gastrocnemius and soleus muscles were perfused in the isolated hemicorpus preparation with buffer containing 4.5% bovine serum albumin (Fraction V), 15 mM glucose, 0.4 mM [^{14}C]phenylalanine and normal plasma levels of the other amino acids. Hearts were perfused with the buffer described above with the exception that 3% bovine serum albumin (Fraction V) was used. Forty-eight hours prior to the experiment, rats (200-300 g) were injected intravenously with 0.9% NaCl (Control) or 60 mg/kg of body weight of alloxan monohydrate (Diabetic) after an overnight fast. The effect of insulin in vitro was tested by the inclusion of insulin in the initial perfusion medium at a concentration of 1 µg/ml (shaded bars).

It would be expected, therefore, that the reactions of peptide-chain initiation would also be limiting in muscle taken from diabetic animals. To test this possibility, the rate of protein synthesis in different muscles from diabetic rats were measured during the first hour of perfusion prior to the onset of perfusion-induced insulin deficiency. The results of synthesis measurements in heart and in two skeletal muscles, the gastrocnemius and soleus, are shown in Figure 3. In gastrocnemius, an example of muscle with mixed red and white fiber distributions, diabetes of 2 days' duration resulted in the anticipated reduction in the rate of protein synthesis. This decrease was due to both loss of tissue RNA and reduced translational efficiency, i.e. synthesis per unit RNA (25). In contrast, in the soleus, a muscle with primarily red fibers, protein synthesis was much less affected by diabetes. The reduction in protein synthesis in this muscle was due entirely to reduced concentrations of RNA, as no decrease in translational efficiency was detected. Similarly, the rate of protein synthesis in heart was not altered in animals with diabetes of 2 days' duration. In gastrocnemius, addition of insulin to the perfusion medium restored protein synthetic efficiency to the same level as that observed in muscle of normal animals. The tissue RNA concentration was not changed by the presence of insulin in the medium. Therefore, only the gastrocnemius could acutely respond to insulin, since this was the only muscle of the three in which translational efficiency was impaired in diabetes.

To gain a better understanding of the differential responses of these muscles to diabetes, the _in vivo_ levels of ribosomal subunits were determined (Figure 4). In gastrocnemius and psoas, muscles with a mixed fiber distribution, ribosomal subunit levels indicated that polysomal disaggregation occurred in the diabetic state. Development of a restraint on peptide-chain initiation, then, was the cause of the reduced translational efficiency determined _in vitro_ (Figure 3). However, the subunit levels in heart muscle from diabetic rats were not elevated and, therefore, did not indicate an impairment in peptide-chain initiation. Subunit levels in soleus were indicative of only a small degree of polysomal disaggregation. Therefore, the _in vitro_ data indicating no impairment in translational efficiency in heart and soleus of 2-day diabetic rats (Figure 3) were accurately reflected in the state of polysomal aggregation in these muscles. It appears that muscles with more red fibers are less susceptible to development of a restraint on peptide-chain initiation caused by the insulin deficiency of diabetes.

These studies indicate that there is some process or factor that protects red muscle in short-term diabetes from the inhibition of protein synthesis caused by insulin deficiency. We proposed previously that the high levels of circulating fatty acids seen in diabetic animals are responsible for maintaining the translational efficiency in heart muscle of rats with diabetes of 2 days' duration (26). Indeed, in heart muscle perfused for 3 hours, the presence of a fatty acid (e.g. palmitate) in the perfusate was as effect-

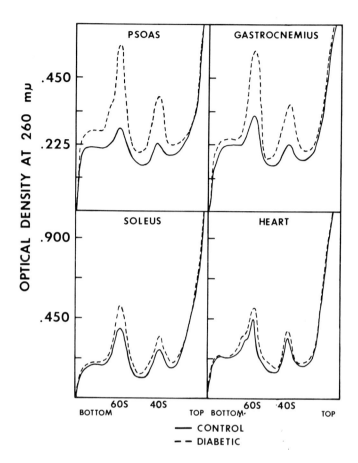

FIGURE 4. Effect of diabetes on the in vivo levels of ribosomal subunits in muscles of the rat. 60S and 40S ribosomal subunits were isolated by sucrose-density gradient centrifugation as described in Figure 1. Control and Diabetic rats were prepared as described in Figure 3. Tissue RNA concentrations (mg RNA per g muscle) in control rats were approximately 1.4 in psoas and gastrocnemius and 1.7 in soleus and heart. RNA concentrations in diabetic rats were approximately 1.1 in psoas and gastrocnemius and 1.5 in soleus and heart.

ive as insulin in maintaining the rate of protein synthesis (Figure 5, left panel). Other studies showed that in addition to palmitate, provision of lactate, pyruvate, acetoacetate, β-hydroxybutyrate, or oleate was also effective in maintaining protein synthesis (27).

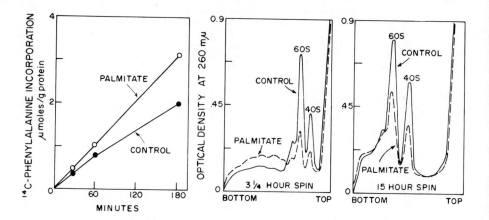

FIGURE 5. Effects of palmitate on protein synthesis and ribosomal aggregation in perfused rat heart. Hearts were perfused for 30-180 min with buffer containing 0.4 mM [^{14}C]-phenylalanine and normal plasma levels of the other amino acids, 15 mM glucose, 3% albumin and, where indicated, 1.5 mM palmitate. Protein synthesis was estimated by measuring phenylalanine incorporation (left panel) as described in Figure 1. Sucrose gradient analysis of polysomes and ribosomal subunits was conducted as described in Figure 1 using hearts that had been perfused for 180 min (center and right panels).

The mechanism of the stimulatory effect of non-carbohydrate substrates on protein synthesis was investigated by measuring intracellular levels of free amino acids, rates of entry of amino acids into the cell, and levels of adenine and guanine nucleotides and creatine phosphate (27). When hearts were perfused for 3 hours with buffer containing normal plasma levels of amino acids and glucose, intracellular levels of all amino acids (except glutamic acid) remained at or above the concentration found in vivo (27). After 3 hours of perfusion with buffer containing glucose and palmitate, only the intracellular level of alanine was lower than that found in vivo (27). Furthermore, palmitate did not appear to supply increased substrate through amino acid transport, since accumulation of α-aminoisobutyric acid was inhibited and entry of aspartic acid, glutamic acid, glycine, isoleucine, lysine, methionine, and tryptophan were unaffected. Increased intracellular levels of glutamic acid and glutamine in hearts given palmitate were probably due to increased supply of carbon to the citric acid cycle. However, glutamic acid was present at very high levels within the cell (7-10 mM) and was not likely to have been rate-limiting for synthesis in control hearts. Intracellular levels of leucine were elevated after 3 hours of perfusion with buffer containing palmitate, perhaps as a result of decreased oxidation. Therefore, increased intracellular availability of amino acids does not appear to account for the stimulatory effect of non-carbohydrate substrates on protein synthesis.

Levels of creatine phosphate were lower in hearts supplied with glucose as substrate, as compared to glucose plus palmitate, whereas ATP levels were unchanged. Energy charge (28) of the adenylate system was 0.88 in hearts given glucose and 0.92 in those supplied with palmitate (27). The role of the somewhat higher energy charge of the adenylate system and the higher levels of creatine phosphate in the effect of fatty acids on protein synthesis is difficult to evaluate. Intracellular levels of ATP and GTP remained well above the K_m values for amino acid activation and peptide-chain elongation (29-31).

As was the case with insulin, addition of palmitate to the perfusate not only maintained the initial rate of protein synthesis, but also maintained polysomes and ribosomal subunits at levels found in vivo during 3 hours of perfusion (Figure 5, center and right panels). In other experiments, the fatty acid was shown to reverse the perfusion-induced impairment in initiation. Provision of physiological levels of oleate, lactate, pyruvate, acetoacetate, β-hydroxybutyrate, or acetate had a similar effect in preventing the restraint on peptide-chain initiation from developing in hearts provided glucose alone (27). These findings suggest that a metabolite common to all of these substrates may be responsible for the stimulation of the synthetic pathway. Maintenance of peptide-chain initiation in heart muscle of diabetic animals, therefore, may be due to a protective effect of fatty acids.

In contrast to the situation in heart muscle, palmitate had no effect on protein synthesis or ribosomal aggregation in perfused skeletal muscle of mixed fiber distribution (5). Absence of an effect of fatty acids in mixed skeletal muscle allowed a reduction in the rate of protein synthesis in insulin deficient states (e.g. diabetes and starvation), and contributed to a net loss of protein from these tissues. Since fatty acids are effective in protecting the heart from the development of an impairment in peptide-chain initiation during insulin deficient states, they may also exert this effect in red skeletal muscle. However, the effects of fatty acids on the rate of protein synthesis and the levels of ribosomal subunits in red skeletal muscle remain to be deter-

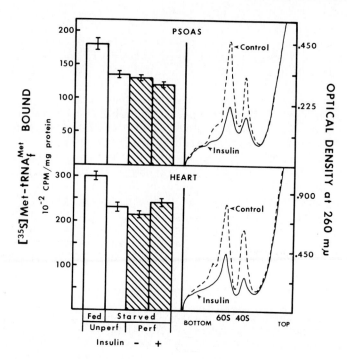

FIGURE 6. Effects of perfusion and insulin on levels of the initiation factor eIF-2 and ribosomal subunits in muscle of the rat. Formation of the ternary complex (met-tRNA$_f^{Met}$ · GTP · eIF-2) was determined by measuring the binding of radioactivity to nitrocellulose filters after a 15 minute incubation of postribosomal supernatant (10-50 µl) from psoas or heart muscle with ^{35}S-Met-tRNA$_f^{Met}$ in a total volume of 150 µl containing 90 mM KCl, 20 mM Tris, pH 7.5, 3 mM dithiothreitol, and 1.33 mM GTP. Muscles were taken from rats (250-350 g) which were fed (Fed) or were fasted for 48 hours (Starved). Heart and psoas muscles were taken from 48-hr fasted rats and were perfused as described in Figures 1 and 3 with buffer containing normal plasma levels of amino acids and with (+) or without (-) the addition of 1 µg/ml insulin. Results are expressed as cpm of ^{35}S-Met-tRNA$_f^{Met}$ bound per mg protein; means ± SEM are shown. Ribosomal subunits from psoas or heart muscle perfused with (Insulin) or without (Control) the addition of 1 µg/ml insulin were isolated on sucrose-density gradients as described in Figure 1. Levels of RNA in psoas and heart were approximately 1.4 and 1.7 mg per g muscle, respectively.

mined. In diabetes of longer duration, the fat stores of the animal would be depleted and peptide-chain initiation in heart may not be maintained.

Because insulin, fatty acids, and non-carbohydrate substrates act to stimulate protein synthesis at the level of peptide-chain initiation, recent investigations have focused on the ability of these factors to affect the partial reactions of this process. Since the formation of a ternary complex of met-tRNA$_f^{Met}$, GTP, and the initiation factor eIF-2 is an early step in the pathway (16,17), possible alterations at this step were first studied. The ability of postribosomal supernatants from both psoas and heart to form this ternary complex was reduced in muscles removed from 48-hour fasted rats (Figure 6). The fasted condition represents an insulin deficient state and levels of ribosomal subunits isolated from psoas are indicative of an impairment in peptide-chain initiation, whereas initiation in heart is not affected under these conditions. Therefore, alterations in the amount of complex formed did not parallel changes in initiation. Furthermore, the formation of ternary complex by postribosomal supernatants taken from perfused muscles was not increased by the presence of insulin in the perfusion medium even though peptide-chain initiation was stimulated under the same conditions (Figure 6). These experiments do not support the concept that the effect of insulin on initiation involved an alteration in ternary complex formation. However, an effect of insulin on this reaction cannot be ruled out, since the binding assay used to determine activity measured total eIF-2 and may not distinguish between _in_ _vivo_ active and inactive forms of the factor. Furthermore, interconversion of active and inactive forms during tissue preparation and assay may occur. It is interesting to note that eIF-2 is subject to phosphorylation and dephosphorylation (32-34). While opinions are divided as to the physiological significance of this covalent modification (35-41), it may represent a means by which the activity of the factor could be regulated by insulin and other agents.

The partial reactions of peptide-chain initiation have been studied most extensively in cell-free systems derived from reticulocyte lysates. In the lysate systems, the initiation step was identified as the rate-limiting step in protein synthesis under a number of conditions including the following: 1) in the absence of added hemin (42-44); 2) in the presence of oxidized glutathione (45,46) or low levels of double-stranded RNA (47); 3) in lysates depleted of low molecular weight components by gel filtration (48), and 4) in the presence of inhibitors of oxidative phosphorylation (49-51). In contrast, a variety of phosphorylated sugars

TABLE 1
CONCENTRATIONS OF GLUCOSE-6-PHOSPHATE IN HEART AND
SKELETAL MUSCLE

Animal	Tissue	Glucose-6-Phosphate μmoles/g dry weight
Normal Rat	Heart	0.80 ± 0.09(4)
Diabetic Rat	Heart	2.23 ± 0.33(4)[a]
Normal Rat	Gastrocnemius	1.15 ± 0.09(4)
Diabetic Rat	Gastrocnemius	1.42 ± 0.19(4)

Effect of diabetes on glucose-6-phosphate concentrations in heart and gastrocnemius muscle. Determinations were made 2 days after induction of diabetes with alloxan (60 mg/kg, IV). The average body weight of the rats was 260 g and 315 g for diabetics and controls, respectively. Following induction of anesthesia (Nembutal, 50 mg/kg) muscles were quick-frozen and pulverized at the temperature of liquid nitrogen. Glucose-6-phosphate concentrations in acid-soluble extracts were determined as described previously (53).
[a] $p < 0.001$ vs normal rat heart

plus $NADP^+$ stimulated initiation in lysates prepared from glucose-starved or ATP-depleted cells (50-51), while a similar stimulatory effect was observed with glucose, citrate, and fructose-1,6-diphosphate in lysates depleted of low molecular weight components by gel filtration (48). These observations implicated a role for phosphorylated sugars, and possibly NADPH production, in the maintenance of the initiation step in reticulocyte lysates. A recent study demonstrated a direct role for glucose-6-phosphate in the regulation of peptide-chain initiation in reticulocyte lysates (52). This report presented evidence that the effect of glucose-6-phosphate was unrelated to its ability to generate NADPH or to its metabolism in glycolysis or the pentose phosphate pathway, suggesting that the compound may act as an allosteric regulator of some step in the initiation process.

A similar role for glucose-6-phosphate in the regulation of peptide-chain initiation may exist in heart and skeletal muscle. While this proposal is only speculative at this time, it should be noted that the levels of this intermediate in muscle are known to change in a manner consistent with

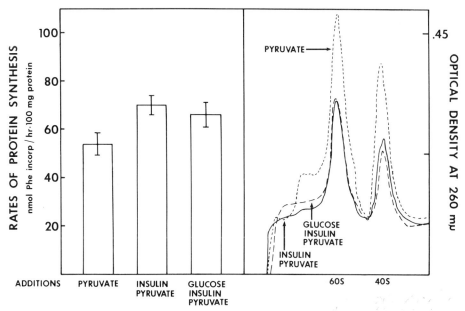

FIGURE 7. Effects of insulin on rates of protein synthesis and levels of ribosomal subunits in hearts perfused with pyruvate and pyruvate plus glucose as substrates. Hearts were perfused for 70 min with non-recirculating buffer containing 15 mM glucose, 0.4 mM phenylalanine, and normal plasma levels of the other amino acids. Hearts were then switched to a recirculating medium containing amino acids as before but with 0.1 µCi/ml [^{14}C]phenylalanine added, and 5 mM pyruvate alone or in combination with 1 µg/ml insulin and 15 mM glucose. Rates of protein synthesis and levels of ribosomal subunits were determined as described in Figure 1 on hearts taken after an additional 60 min of perfusion. In hearts perfused with only glucose (15 mM) in the recirculating medium, the rate of protein synthesis was 44 ± 2 nmoles phenylalanine incorporated per h per 100 mg protein.

such a role. For example, the maintenance of initiation in hearts of diabetic rats was associated with elevated levels of glucose-6-phosphate (Table 1). In contrast, the impairment in initiation that developed in skeletal muscle of diabetic rats was associated with no increase in levels of glucose-6-phosphate (Table 1). Hearts perfused in the presence of glucose and palmitate contained higher glucose-6-phosphate levels than those perfused in the presence of glucose alone (54), suggesting that the effect of fatty acids on protein synthesis may be related to their ability to cause an elevation in the concentration of this metabolic intermediate. Fatty acids were identified as the major oxidative substrates of heart muscle (55), accounting for 75-90% of cardiac oxygen consumption. Provision of fatty acids decreased glucose utilization in heart by inhibiting phosphofructokinase (56), leading to an accumulation of glucose-6-phosphate which, in turn, inhibited hexokinase (57) and membrane transport of glucose (58). Attempts to demonstrate a similar relationship between fatty acid utilization and glucose metabolism in mixed skeletal muscle were unsuccessful (2,59), although findings in red skeletal muscle (60) suggested that it may be more like heart in this respect. Thus, the abilities of various muscles to utilize fatty acids and elevate glucose-6-phosphate levels may determine the extent to which an impairment in peptide-chain initiation develops in diabetes and starvation.

Glucose-6-phosphate levels were also elevated in heart and skeletal muscle perfused with glucose plus insulin compared to muscles perfused with glucose alone (61), raising the possibility that glucose-6-phosphate may mediate the effects of insulin on protein synthesis. However, insulin stimulated peptide-chain initiation in skeletal muscle perfused in the absence of substrate (5). Insulin was also effective at stimulating initiation but did not elevate glucose-6-phosphate levels (unpublished data) in hearts perfused with pyruvate as substrate (Figure 7). Therefore, insulin appears capable of regulating initiation by a mechanism independent of glucose-6-phosphate levels.

On the basis of these findings, regulation of peptide-chain initiation appears to be analogous to the model proposed for the control of glycogen synthase (62). In this model, glycogen synthase is activated by glucose-6-phosphate via a mechanism involving the activation of a phosphoprotein phosphatase. Insulin is proposed to act independently of glucose-6-phosphate, by a mechanism involving an inhibition of a protein kinase. Whether glucose-6-phosphate and insulin are acting to control the state of phosphorylation of a component of the peptide-chain initiation process is yet to be established.

REFERENCES

1. Morgan, H. E., Earl, D. C. N., Broadus, A., Wolpert, E. B., Giger, K. E., and Jefferson, L. S. (1971) J. Biol. Chem. 246, 2152.
2. Jefferson, L. S., Koehler, J. O., and Morgan, H. E. (1972) Proc. Natl. Acad. Sci. USA 69, 816.

3. Rannels, D. E., McKee, E. E., and Morgan, H. E. (1977) In "Biochemical Actions of Hormones" (G. Litwack, ed.) Vol. 4, pp. 135-195. Academic Press, New York.
4. McKee, E. E., Cheung, J. Y., Rannels, D. E., and Morgan, H. E. (1978) J. Biol. Chem. 253, 1030.
5. Jefferson, L. S., Li, J. B., and Rannels, S. R. (1977) J. Biol. Chem. 252, 1476.
6. Chance, B., and Williams, G. R. (1956) Advan. Enzymol. 17, 65.
7. Jackson, R. J. (1975) In "Synthesis of Amino Acids and Proteins" (Intern. Rev. Science, Biochemistry Series One) (H. Arnstein, ed.) pp. 89-135. University Park, Baltimore.
8. Martin, T. E., Rolleston, F. S., Low, R. B., and Wool, I. G. (1969) J. Mol. Biol. 43, 135.
9. Reader, R. W., and Stanners, C. P. (1967) J. Mol. Biol. 28, 211.
10. Martin, T. E., and Hartwell, L. H. (1970) J. Biol. Chem. 245, 1504.
11. Zylber, E. A., and Penman, S. (1970) Biochim. Biophys. Acta 204, 221.
12. Botts, J., Chaskin, A., and Schmidt, L. (1966) Biochemistry 5, 1360.
13. Polimeni, P. I., and Page, E. (1973) Circulation Res. 33, 367.
14. Rose, I. A. (1968) Proc. Natl. Acad. Sci. USA 61, 1079.
15. Morgan, H. E., Jefferson, L. S., Wolpert, E. B., and Rannels, D. E. (1971) J. Biol. Chem. 246, 2163.
16. Trachsel, H., Erni, B., Schreier, M. H., and Staehelin, T. (1977) J. Mol. Biol. 116, 755.
17. Benne, R., and Hershey, J. W. B. (1978) J. Biol. Chem. 253, 3078.
18. Rannels, D. E., Pegg, A. E., Rannels, S. R., and Jefferson, L. S. (1978) Am. J. Physiol. 235(2), E126.
19. Kipnis, D. M., and Noall, M. W. (1958) Biochim. Biophys. Acta 28, 226.
20. Wool, I. G., and Krahl, M. E. (1959) Am. J. Physiol. 196, 961.
21. Davey, P. J., and Manchester, K. L. (1969) Biochim. Biophys. Acta 182, 85.
22. Manchester, K. L. (1970) Biochem. J. 117, 457.
23. Rannels, D. E., Kao, R., and Morgan, H. E. (1975) J. Biol. Chem. 250, 1694.
24. Neely, J. R., Liebermeister, H., Battersby, E. J., and Morgan, H. E. (1967) Am. J. Physiol. 212, 804.
25. Millward, D. J., Garlick, P. J., Stewart, R. J. C., Nnanyelugo, D. O., and Waterlow, J. C. (1975) Biochem. J. 150, 235.

26. Jefferson, L. S., Rannels, D. E., Munger, B. L., and Morgan, H. E. (1974) Fed. Proc. 33, 1089.
27. Rannels, D. E., Hjalmarson, A. C., and Morgan, H. E. (1974) Am. J. Physiol. 226, 528.
28. Atkinson, D. E. (1968) Biochemistry 7, 4030.
29. Bergmann, F. H., Berg, P., and Dieckmann, M. (1961) J. Biol. Chem. 236, 1735.
30. Lin, S. Y., McKeehan, W. L., Culp, W., and Hardesty, B. (1969) J. Biol. Chem. 244, 4340.
31. Siler, J., and Moldave, K. (1969) Biochim. Biophys. Acta 195, 138.
32. Kramer, G., Cimadevilla, J. M., and Hardesty, B. (1976) Proc. Natl. Acad. Sci. USA 73, 3078.
33. Levin, D. H., Ranu, R. S., Ernst, V., and London, I. M. (1976) Proc. Natl. Acad. Sci. USA 73, 3112.
34. Farrell, P. J., Balkow, K., Hunt, T., Jackson, R. J., and Trachsel, H. (1977) Cell 11, 187.
35. Datta, A., de Haro, C., Sierra, J. M., and Ochoa, S. (1977) Proc. Natl. Acad. Sci. USA 74, 1463.
36. Lenz, J. R., and Baglioni, C. (1978) J. Biol. Chem. 253, 4219.
37. Clemens, M. J., Henshaw, E. C., Rahamimoff, H., and London, I. M. (1974) Proc. Natl. Acad. Sci. USA 71, 2946.
38. Pinphanichakarn, P., Kramer, G., and Hardesty, B. (1976) Biochem. Biophys. Res. Commun. 73, 625.
39. de Haro, C., Datta, A., and Ochoa, S. (1978) Proc. Natl. Acad. Sci. USA 75, 243.
40. de Haro, C., and Ochoa, S. (1978) Proc. Natl. Acad. Sci. USA 75, 2713.
41. Ranu, R. S., London, I. M., Das, A., Dasgupta, A., Majumdar, A., Ralston, R., Roy, R., and Gupta, N. K. (1978) Proc. Natl. Acad. Sci. USA. 75, 745.
42. Adamson, S. D., Herbert, E., and Godchaux, W., III. (1968) Arch. Biochem. Biophys. 125, 671.
43. Zucker, W. V., and Schulman, H. M. (1968) Proc. Natl. Acad. Sci. USA 59, 582.
44. Rabinovitz, M., Freedman, M. L., Fisher, J. M., and Maxwell, C. R. (1969) Cold Spring Harbor Symp. Quant. Biol. 34, 567.
45. Kosower, N. S., Vanderhoff, G. A., Benerofe, B., Hunt, T., and Kosower, E. M. (1971) Biochem. Biophys. Res. Commun. 45, 816.
46. Kosower, N. S., Vanderhoff, G. A., and Kosower, E. M. (1972) Biochim. Biophys. Acta 272, 623.
47. Ehrenfeld, E., and Hunt, T. (1971) Proc. Natl. Acad. Sci. USA 68, 1075.

48. Hunt, T. (1976) Br. Med. Bull. 32, 257.
49. Freudenberg, H., and Mager, J. (1971) Biochim. Biophys. Acta 232, 537.
50. Giloh (Freudenberg), H., and Mager, J. (1975) Biochim. Biophys. Acta 414, 293.
51. Giloh (Freudenberg), H., Schochot, L., and Mager, J. (1975) Biochim. Biophys. Acta 414, 309.
52. Ernst, V., Levin, D. H., and London, I. M. (1978) J. Biol. Chem. 253, 7163.
53. Lang, G., and Michal, G. (1974) In "Methods of Enzymatic Analysis" (H. U. Bergmeyer, ed.), Vol. 3, pp. 1238-1242. Academic Press, New York.
54. Neely, J. R., Whitfield, C. F., and Morgan, H. E. (1970) Am. J. Physiol. 219, 1083.
55. Bing, R. J., Siegal, A., Ungar, I., and Gilbert, M. (1954) Am. J. Med. 16, 504.
56. Newsholme, E. A., Randle, P. J., and Manchester, K. L. (1962) Nature 193, 270.
57. England, P. J., and Randle, P. J. (1967) Biochem. J. 105, 907.
58. Neely, J. R., Bowman, R. H., and Morgan, H. E. (1969) Am. J. Physiol. 216, 804.
59. Goodman, M. N., Berger, M., and Ruderman, N. B. (1974) Diabetes 23, 881.
60. Rennie, M. J., and Holloszy, J. O. (1977) Biochem. J. 168, 161.
61. Regen, D. M., Davis, W. W., Morgan, H. E., and Park, C. R. (1964) J. Biol. Chem. 239, 43.
62. Lawrence, J. C., Jr., and Larner, J. (1978) J. Biol. Chem. 253, 2104.

EFFECT OF PHOSPHORYLATION ON eIF-2

William C. Merrick

Department of Biochemistry, Case Western Reserve University, School of Medicine, Cleveland, Ohio 44106

ABSTRACT The molecular structure of eIF-2 has been determined to be an $\alpha\beta\gamma$ trimer of molecular weight 122,000. The GTP initially bound by eIF-2 to form a soluble ternary complex appears to be the same GTP that is hydrolyzed as a prerequisite for subunit joining. Using the above information, experiments were performed to test the effect of HCR mediated phosphorylation of eIF-2 on the biologic function of eIF-2. In every assay tested, whether rate or extent of reaction was measured, there appeared to be no difference in biologic activity between phosphorylated and non-phosphorylated eIF-2.

INTRODUCTION

Within the past five years, three different laboratories have purified most, if not all, of the protein synthesis initiation factors from rabbit reticulocytes (1-3). The preparation of these factors has allowed an extensive investigation into the stepwise utilization of these factors in the formation of an 80S initiation complex with natural (4,5) or synthetic (6) mRNA templates. The overall agreement between these three laboratories formed the corner stone for the foundation of a new nomenclature for eukaryotic initiation factors (7). In addition to the three laboratories cited above, many other laboratories have been instrumental in initially identifying initiation factor activities and elucidating steps in the initiation complex reaction sequence (see reviews 8,9). With an increased understanding of the mechanism of protein synthesis initiation, a new area in protein synthesis has begun to be examined at the molecular level, the area of protein synthesis regulation. Following the initial identification of eIF-2 in the regulation of protein synthesis in hemin deficient lysates (10), many investigators have examined the events which lead to the ceasation of protein synthesis initiation (8,9). The result of this work has indicated that a covalent modification of eIF-2 by a highly specific protein kinase (termed HCR for historical reasons (11)) may be the single event responsible for the inactivation of the initiation complex pathway. As will become evident later, the exact site of inactivation is not clear. There have been reports on several different possibilities and it is also possible that in fact several sites may be effected, not just one. In an attempt to more fully understand how phosphorylation of eIF-2 might alter its biologic activity, studies were undertaken to physically characterize eIF-2 and it's subunits and to better define several simple assay systems to test for biologic differences between phosphorylated and non-phosphorylated eIF-2.

RESULTS

Previous reports in many laboratories have indicated that eIF-2 is composed of three non-identical subunits (1-3, 12-17). However, the stoichiometry of these subunits has been in doubt based upon the wide range of molecular weights reported (90,000 to 180,000) for native eIF-2 (17-22). Recent studies in my laboratory, primarily by Dr. Michele Lloyd, have determined that eIF-2 is composed of one each of three non-identical polypeptides (23). Purification of each of the subunits allowed a variety of correlations between native eIF-2 and reconstitutions of the $\alpha\beta\gamma$ oligomer based upon individual subunit properties. In particular, studies using sedimentation equilibrium analysis, protein cross-linking, and amino acid analysis were consistent only with an $\alpha\beta\gamma$ oligomer. The major curiousity of these studies was that the β subunit which behaves as a 50,000 dalton polypeptide in SDS gel electrophoresis is in fact a 35,000 dalton polypeptide (see Table I). The error in apparent molecular weight by SDS gel electrophoresis may reflect the high content of basic amino acids (20.1%, ref. 23), but other possibilities have not been excluded. Additional characteristics of both eIF-2 and its subunits are listed in Table I.

Having established the molecular structure of eIF-2, the apparent target for translational control, suitable assay systems were then required to attempt to monitor changes in eIF-2 activity as a result of phosphorylation. One of the simplest assays is the AUG dependent synthesis of methionyl-puromycin (24). Use of the assay had previously indicated that two separate pathways existed for 80S initiation complex formation (or methionyl-puromycin synthesis): one pathway used eIF-2 to direct the binding of Met-tRNA$_f$ to 40S subunits; the other used eIF-2A to direct the AUG-dependent binding of Met-tRNA$_f$ to 40S subunits (Figure 1; ref. 24,25). The existance of this dual pathway has recently allowed us to better understand both the function of some of the initiation

TABLE I
CHARACTERISTICS OF NATIVE eIF-2 AND IT'S SUBUNITS

Characteristic	eIF-2 Subunits			Native eIF-2
	α	β	γ	αβγ
M_r	32,000	35,000	55,000	122,000
pI	5.1	5.4	8.9	6.4
Ligand binding	GDP	mRNA Met-tRNA$_f$	—	GTP, GDP Met-tRNA$_f$ mRNA
Phosphorylation	(+)	(+)	—	+

Molecular weights (M_r) were determined by sedimentation equilibrium analyses in 6M guanidine hydrochloride (subunits) or buffered 100 mM KCl (native eIF-2). Isoelectric points (pI) were determined by isoelectric focusing in polyarcylamide gels using denaturing (8 M urea, subunits) or non-denaturing (native eIF-2) Ampholine buffers. Ligand binding characteristics for eIF-2 subunits were as reported by Barrieux and Rosenfeld (12). The ligand binding and phosphorylation of eIF-2 have been determined by several laboratories (8,9). The (+) sign indicates which subunits in native eIF-2 have been reported to be phosphorylated.

factors and the fate of the GTP initially bound to the 40S subunit as part of the ternary complex.

One of the initial questions was whether or not the GTP in the ternary complex was the same one hydrolyzed to allow 80S complex formation (Figure 1, left side) or whether both pathways utilized a separate GTP to allow 60S subunit joining. Therefore, we examined GTP utilization in methionylpuromycin synthesis to determine whether the Km for GTP for both pathways was the same and reflected the Km for eIF-5 directed GTP hydrolysis (~ 10 µM; ref. 26) or whether the pathway using eIF-2 might reflect the Km for GTP for ternary

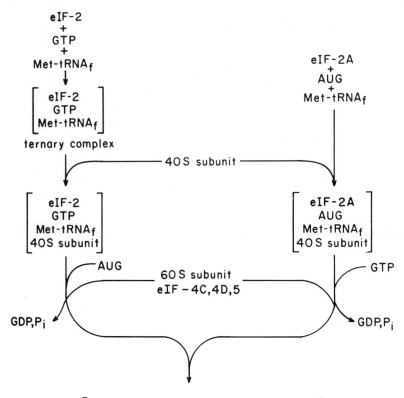

FIGURE 1. Eukaryotic Initiation Complex Formation
See text and ref. 25 for details.

complex formation (~ 0.5 µM; ref. 18). The results are presented in Figure 2. As may be evident, the reaction with eIF-2 is more rapid and requires less GTP. Analysis of the data by plots of 1/V vs. 1/S indicate that the Km for GTP for the eIF-2 pathway was 1.1 µM and for the eIF-2A pathway was 12.6 µM (the relative V_{max} values were 2.5 to 1 respectively). These differences in Km value for GTP are consistent with the interpretation that the GTP bound in ternary complex to the 40S subunit is the GTP hydrolyzed to allow 80S initiation complex formation.

This interpretation has been confirmed using a much different approach which required the isolation of 40S subunit · eIF-2 · GTP · Met-tRNA$_f$ complexes by Sepharose 6B

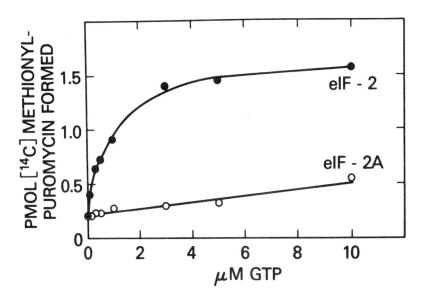

FIGURE 2. GTP Dependent Synthesis of Methionyl-Puromycin using either eIF-2 or eIF-2A. The AUG-dependent synthesis of methionyl-puromycin using [^{14}C] Met-tRNA$_f$, eIF-4C, eIF-4D, eIF-5, and sucrose cushion ribosomes was as described previously (24). The GTP concentration was varied from 0.1 μM to 100 μM in the presence of either eIF-2 or eIF-2A as indicated in the figure.

chromatography. The isolation of such complexes eliminates unbound GTP and Met-tRNA$_f$ so that subsequent reactions are dependent on the presence of these components on the 40S subunit. The utilization of 40S complexes is presented in Table II. The initial 40S complexes usually contain a slight excess of Met-tRNA$_f$ over GTP. The addition of eIF-4C, eIF-4D, eIF-5, 60S subunits and puromycin leads to the conversion of about 50% of the [^{14}C] Met-tRNA$_f$ present in the 40S complexes to [^{14}C] methionyl-puromycin. At the same time there is a similar, although somewhat smaller, amount of [^{32}P]PO$_4$ released from the bound [γ-^{32}P]GTP. This experiment offers direct proof that the GTP initially bound as part of the ternary complex is hydrolyzed during the formation of an 80S initiation complex.

With this knowledge in hand, attempts were then made to evaluate the effect the phosphorylation of eIF-2 might have

TABLE II
REACTION OF 40S COMPLEXES TO YIELD METHIONYL-PUROMYCIN AND GTP HYDROLYSIS

	40S Complex		Product		Time of Incubation
	Met-tRNA$_f$	GTP	Methionyl-puromycin	PO$_4$ Released	(at 37°)
	(pmol)		(pmol)		
Experiment 1	1.84	1.30	0.92	0.88	30 min.
Experiment 2	1.73	1.44	0.94	0.65	30 min.
Experiment 3	1.68	1.04	0.50	0.40	10 min.

For the isolation of 40S preinitiation complexes, 1 ml reaction mixtures were incubated at 37° for 15 min and contained: 20 mM Hepes-KOH, pH 7.5; 2 mM MgCl$_2$; 100 mM KCl; 1 mM dithiothreitol; 3.0 A$_{260}$ units of AUG; 320 pmol eIF-2; 400 pmol 40S subunits; 400 pmol [^{14}C]Met-tRNA$_f$ (sp. act. 260 mCi/mmol); and 80 µM [γ-^{32}P]GTP (sp. act. 403 mCi/mmol). Following incubation, the reaction mixture was applied to a 0.9 x 20 cm column of Sepharose 6B equilibrated with 20 mM Hepes-KOH, pH 7.5, 100 mM KCl, 3 mM MgCl$_2$ and 1 mM dithiothreitol. The column was maintained at 4° and 40S preinitiation complexes were eluted in the void volume at a flow rate of 20 ml per hr. To test for methionyl-puromycin synthesis, 100 µl of column fractions were mixed with an equal volume of buffer containing eIF-4C, eIF-4D, eIF-5, and 60S subunits. The final concentrations of all components were the same as those optimal for methionyl-puromycin synthesis (24) except there was no added GTP, phosphoenolpyruvate or pyruvate kinase. Following a second incubation at 37°, [^{14}C]methionyl-puromycin and [^{32}P]PO$_4$ were quantitated by extraction into an organic phase (24) and subsequent scintillation spectroscopy of aliquots of the organic phase.

on eIF-2 biologic activity. These studies were greatly
facilitated by the generous collaborative efforts of Dr.
Jolinda Traugh. The general outline of the experiments was
to treat eIF-2 with HCR in the presence or absence of $[\gamma\text{-}^{32}P]$
ATP, repurify the eIF-2 by phosphocellulose chromatography
and then test both preparations for biologic activity. The
eIF-2 was quantitated by A_{280} and by radioactivity as the

TABLE III
EFFECT OF HCR PHOSPHORYLATION ON eIF-2 ACTIVITY

Assay	pmol [^{3}H]Met-tRNA$_f$ incorporated into complex (1,2,3) or product (4)	
	eIF-2	eIF-2~PO$_4$
1. Ternary complex formation	8.8	8.5
2. Met-tRNA$_f$ binding to 40S subunits	7.9	7.5
3. Met-tRNA$_f$ binding to 80S ribosomes	3.8	3.3
4. Methionyl-puromycin synthesis	4.2	4.3

Radiolabeled eIF-2, prepared by reductive
methylation with [^{14}C]HCHO (6,27), was treated
with HCR in the presence and absence of $[\gamma\text{-}^{32}P]$ATP.
The treated eIF-2 was then repurified by phosphocellulose column chromatography and both preparations were tested for biologic activity using
previously established assay conditions (24).
Ternary complex formation was determined by retention on Millipore filters; Met-tRNA$_f$ binding
to 40S subunits or 80S ribosomes was determined
by sucrose density gradient centrifugation;
methionyl-puromycin synthesis was determined as
the transfer of [^{14}C]methionine from [^{14}C]Met-tRNA$_f$
to organic extractable [^{14}C]methionyl-puromycin.
For each experiment equivalent amounts of eIF-2
were used and the phosphorylated eIF-1 contained
approximately 2 mol of [^{32}P]PO$_4$ per mol of eIF-2α
subunit.

preparation of eIF-2 used had previously been made radioactive by reductive methylation using [^{14}C]HCHO and sodium borohydride (6,27). For the studies reported here, between 1.8 and 2.1 pmol of [^{32}P]PO$_4$ were incorporated per mol of eIF-2. The two types of eIF-2 were tested in the several different assays as indicated in Table III. As can be seen, little difference between the two eIF-2 preparations was noted whether the assay measured the extent of reaction (experiments 1,2,3) or the rate of the reaction (experiment 4; ternary complex formation, data not shown). From these experiments, it would appear that there has been no alteration of eIF-2 activity as a result of phosphorylation of eIF-2. A similar conclusion was reported by Trachsel and Staehelin (28).

In as much as there appear to be little or no change in the ability of eIF-2 to catalyze the synthesis of an 80S initiation complex, the question arose where else might eIF-2 activity be effected. Two pieces of information came to mind. First, preliminary results indicated that when eIF-2 was added to hemin deficient lysates to allow recovery of protein synthesis, there was only a stoichiometric utilization of eIF-2 (i.e., one mol of globin chain synthesized per mol of eIF-2 added) (29). Secondly, it had been reported earlier that GDP was a very potent inhibitor of ternary complex formation and that perhaps the ratio of GDP to GTP might be involved in the regulation of protein synthesis initiation (30). These two observations suggested to us the possibility that a eIF-2 · GDP complex might be a natural part of the cyclic utilization of eIF-2 as has been observed for the prokaryotic elongation factor Tu (31). It might then be possible that phosphorylation of eIF-2 would lead to an eIF-2 ~ PO$_4$ · GDP complex which might not readily dissociate into eIF-2 ~ PO$_4$ and GDP (either freely or aided by some unknown protein whose function would be similar to EF - Ts (31)). To test this possibility, both phosphorylated and non-phosphorylated eIF-2 were incubated with [8-^{3}H]GDP for 5 min at 4° and then a 10-fold excess of unlabeled GDP was added to the reaction mixture and [8-^{3}H]GDP binding was monitored as a function of time. The results are presented in Figure 3. At zero time the level of [8-^{3}H]GDP binding was the same for both eIF-2 preparations and subsequently, the loss of [8-^{3}H]GDP from both eIF-2 preparations occured at the same rate. This result indicated that there was no change in the eIF-2 binding affinity for GDP as a result of HCR mediated phosphorylation of eIF-2.

During the period of time that these studies were being performed, many other laboratories were also examining the effect of HCR phosphorylation on eIF-2 activity. Three

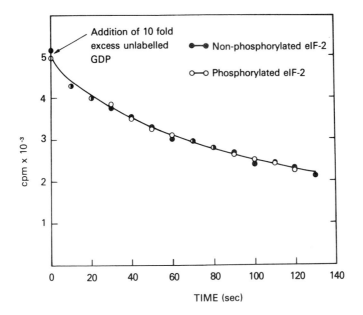

FIGURE 3. [8-^{3}H]GDP Binding to eIF-2 - eIF-2 was incubated in 10 µM [8-^{3}H] in a reaction mixture which contained: 20 mM Tris·HCl, pH 7.5; 100 mM KCl; 1 mM dithiothreitol; and 3 mM MgCl$_2$. Incubation was for 5 min at 4°. Subsequently a 10-fold molar excess of unlabeled GTP was added to the reaction mixture. At time zero and every 10 sec. after that aliquots were tested for the binding of [8-^{3}H]GDP. Samples were diluted with 2 ml cold 100 mM KCl, 20 mM Tris-HCl, pH 7.5, and 5 mM MgCl$_2$ and bound [8-^{3}H]GDP was determined by retention on Millipore filters after vacuum filtration.

different laboratories were able to demonstrate a change in eIF-2 biologic activity accompanying HCR directed phosphorylation (32-34). These studies measured changes in the extent of formation of the ternary complex (33) or in the extent of 40S subunit · ternary complex formation (32,34). A careful examination of the conditions used to demonstrate these differences indicated that the properties of their non-phosphorylated eIF-2 were different from ours. A simplified comparison of "our" eIF-2 (form 1) with "their" eIF-2 (Form 2) is presented in Table IV. While most of the table is self-explanatory in general terms, a few specific points should be made. First, considering the temperature of incubation and the measured rate of ternary complex formation,

TABLE IV
CHARACTERISTICS OF "DIFFERENT" FORMS OF eIF-2

Characteristic	"Form 1"	"Form 2"
1. Rate of ternary complex formation	rapid at 4°	slow at 37°
2. Extent of ternary complex formation	10-40%	less than 10%
3. Stimulation of ternary complex formation by additional protein	−	+
4. Inhibition of ternary complex formation by Mg^{++} ion	±	++
5. Direct binding of ternary complex to 40S subunits	+	−

The characteristics of Form 1 eIF-2 are consistent with and derived from ref. 4-6. The characteristics of Form 2 eIF-2 are from ref. 22, 32-34. In as much as there are considerable differences in assay technique, it is not possible to make quantitative correlations between the two forms of eIF-2. However, a − vs. + rating (line 3) indicates a several fold stimulation. A ± vs. ++ rating (line 4) would indicate a 40% reduction compared to a 95% reduction. The differences in line 5 would be similar to those in line 3.

Form 1 eIF-2 reacts about two orders of magnitude more rapidly than Form 2 eIF-2. Secondly, Form 1 eIF-2 directed ternary complex formation is not sensitive to Mg^{++} at 4°, but there is about 40% reduction in ternary complex formation when measured in the presence of 5 mM $MgCl_2$ at 37° (Merrick, unpublished work). By comparison, ternary complex formation with Form 2 is highly sensitive to Mg^{++} ion (> 90% reduction at 0° in the presence of 5 mM $MgCl_2$, ref. 32). Finally, it should be noted that those reports which have observed differences between phosphorylated and non-phosphorylated eIF-2 have measured differences in extent of reaction (32-34). From these reports it is not clear that a change in the rate of protein synthesis will accompany these observed changes in extent of reaction.

DISCUSSION

Our failure to detect any difference in the biologic activity of eIF-2 following phosphorylation by HCR lead us to consider the following question: "Is the phosphorylation of eIF-2 a necessary, but not sufficient condition for the inhibition of protein synthesis initiation"? If the answer to this question were yes, then some other event must occur concomitant with the phosphorylation of eIF-2. In as much as highly purified HCR is capable of inducing protein synthesis shut-off in reticulocyte lysates (35-37), the most reasonable possibility would appear to be that there is at least one additional substrate for HCR which can effect protein synthesis initiation. Even this possibility seems remote as the HCR specificity is such that eIF-2 is the only known substrate for this protein kinase.

An alternate possibility is that during the normal initiation cycle, eIF-2 must interact with some protein or factor which is sensitive to the phosphorylation of eIF-2 by HCR. In this regard, the AUG directed synthesis of an 80S initiation complex represents a great simplification from the translation of "normal" mRNAs. As can be seen in Figure 4, there are many additional requirements for the initiation of a natural mRNA. The additional initiation factors (eIF-1, eIF-3, eIF-4A, and eIF-4B) represent an additional 900,000 daltons of protein in about 14 polypeptide chains (compare with the AUG system which requires 6 polypeptide chains and 280,000 daltons of protein). This additional protein does not include any protein which may normally be bound to mRNA (as mRNPs) or any factors which may positively or negatively effect the rate of protein synthesis initiation which have yet to be identified. This additional protein, much of which must bind to the 40S subunit at some point, could very easily be responsible for differentiating between phosphorylated and non-phosphorylated eIF-2. It should be noted that a complication arises in that several of the initiation factors appear to exist normally with some degree of phosphorylation (38-39). Preliminary studies with several of the reticulocyte kinases and initiation factors would seem to indicate that the purified factors contain relatively little phosphate (Traugh and Merrick, unpublished data). Therefore, it is possible that several initiation factors must be phosphorylated before the phosphorylation of eIF-2 becomes inhibitory.

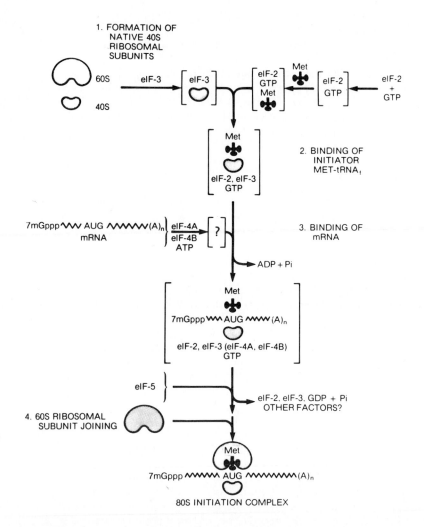

FIGURE 4 Possible Pathway for Natural mRNA-Directed Synthesis of 80S Initiation Complexes. The flow diagram pictured above is a composite from the studies reported in ref 4-6. Factors eIF-1, eIF-4C, and eIF-4D are not included as they tend to stimulate several steps in the pathway. However, the major effect of eIF-4C and eIF-4D seems to be at step 4.

There is one final word of caution that should be presented. Most, if not all, of the assays with purified initiation factors fail to demonstrate catalytic utilization of the factors except for eIF-5. Also, the reconstituted protein synthesizing systems only function at about 5% of the efficiency of the rabbit reticulocyte lysate. Therefore, it is quite probable that it will be difficult to demonstrate a quantitative change in eIF-2 activity that corresponds to the loss of synthetic activity in hemin-deficent lysates. This increases the need for the careful evaluation of minor changes (2-3 fold) in eIF-2 activity until they have been verified in a highly catalytic assay system such as the reticulocyte lysate.

ACKNOWLEDGEMENTS

Much of the work reported here was the dissertation project of Dr. Michele Lloyd. Additional help and discussion was provided by Drs. Rosemary Jagus, James Osborne, Jr., Daniel Peterson, Brian Safer and Jolinda Traugh. Excellent technical assistance was provided in part by Ms. Karen Berry, Ms. Kathy Li, and Mr. Wayne Kemper. Expert editorial assistance was patiently provided by Ms. Yvonne Coleman.

REFERENCES

1. Benne, R. and Hershey, J.W.B. (1979) Methods Enzymol. 60, (in press).
2. Schrier, M., Erni, B., and Staehelin, T. (1977) J. Mol. Biol. 116, 727-753.
3. Merrick, W.C. (1979) Methods Enzmol. 60, 101-108.
4. Trachsel, H., Erni, B., Schrier, M.H., and Staehelin, T. (1977) J. Mol. Biol. 116, 755-767.
5. Benne, R. and Hershey, J.W.B. (1978) J. Biol. Chem. 253, 3078-3087.
6. Peterson, D.T., Merrick, W.C., and Safer, B. (1979) J. Biol. Chem. (in press).
7. Anderson, W.F., Bosch, L., Cohn, W.E., Lodish, H., Merrick, W.C., Weissbach, H., Wittman, H.G., and Wool, I.G. (1977) FEBS lett. 76, 1-10.
8. Safer, B., and Anderson, W.F. (1979) Crit. Rev. Biochem. (in press).
9. Revel, M., and Groner, Y. (1978) Ann. Rev. Biochem. 47, 1079-1126.
10. Clemens, M.J., Safer, B., Merrick, W.C., Anderson, W.F., and London, I.M. (1975) Proc. Natl. Acad. Sci. USA 72, 1286-1290.
11. Maxwell, C.R. and Rabinovitz, M. (1969) Biochem. Biophys. Res. Commun. 35, 79-85.

12. Barrieux, A. and Rosenfeld, M.G. (1977) J. Biol. Chem. 252, 3843-3847.
13. Kramer, G., Cimadevilla, J.M., and Hardesty, B. (1976) Proc. Natl. Acad. Sci. U.S.A. 73, 3078-3082.
14. Dasgupta, A., Majumder, A., Caldiroli, E., Chatterjee, B., Palmieri, S., Das, A., and Gupta, N.K. (1977) Fed. Proc. 36, 869.
15. Farrell, P.J., Balkow, K., Hunt, T., Jackson, R.J., and Trachsel, H. (1977) Cell 11, 187-200.
16. Kaempfer, R., Hollender, R., Abrams, W., and Israeli, R. (1978) Proc. Natl. Acad. Sci. USA 75, 209-213.
17. Harbitz, I. and Hauge, J.G. (1976) Arch. Biochem. Biophys. 176, 766-778.
18. Safer, B., Adams, S.L., Anderson, W.F., and Merrick, W.C. (1975) J. Biol. Chem. 250, 9076-9082.
19. Benne, R., Wong, C., Leudi, M., and Hershey, J.W.B. (1976) J. Biol. Chem. 251, 7675-7681.
20. Staehelin, T., Trachsel, H., Erni, B., Boshcetti, A., and Schrier, M.H. (1975) Proc. 10th FEBS Meeting, 309-323.
21. Cashion, L.M., and Stanley, W.B., Jr. (1974) Proc. Natl. Acad. Sci. USA 71, 436-440.
22. Gupta, N.K., Chatterjee, B., Chen, Y., and Majumdar, A. (1974) Biochem. Biophys. Res. Commun. 58, 699-706.
23. Lloyd, M.A., Ph.D. dissertation, George Washington University, 1978.
24. Merrick, W.C. (1979) Methods Enzymol. 60, 108-123.
25. Adams, S.L., Safer, B., Anderson, W.F., and Merrick, W.C. (1975) J. Biol. Chem. 250, 9076-9082.
26. Merrick, W.C., Kemper, W.M., and Anderson, W.F. (1975) J. Biol. Chem. 250, 5556-5562.
27. Ottesen, M. and Svenson, B. (1971) C.R. Trav. Lab. Carlsburg 38, 445-456.
28. Trachsel, H. and Staehelin, T. (1978) Proc. Natl. Acad. Sci. USA 75, 204-208.
29. Safer, B., Peterson, D.T., and Merrick, W.C. (1977) in Proceedings of the International Symposium on Translation of Synthetic and Natural Polynucleotides (Legocki, A.B., ed) pp. 24-31, Poznan.
30. Walton, G.M. and Gill, G.N. (1975) Biochim. Biophys. Acta 390, 231-245.
31. Miller, D.L. and Weissbach, H. (1977) in Molecular Mechanisms of Protein Biosynthesis (Weissbach, H. and Pestka, S., ed) pp. 324-374, Academic Press, New York.
32. Ranu, R.S., London, I.M., Das, A., Dasgupta, A., Majumdar, A., Ralston, R., Roy, R. and Gupta, N.K. (1978) Proc. Natl. Acad. Sci. USA 75, 745-749.

33. deHaro, C., Datta, A., and Ochoa, S. (1978) Proc. Natl. Acad. Sci. USA 75, 243-247.
34. Henderson, A.B., Kramer, G., Pinphanichakarn, P., Wallis, M.H., and Hardesty, B. (1977) Fed. Proc. 36, 730.
35. Ranu, R.S. and London, I.M. (1976) Proc. Natl. Acad. Sci. USA 73, 4349-4353.
36. Traugh, J. (personal communication).
37. Hunt, T. (personal communication).
38. Floyd, G.A., Merrick, W.C., and Traugh, J.A. (1979) Eur. J. Biochem. (in press).
39. Benne, R., Edman, J., Traut, R.R., and Hershey, J.W.B. (1978) Proc. Natl. Acad. Sci. U.S.A. 75, 108-112.

MODULATION OF PROTEIN SYNTHESIS AND EUKARYOTIC
INITIATION FACTOR 2 (eIF-2) FUNCTION DURING
NUTRIENT DEPRIVAL IN THE EHRLICH ASCITES TUMOR CELL

Edgar C. Henshaw, Walter Mastropaolo,
and A. R. Subramanian

University of Rochester School of Medicine and
Dentistry, Rochester, New York, U.S.A., and the
Max-Planck-Institut fur Molekulare Genetik,
Berlin, West Germany

ABSTRACT We have shown previously that initiation is inhibited in Ehrlich cells deprived of an essential amino acid or of glucose, and that an inhibited step in initiation is the binding of Met-tRNA$_f$ to the native 40s ribosomal subunit. This step is mediated by eukaryotic initiation factor 2 (eIF-2). In the present work we have studied the amount and degree of phosphorylation of eIF-2 in the ribosome-bound (KCl wash) and soluble (S-100) fractions of fed, glucose-deprived, and amino acid-deprived cells. The amount of eIF-2, measured by the Millipore filter binding assay, is similar in fed and deprived cells, in both the KCl wash and the S-100 fractions. In cells exposed to ^{32}Pi for two hours, the 38,000 dalton subunit of eIF-2 is labeled in eIF-2 purified from total ribosomes or from native 40s subunits, but not in eIF-2 purified from the soluble fraction. Function of eIF-2 requires cycling from the free to the ribosome-bound state, and these data suggest that phosphorylation of eIF-2 may play a role in its cycling. The degree of labeling of the 38,000 dalton subunit is similar in eIF-2 of fed compared to nutrient-deprived cells, whether extracted from the total ribosomal KCl wash fraction or from the native 40s subunits. The 48,000 dalton subunit is also phosphorylated in vivo and differences were not seen between fed and deprived cells. If the degree of labeling reflects the extent of phosphorylation of the proteins, these data indicate that phosphorylation of eIF-2 is not the mechanism regulating initiation in glucose- or amino acid-deprived Ehrlich cells. A ribosomal protein of molecular weight 36,000, presumably S6, is much more highly labeled by ^{32}Pi in ribosomes of fed than of deprived cells. Monomeric ribosomes and polyribosomes are equally labeled.

INTRODUCTION

Nutrient deprival of an animal, by fasting or by a protein-free diet, leads to marked inhibition of protein synthesis in several tissues, notably skeletal muscle and liver. Since the plasma glucose and amino acid concentrations are maintained by homeostatic mechanisms, these tissues are probably not deprived of nutrients themselves, but are responding to hormonal signals. In addition, mammalian tissues have mechanisms, presumably protective, for dealing with direct tissue stresses and injuries. In particular, many tissues in vivo respond with inhibition of protein synthesis to direct tissue nutrient deprivals such as occur in ischemia, and to physical stresses such as high or low temperature or pH.

We have studied the response of protein synthesis in the Ehrlich ascites tumor cell in culture to direct nutrient deprival. The cells are grown in suspension culture, where nutrient conditions can be accurately controlled. It has been shown that in Ehrlich cells and in several other cell lines deprival of glucose or of any one of 13 essential amino acids leads to a decrease in the size and number of polyribosomes and the accumulation of monomeric ribosomes, that is, leads to "polyribosome runoff" (1-5). Since polypeptide chain elongation is itself slowed rather than speeded up under these conditions, the runoff indicates an even greater inhibition of the reactions whereby the monomeric ribosomes re-associate with messenger RNA (in polyribosomes), i.e. an inhibition of polypeptide chain initiation.

The reactions of initiation begin with the dissociation of monomeric ribosomes into free 40s and 60s subunits, which are referred to as "native" subunits because they are the only naturally occurring free subunits in the cell, the others being incorporated into ribosomes. eIF-3 then binds to the native 40s subunit, followed by Met-tRNA$_f$, the initiator aminoacyl tRNA. The binding of Met-tRNA$_f$ to the native 40s subunit depends upon eIF-2 and requires two sequential reactions:

The non-covalent formation of a ternary complex:
(1) Met-tRNA$_f$ + eIF-2 + GTP $\rightleftarrows$ MET-tRNA$_f$·eIF2·GTP
(2) The non-covalent binding of this complex to the native 40s subunit. The 40s subunit complex then binds to the initiation site of the mRNA, followed by the 60s subunit.

We have shown that in the Ehrlich cell, deprival of

glucose or an amino acid results in a 50-75% decrease in the amount of Met-tRNAf bound to native 40s subunits, implying that either formation of the ternary complex or binding of this complex to 40s subunits is inhibited (6). In the present work we are examining the components of these two reactions in an effort to elucidate the mechanism of the inhibition of Met-tRNAf binding. We are studying in particular the possibility that eIF-2 activity is modulated by phosphorylation of one of its subunits. We have also examined the phosphorylation of the ribosomal proteins.

eIF-2

The eIF-2 of the Ehrlich cell resembles the rabbit reticulocyte factor. On DEAE-cellulose it elutes at about 0.18 $\underline{M}$ KCl; on phosphocellulose it elutes at 0.5 $\underline{M}$ KCl. Analyzed on Sephadex G-200 after purification, its molecular weight is 151,000. It consists of 3 subunits of molecular weight 38,000; 48,000; and 52,000 according to SDS polyacrylamide gel electrophoresis (PAGE).

The activity of eIF-2 has been compared in the ribosomal KCl-wash fraction of fed cells and of cells deprived of glutamine for 1 hour, of 13 amino acids for 1 hour, and of cells allowed to enter plateau phase of growth through depletion of the medium. (This appears to be due principally to depletion of glucose, and of serum growth factors (7).) Activity was measured using the Millipore filter binding assay for formation of ternary complex (8). Yield of KCl-wash protein was similar in fed and deprived cells, and no difference between the activity per unit of protein was detected. Because there are interfering components in the crude KCl-wash fraction, activity was also measured after partial purification of eIF-2 from the KCl-wash on DEAE-cellulose (Table 1). Again, no significant differences were detected.

Activity is also present in the soluble fraction (S-100) and is equal in fed and glutamine-deprived cells (data not shown). We therefore have evidence against important changes in amount of eIF-2. Since the inhibition of protein synthesis occurs within 15 minutes after amino acid deprival and within two hours of glucose deprival, and is reversed within 10 minutes upon refeeding, changes in amount of eIF-2 are perhaps not to be expected. As described by others in this Symposium, changes in degree of phosphorylation of eIF-2 would probably not be detected by the Millipore filter assay used here.

Because of the evidence in reticulocytes for the regulatory function of a kinase which phosphorylates the 38,000 dalton subunit, we have studied this possibility in Ehrlich

TABLE I
eIF-2 ACTIVITY IN FED AND NUTRIENT-DEPRIVED EHRLICH CELLS

Cells	µg Protein in DEAE Fraction Per 10^6 Cells[a]	eIF-2 Activity[b]	
		per µg Protein	per 10^6 Cells
Fed	2.25	5.58	12.6
Minus AA	2.05	4.93	10.1
Non-replenished	2.98	4.33	13.0

[a]KCl wash was prepared as described (8) from cultures of cells growing in complete medium; from cells deprived of 13 essential amino acids for 1 hour; and from cells suspended in complete medium at 5 x 10^5 cells/ml and grown for 24 hours without replenishment (7). Fed and deprived state was always confirmed by sucrose gradient analysis of ribosome patterns in cytoplasmic extracts (9). Preparations were precipitated with ammonium sulfate and the 30-80% fraction was analyzed by column chromatography on DEAE-cellulose, with gradient elution from 0.1 to 0.3 $\underline{M}$ KCl. The activity of eIF-2 was measured in a Millipore filter binding assay (8), and fractions containing eIF-2 were pooled and dialyzed.

[b]Preparations from fed and deprived cells were compared in the same assay. Activity versus concentration is not entirely linear, and specific activity was compared at a constant activity level.

cells. In collaboration with Dr. London's group, it was found that a crude soluble extract of Ehrlich cells (S-100) contained an inhibitor of the reticulocyte lysate cell-free protein-synthesizing system which inhibits initiation with a time course similar to that of the reticulocyte eIF-2 kinase, and is reversed by added eIF-2 (10). Using inhibition of the reticulocyte lysate system as an assay, we have attempted to purify the inhibitor. It is precipitated by 40% ammonium sulfate; is eluted from DEAE-cellulose at 0.25 $\underline{M}$ KCl; and is eluted from phosphocellulose (pH 7.0) at below 0.05 $\underline{M}$ KCl. Activity is lost during these steps and is destroyed by freezing; activity also diminishes during storage on ice. We have therefore not succeeded in purifying the protein responsible for the inhibitory activity. However, in the presence of [gamma-^{32}P] (ATP^{32}) the 38,000 dalton subunit is labeled

MODULATION OF PROTEIN FUNCTION

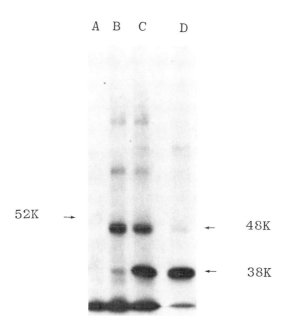

FIGURE 1. eIF-2 Kinase Activity of the Ehrlich Cell Inhibitor of Reticulocyte Lysate Protein Synthesis. Ehrlich cell inhibitor was purified as described in the text. Ehrlich cell eIF-2 was purified from a 0.5 M KCl wash of the ribosomes by ammonium sulfate precipitation (35-70% fraction), phosphocellulose chromatography (gradient elution with 0.4-0.8 M KCl) and DEAE-cellulose chromatography (gradient elution with 0.1-0.25 M KCl). Rabbit reticulocyte heme regulated inhibitor (HRI) was purified from the postribosomal supernatants of lysed reticulocytes. The HRI in these supernatants was activated by N-ethylmaleimide treatment (10) then purified by ammonium sulfate precipitation (0-30% fraction), DEAE-cellulose chromatography (gradient elution with 0.1-0.4 M KCl), and phosphocellulose chromatography (elution at 0.04 M KCl).

Kinase activity was assayed in 50 μl of solution containing 0.1 mM ATP^{32} (0.5 Ci/mMol), 3.0 mM magnesium acetate, 50 mM MOPS pH 7.2. The assay solutions were incubated 15 minutes at 28°. ^{32}P-labeled proteins from these incubations were precipitated with 10% trichloroacetic acid(TCA), centrifuged, and the precipitate dissolved in 30 μl of SDS-PAGE sample buffer(11). 25 μl of each sample was analyzed by SDS-PAGE according to Laemmli (11) on 9% acrylamide gels. After electrophoresis the gels were stained with Coomassie
(Legend continued on next page.)

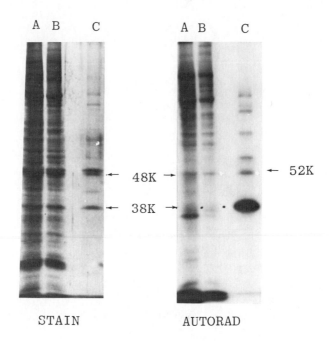

FIGURE 2. ^{32}P-labeled KCl Wash Proteins from Fed and Glutamine-Deprived Cells. Cells at a density of 10 x 10^6 per ml were incubated for 2 hours at 37° in 100 ml of complete or glutamine-deprived media containing 20 μCi/ml of ^{32}Pi (40 mCi/mMol). Cells were then harvested and KCl wash prepared. eIF-2 was partially purified by phosphocellulose chromatography with batchwise elution between 0.4-0.8 M KCl. The ^{32}P-labeled proteins in the eIF-2 fractions were precipitated with 10% TCA and analyzed by SDS-PAGE as described in Fig. 1. Samples containing equal amounts of Met-tRNAf binding activity from fed or deprived cells were applied to each lane. <u>Lane A</u> KCl wash proteins from fed cells. <u>Lane B</u> KCl wash proteins from glutamine-deprived cells. <u>Lane C</u> Purified eIF-2 labeled with ATP^{32} and HRI as described in Fig. 1.

Figure 1 legend continued from previous page

blue, dried and autoradiographed for 8 days with Kodak XRP-6 film.

<u>Lane A</u> 18 μg of Ehrlich inhibitor. <u>Lane B</u> 18 μg of Ehrlich inhibitor, 1.1 μg of eIF-2. <u>Lane C</u> 18 μg of Ehrlich inhibitor, 1.1 μg of eIF-2, 0.2 μg of HRI. <u>Lane D</u> 1.1 μg of eIF-2, 0.2 μg of HRI.

by crude inhibitor preparations purified on phosphocellulose or DEAE cellulose, as shown in Fig. 1, lane B. It is also evident that there is an even more active kinase for the 48,000 dalton subunit in the crude preparations. We have no evidence that in Ehrlich cells either the 48,000 or 38,000 dalton kinase activity causes the inhibition of the retic system. The data do demonstrate the existence in Ehrlich cells of an enzyme capable of phosphorylating the 38,000 dalton subunit of eIF-2.

To assess the importance of phosphorylation of eIF-2 in regulating protein synthesis during nutrient deprival, we measured the labeling by ^{32}P-orthophosphate(^{32}Pi) of eIF-2 in the cell under fed and deprived conditions. Fed cells and cells deprived of glutamine were labeled for 2 hours with ^{32}Pi; KCl-wash was prepared; and a fraction enriched for eIF-2 was isolated by stepwise elution from phosphocellulose (the 0.4 to 0.8 M KCl fraction). The latter step is important because there are phosphorylated proteins in the Ehrlich cell which migrate very near to the 38,000 dalton subunit of eIF-2 on one-dimensional SDS gels and can confuse the interpretation of bands in this region. Fig. 2 shows the SDS-PAGE analyses of these preparations. In the stained gel the three bands of eIF-2 can be located by reference to the purified eIF-2 lane. There are similar amounts of eIF-2 in each preparation. In the autoradiogram the location of the 38,000 dalton subunit is indicated by the prominent phosphorylated band in eIF-2 incubated with ATP^{32} and HRI (lane C). In vivo phosphorylation of the 38,000 dalton subunit of eIF-2 is faint, but it is equal in the fed and deprived cells (lanes A & B). In Fig. 3 fed and glutamine-deprived, ^{32}Pi-labeled cells have been fractionated into a magnesium-precipated ribosomal pellet and a soluble protein fraction. This method separates soluble proteins from the ribosomes better than centrifugation alone. eIF-2 was then partially purified from each fraction. Reference to the stained gel indicates that eIF-2 is present in about equal amounts in the lanes from the fed (lane F) and deprived(-AA) KCl-washes, and the autoradiogram shows that phosphorylation is also similar. Finally, we have also investigated the eIF-2 of the native 40s subunits. Subunits were isolated by sucrose gradient analysis after a two hour ^{32}Pi labeling period; eIF-2 was partially purified and was analyzed by SDS-PAGE (Fig. 4). The relevant lanes (-AA, -G, and F) contain similar amounts of eIF-2 (or slightly less in the fed) and there is no difference in the labeling of the 38,000 dalton subunit on the autoradiograms (lanes -AA, -G, and F) among fed, glucose-deprived and glutamine-deprived cells. Thus, although we find an enzyme which can phosphorylate eIF-2, and evidence for in vivo phosphoryl-

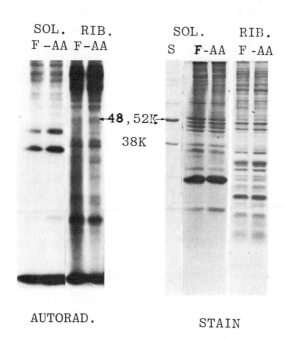

FIGURE 3. ^{32}P-labeled Soluble and Ribosomal Proteins from Fed and Glutamine-Deprived Ehrlich Cells. Cells were labeled with ^{32}Pi as described in Fig. 2, lysed in hypotonic buffer containing 0.4% deoxycholate and 0.4% Triton X-100, and the ribosomes purified from the post mitochondrial supernatant of the lysate by precipitation with 0.05 M magnesium acetate. eIF-2 was partially purified from soluble and ribosomal fractions by phosphocellulose chromatography as described in Fig. 2 except that in the case of the soluble fraction eIF-2 was purified further by DEAE-chromatography with batch elution between 0.1 and 0.25 M KCl. eIF-2 was then analyzed by SDS-PAGE as described in Fig. 1 except that the acrylamide concentration was 12%. Lane F Fed cell proteins. Lane -AA Proteins from glutamine-deprived cells. Lane S 1.1 µg of purified eIF-2 standard.

ation, we do not find detectable differences between phosphorylation of eIF-2 in fed and nutritionally deprived Ehrlich cells, either in the total KCl-wash or in the eIF-2 on the native 40s subunits, and therefore we find no indication that this mechanism is important in the nutritional response of these cells.

We have also studied the eIF-2 in the soluble cell fraction (S-100). eIF-2 is released from the initiation complex, with the hydrolysis of GTP, when the 60s subunit joins the 40s subunit on the initiation site on the mRNA. Thus, eIF-2 must exist as unbound or "soluble" protein during a portion of its cycle in protein synthesis, and therefore some fraction of total eIF-2 must be in the unbound state at any time. The

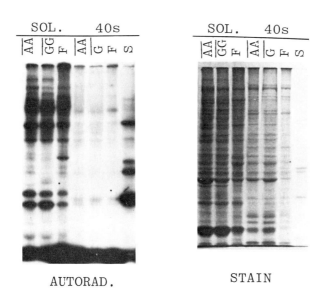

FIGURE 4. ^{32}P-labeled Soluble and 40s Subunit-Bound Proteins from Fed and Nutrient-Deprived Ehrlich Cells. Cells were labeled with ^{32}P and lysed as described in Figs. 2 and 3. Postmitochondrial supernatants from the lysates were layered on 20-40% (w/v) sucrose gradients and centrifuged at 25,000 rpm for 3.5 hours in a SW 25.1 rotor. The gradients were fractionated into soluble, 40s, 80s, and polyribosomal fractions. eIF-2 was partially purified and analyzed by SDS PAGE as described in Figs. 1 and 2. Lane F Proteins from fed cells. Lane -AA Proteins from glutamine-deprived cells. Lane -G Protein from glucose-deprived cells. Lane S Purified eIF-2 treated with HRI and ATP^{32} as described in Fig. 1.

released eIF-2 must also re-associate with GTP in order to function, and it may have to eject GDP to do so, as is the case with EF-Tu in bacteria, so that differences related to cycling may exist between free and bound eIF-2. In recent experiments we have found as much as 50% of Ehrlich cell eIF-2 in the soluble fraction. In Fig. 3 the soluble and total bound eIF-2 are compared. Although there is, if anything, more eIF-2 in the soluble fraction lanes (see the stained gel), there is no detectable radioactivity, whereas there is definitely label in the bound fraction. In Fig. 4 soluble eIF-2 is compared with that bound specifically to 40s subunits. The distinction is again evident: the unbound eIF-2 is not detectably labeled while the bound is.

This surprising finding is subject to several interpretations. We have not yet excluded the possibility that phosphatases in the soluble fraction are responsible for the apparent absence of labeling. However, we have carried out the steps rapidly and in the cold. If the finding is not artefactual, it might mean either that soluble eIF-2 is less phosphorylated than bound, or that it is phosphorylated but that the phosphate turns over very slowly. We have used only 1 and 2 hour labeling periods and cannot choose rigorously between these two possibilities. However, the rapid turnover of phosphate on this subunit in reticulocytes favors the hypothesis that unbound eIF-2 is non-phosphorylated. From our labeling data we cannot estimate what proportion of bound eIF-2 is phosphorylated, because we do not know the specific activity of the ATP^{32} pool, or the amount of protein in each band. Visual comparison of the intensity of labeling of different protein bands and the intensity of dye staining suggests that the specific activity of bound eIF-2 is not as high as that of some of the other phosphoprotein bands, which might mean that only a portion of the bound molecules are labeled. In collaboration with Bruce Voris and Donald Young we have recently used the O'Farrell two-dimensional PAGE technique (12), which resolves the phosphorylated and non-phosphorylated forms of the 38,000 dalton subunit (13), to examine eIF-2. Preliminary results indicate that eIF-2 purified from the KCl wash contains both phosphorylated and non-phosphorylated forms, and that the phosphorylated form is about 25% of the total. The soluble eIF-2 has yet to be analyzed using this technique.

A difference between the phosphorylation of free and bound eIF-2 suggests a role for phosphorylation in eIF-2 cycling. On the basis of entirely different evidence, London et al. have previously suggested that the inhibitory effect of phosphorylation of eIF-2 in the reticulocyte system may be on cycling of the factor (14). Phosphorylation does not appear to be a regulatory mechanism in the Ehrlich cells under our conditions, and a large pool of non-phosphorylated eIF-2 exists in the soluble fraction and in the bound fraction.

Phosphorylation of Ribosomal Proteins. We have also studied the degree of phosphorylation of ribosomal proteins, by labeling fed and nutritionally deprived cells for 2 hours with ^{32}Pi, as described above, and analyzing the ribosomal proteins by SDS-PAGE. Fig. 5 shows the autoradiogram of a gel analysis of the proteins of polyribosomes from fed, glucose-deprived, and glutamine-deprived cells, and of mono-

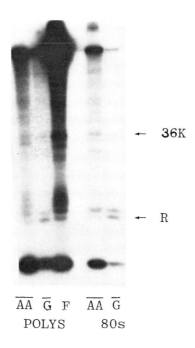

FIGURE 5. ^{32}P-labeled Ribosomal Proteins from Fed and Nutrient-Deprived Ehrlich Cells. Cells were labeled with ^{32}Pi and polyribosomal and 80s subunit fractions were purified as in Fig. 4. The ribosomes in both fractions were then concentrated by centrifugation at 30,000 rpm for 10 hours in a 70 Ti rotor. A portion of the ribosomal pellet was dissolved directly in SDS-PAGE sample buffer and 0.5 A$_{260}$ unit was analyzed by SDS-PAGE electrophoresis on 12% acrylamide gels and autoradiographed as described in Fig. 1. Lane F Protein from fed cells. Lane -G Protein from glucose-deprived cells. Lane -AA Protein from glutamine deprived cells.

meric ribosomes from the two deprived cultures. (Monomeric ribosomes are very low in well-fed cells.) The stained gel indicates equal amounts of protein in each lane. RNA was not removed from these preparations and the labeled ribosomal RNA complicates especially the fed-cell analysis, but can be corrected for in the region of a particular band by measuring a neighboring background area. Comparison of the radioactivity of a band in question with that of a reference band in the lower-molecular-weight region will correct for possible differences in ATP pool specific activities in fed and deprived cells. Visual inspection suggests that the band in the 36,000 dalton region of the polysomes of fed cells is much more intense than the reference band, while the 36,000 dalton band of both the polyribosomes and monomeric ribosomes of deprived cells is less radioactive than the reference band. To quantitate this, the 36,000 dalton band, several reference bands, and neighboring background regions were cut from stained gels and the radioactivity was measured (Table 2). The 36,000 dalton band was far more radioactive

TABLE II
PHOSPHORYLATION OF A 36,000 DALTON PROTEIN IN POLYRIBOSOMES AND 80s SUBUNITS FROM FED AND NUTRIENT-DEPRIVED EHRLICH CELLS[a]

Gel Lane	36K cpm / R cpm	protein[b] (dye units)	36K cpm / protein
Polys F	1.4	218	0.57
Polys -G	0.25	289	0.045
Polys -AA	0.27	391	0.020
80s -G	0.25	162	0.031
80s -AA	0.17	189	0.042

[a]Lanes were sliced from a slab gel that was processed in parallel to the gel described in Fig. 5. Protein bands cooresponding to the ^{32}P-labeled bands (36K and R arrows in the autoradiogram) were cut out, dried and counted. Background counts were determined by cutting out a portion of the gel with identical dimensions adjacent to the labeled band. The counts in these slices were subtracted from the counts in the labeled bands.
[b]Protein was estimated from the area of the peak corresponding to the labeled band in a densitometric scan of the gel lane before processing for counting.

with respect to the reference band in polyribosomes of fed cells than in monomers or polyribosomes of deprived cells. The relative phosphorylation of this band also may be compared by dividing the radioactivity by the protein in the band. The results of this analysis (Table 2) indicate more than 10-fold greater phosphorylation of the 36,000 dalton band in fed than in deprived cells. The difference is so large that it could hardly be artefactual, despite the high RNA background in the fed cells. Although the radioactivity in some bands was low, the data also suggest that there are not large differences between the polyribosomes and the monomers in the phosphorylation of this protein, in either type of deprived cell.

On one-dimensional gels ribosomal structural proteins are usually not completely resolved from each other, and identification requires 2 dimensional techniques. Two-dimensional gels are shown in Fig. 6. These gels are consistent with the interpretation that the highly labeled protein in the fed cells is S6 (by reference to authentic S6), and demonstrate the striking reduction in intensity of phosphorylation of S6 in the cells deprived either of glucose or of glutamine. These gels also show a difference in the relative intensity of labeling of two acidic proteins indicated by the two carets at the bottom of Fig. 6. These proteins are altered in different ways by amino acid deprivation and by glucose deprivation. This finding has not yet been confirmed. Alterations in the labeling of an acidic protein in Krebs II cells have been reported during glucose and amino acid deprival (15).

At the present time any relationship between de-phosphorylation of S6 and the regulation of protein synthesis is purely speculative. A number of recent studies have shown increased phosphorylation of S6 under physiological circumstances in which protein synthesis is stimulated. Thus, phosphorylation is increased in growing compared to stationary or confluent HeLa and BHK cells (16,17); in regenerating liver; and in reticulocytes compared to mature red blood cells (18,19). However, the opposite occurs in certain other circumstances. Phosphorylation of S6 is decreased in growing Tetrahymena compared to cells which are starved and cease to grow(20); it is increased in the liver of diabetic rats and is decreased by insulin (21), whereas protein synthesis is stimulated by insulin. Thus, the relationship of phosphorylation of S6 to the rate of protein synthesis must be considered, at best, complex. It is evident that phosphorylation is subject to extensive modulation.

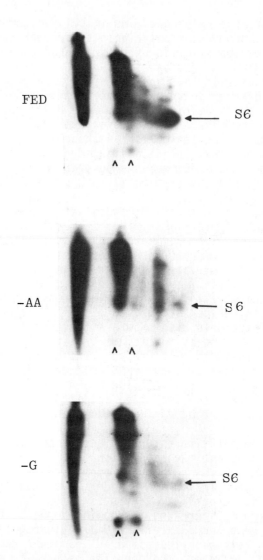

FIGURE 6. Two Dimensional PAGE Analysis of ^{32}P-labeled Ribosomal Proteins from Fed and Nutrient-Deprived Ehrlich Cells. Cells were labeled with ^{32}Pi as described in Fig. 2 and the ribosomal fraction of the cell obtained as described in Fig. 3. Ribosomal proteins were extracted with acetic acid and analyzed by two dimensional PAGE as described (25). The gels were dried and autoradiographed as described in Fig. 1.

Leader and Coia have recently shown that S6 is more highly labeled in polyribosomes than in monomeric ribosomes of BHK cells, suggesting that phosphorylation might enhance the probability of a ribosome's entering the pathway of initiation (22). However, it is certainly clear that if there is a functional difference in probability of initiation between ribosomes with phosphorylated and non-phosphorylated S6, this difference is not absolute. Kabat found that a 40s subunit phosphoprotein, which is almost surely S6, was phosphorylated on monomeric ribosomes and polyribosomes of reticulocytes and sarcoma 180 cells, although it was more highly labeled in reticulocyte polyribosomes (19,23,24). It was equally labeled in monomers and polyribosomes of regenerating liver (18). In the Ehrlich cells we do not detect large differences between monomers and polyribosomes of deprived cells, in which phosphorylated S6 is comparatively low and initiation is limiting. In this case, if phosphorylation conferred a large advantage in initiation, the difference in S6 phosphorylation between monomers and polyribosomes would be expected to be pronounced. However our data are insufficient to rule out small differences.

In conclusion, we have shown that the phosphorylation of eIF-2 may be important in the cycling of eIF-2 but probably does not play an absolute role in the control of protein synthesis by nutritional means in the Ehrlich cell. On the other hand the phosphorylation of S6 may be important to the regulatory mechanism. It is evident that protein synthesis may be controlled by several mechanisms depending on the stimulus, cell type and other circumstances and each case must be evaluated individually.

REFERENCES

1. van Venrooij, W. J. W., Henshaw, E. C., and Hirsch, C. A. (1972). Biochim. Biophys. Acta 259, 127-137.
2. Vaughan, M. H., Jr., Pawlowski, P. J., and Forchhammer, J. (1971). Proc. Nat. Acad. Sci., U.S.A. 68, 2057-2061.
3. Lee, S. Y., Krsmanovic, V., and Brawerman, G. (1971). Biochemistry 10, 895-900.
4. Eliasson, E., Bauer, G. E., and Hultin, T.(1967). J. Cell Biol. 33, 287-297.
5. Hogan, B. L. M., and Korner, A. (1968). Biochim. Biophys. Acta 169, 129-138.
6. Pain, V. M., and Henshaw, E. C. (1975). Eur. J. Biochem 57, 335-342.
7. van Venrooij, W. J. W., Henshaw, E. C., and Hirsch, C. A. (1970). J. Biol. Chem. 245, 5947-5953.

8. Smith, K. E., and Henshaw, E. C. (1975) Biochemistry 14, 1060-1067.
9. Hirsch, C. A., Cox, M. A., Van Venrooij, W. J. W., and Henshaw, E. C. (1973). J. Biol. Chem 248, 4377-4385.
10. Kramer, G., Cimadevilla, J. M., and Hardesty, B. (1976). Proc. Nat. Acad. Sci., U.S.A. 73, 3078-3082.
11. Laemmli, U. K. (1970). Nature 227, 680-685.
12. O'Farrell, P. H. (1975). J. Biol. Chem 250, 4007-4021.
13. Farrell, P. J., Hunt, T. J., and Jackson, R. J. (1978). Eur. J. Biochem. 89, 517-521.
14. Cherbas, L., and London, I. M. (1976). Proc. Nat. Acad. Sci. U.S.A. 73, 3506-3510.
15. Leader, D. P., and Coia, A. A. (1978). FEBS Lett. 90, 270-274.
16. Lastick, S. M., Nielsen, P. J., and McConkey, E. H. (1977). Molec. Gen. Genet. 152, 223-230.
17. Leader, D. P., Rankine, A. D., and Coia, A. A. (1976). Biochem. Biophys. Res. Commun. 71, 966-974.
18. Gressner, A. M., and Woll, I. G. (1974). J. Biol. Chem. 249, 6917-6925.
19. Kabat, D. J. (1972) Biol. Chem. 247, 5338-5344.
20. Kristiansen, K., and Kruger, A. (1978). Biochim. Biophys. Acta 521, 435-451.
21. Gressner, A. M., and Woll, I. G. (1976). Nature 259, 148-150.
22. Leader, D. P., and Coia, A. A. (1978). FEBS Lett. 90, 270-274.
23. Bitte, L., and Kabat, D. (1972). J. Biol. Chem. 247, 5345-5350.
24. Kabat, D. (1970). Biochemistry 9, 4160-4175.
25. Subramanian, A. R. (1975). Eur. J. Biochem. 45, 541-546.

ACKNOWLEDGEMENT

This work was supported by USPHS grants CA-21663 and CA-11198. We thank Dr. Sie Ting Wong for preparing the Ehrlich cell inhibitor of protein synthesis and Eileen Canfield for excellent technical assistance.

GENETIC DEFECTS OF THE HUMAN RED BLOOD CELL AND HEMOLYTIC ANEMIA[1]

William N. Valentine[2]
and Donald E. Paglia[3]

Departments of Medicine and Pathology,
University of California, Center for the
Health Sciences, Los Angeles, CA 90024

ABSTRACT Lacking a nucleus, intracellular organelles, or any capacity for protein synthesis, the human red cell derives its small energy requirements from the production of lactate from glucose and from salvage pathways for adenine ribonucleotides. Severe heritable deficiencies of enzymes of anaerobic glycolysis usually cause hemolytic anemia, presumably secondary to associated deficiencies in energy generation. Lactate dehydrogenase deficiency is recognized but is not associated with anemia. Deficiencies of both enzymes of glutathione synthesis, of glutathione reductase and possibly glutathione peroxidase, and of glucose-6-phosphate dehydrogenase (G6PD) also produce hemolytic syndromes, primarily due to inability to protect hemoglobin from oxidative denaturation secondary to dysfunction of the pentose phosphate shunt. Severe deficiency of pyruvate kinase (PK) activity is second only to that of G6PD as a cause of hemolytic anemia in man. The deficient red cell isozyme is closely related to the L isozyme of liver, differs immunologically from the M_1 isozyme of muscle and in other ways from the M_2 isozyme of leukocytes and certain other tissues. Fructose-1,6-diphosphate is an allosteric modifier of red cell PK. Phosphoglycerate kinase and G6PD deficiency are X-chromosome linked; all other deficiencies

---

[1] Investigations reported from the authors' laboratory were supported by Grant #HLB-12944 from the National Institutes of Health.
[2] Present address: Department of Medicine, University of California, Center for the Health Sciences, Los Angeles, California, 90024.
[3] Present address: Division of Surgical Pathology, University of California, Center for the Health Sciences, Los Angeles, California, 90024.

are autosomally transmitted, hemolysis most often occurring only when two defective genes are inherited. Genetic polymorphism dictates that most affected subjects are doubly heterozygous for separate mutant genes except where consanguinity exists. An inherited hemolytic syndrome associated with a recessive, autosomally transmitted deficiency of a unique pyrimidine-specific 5'-nucleotidase is defined and is associated with enormous intracellular accumulations of pyrimidine ribonucleotides which are normally undetectable. An acquired lead-induced deficiency of the same nucleotidase results in hemolysis in subjects with severe lead poisoning. Increased red cell adenosine deaminase activity (45-70 fold), inherited as a Mendelian dominant, also causes a hemolytic syndrome. The associated diminished ATP is believed to result from inability of adenosine kinase to salvage adenosine for nucleotide replenishment in the face of the competing massive deamination.

The human erythrocyte has been molded by evolutionary forces to perform a single task, the transportation of oxygen to the tissues and carbon dioxide back to the lungs. Packaged within the cell as about a 33 percent solution, its oxygen-transport pigment, hemoglobin, contributes 95 percent of the dry weight of the erythrocyte. Alone of all the cells of the body, the mature human erythrocyte lacks a nucleus, DNA, RNA, ribosomes, mitochondria or indeed any form of intracellular organelles. It has no residual capacity for protein or lipid synthesis nor for oxidative phosphorylation. Fortunately, oxygen transport requires no expenditure of energy, for only 5 percent of its constituents other than water are available to provide a plasma membrane and to carry out functions essential for its survival.

In normal adult man, the roughly 25 trillion red cells circulate for about 120 days. During this time, while energy requirements are small they are not nonexistent. The biconcave discoidal shape and the ready deformability of the plasma membrane must be preserved. The red cell must pump cations against electrochemical gradients, synthesize glutathione (GSH), replenish losses from its adenine nucleotide pool, maintain the iron of its hemoglobin in the functional ferrous form, and protect its hemoglobin against oxidative denaturation. It is handicapped in accomplishing these tasks by its inability to replace any losses of its initial complement of enzymatic proteins, each of which undergoes denaturation at some characteristic rate under any given set of conditions. The gradual attrition of its catalytic capacities and inevitable slow denaturation of its structural proteins ultimately lead to changes which mark it for destruction. The final event is its engulfment and dissolution by the phagocytic macrophage.

In an adult male of average size, the maintenance of normal erythrocyte numbers requires the production of roughly 10 billion replacements an hour by the bone marrow, the factory whose assembly lines must compensate for the inevitable daily destruction of circulating cells. While early red cell precursors have a nucleus, this is extruded before the maturing erythrocytes leave the marrow compartment as non-nucleated reticulocytes. The latter continue to possess some mitochondria and ribosomes for 1 to 3 days at most, during which time supravital stains reveal the retained organelles and RNA as a fine reticulum. During the brief period before these too are lost, the reticulocyte does retain deteriorating capacities for oxidative phosphorylation and for a modicum of protein synthesis. When the life span of the erythrocyte is significantly shortened, the percentage of reticulocytes rises, the marrow increases its production of red cells up to several fold, and the hematologist designates the resulting syndrome as hemolytic. The term "hemolytic anemia" is usually reserved for situations where hemolysis is marked, where it is prominent in the pathogenesis of the anemia, and where often even the hyperplastic marrow is incapable of maintaining normal erythrocyte numbers and hemoglobin concentrations in the blood. The capacities of normal marrow are such, however, that up to a point even marked shortening of red cell life span can be compensated for by increased marrow production of erythrocytes.

In view of the unique specialization of the erythrocyte and its relative metabolic impoverishment even under normal circumstances, it is not surprising that genetically determined handicaps compromising its already limited metabolic capacities should lead, when severe, to its premature destruction and, by definition, to hemolytic syndromes. Such has proved to be the case.

The mature, akaryote human erythrocyte supports itself energetically by way of what probably is the most primitive and universal pathway of metabolism, conversion of glucose to pyruvate and lactate and storage of a portion of the energy generated in the form of ATP. Severe red cell deficiencies of seven enzymes of the anaerobic Embden-Meyerhof pathway have now been documented (Figure 1) (1-12). Six are associated with frank hemolytic syndromes. The seventh, severe deficiency of lactate dehydrogenase (LDH), is well defined but not associated with hemolysis (10). In part, this may

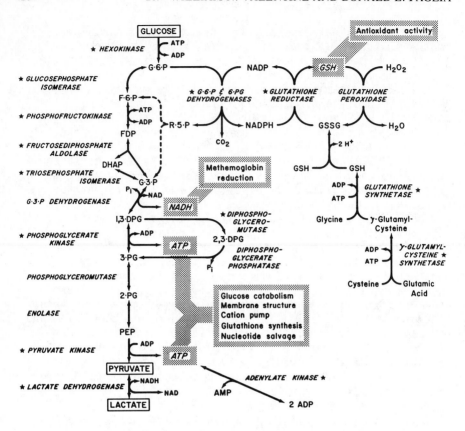

FIGURE 1. Pathways of glycolysis and glutathione metabolism in the human erythrocyte. Stars indicate enzymes for which genetically determined deficiency states have been defined. (Reproduced by courtesy of Williams and Wilkins, Co., Baltimore.)

be due to the fact that both pyruvate and lactate are freely diffusible, and can escape the red cell to be metabolized and excreted elsewhere. One might have presumed that inability to reoxidize NADH adequately at the terminal step in glycolysis would seriously disrupt the metabolism of the glycolytically dependent human erythrocyte. It appears, however, that alternative mechanisms of NADH oxidation are sufficiently compensatory to prevent hemolysis. Except for phosphoglycerate kinase (PGK) (8, 13, 14), all genes coding for Embden-Meyerhof pathway enzymes reside on autosomes. With the exception of a mild hemolytic syndrome present in one form of heterozygous phosphofructokinase (PFK) deficiency (15, 16), individuals heterozygous for a defective gene are with rare exceptions phenotypically and hematologically normal, but a partial deficiency, usually associated with a loss of about fifty percent of catalytic activity, is demonstrable by enzyme assay. Homozygotes and subjects doubly heterozygous for separate defective mutant genes have variably severe hemolytic anemia.

PGK deficiency is X-chromosome linked (8, 13, 14). Hemizygous males have full expression of the hemolytic syndromes associated with severe PGK deficiency; heterozygous females are a mosaic, in accordance with the Lyon-Beutler hypothesis (17, 18) of random functional inactivation of one X-chromosome in each cell. One cell population is just as severely deficient as that of the male haploid for the X-chromosome, the other is entirely normal. Depending on the ratios of the two populations, clinical expression runs the gamut from phenotypic normalcy to severe hemolytic anemia.

We believe that the final common denominator responsible for premature red cell destruction in the syndromes associated with severe deficiency of Embden-Meyerhof pathway enzymes is metabolic depletion secondary to inability to maintain cellular ATP. Unable to synthesize any enzyme protein after the reticulocyte state, the mature erythrocyte that is further handicapped by a severe, genetically determined reduction of already limited resources may survive briefly in the vigor of youth. But shortly it is unable to meet even its own austere needs, and finds an ultimate graveyard in the phagocytic macrophage.

Energy generation in the human red cell is complicated by the presence of the Rapaport-Luebering shunt (Figure 1) (19). The shunt permits the ATP generating step catalyzed by PGK to be bypassed at the same time 1,3-diphosphoglycerate is mutated to the important metabolite, 2,3-diphosphoglycerate (2,3-DPG). The latter may constitute up to 50 percent of the organic phosphate of the red cell (20). In minute catalytic amounts, 2,3-DPG serves as a necessary cofactor for monophosphoglycerate mutase which catalyzes the reversible conversion of 3- and 2-phosphoglycerate. In higher concentrations, 2,3-DPG has a variety of regulatory effects, the most prominent of which is the alteration of the oxygen dissociation curve of hemoglobin (21). In the presence of 2,3-DPG, there is an increase in the P_{50}, the partial pressure of oxygen at which red cell hemoglobin is half-saturated. The lessened avidity of hemoglobin for oxygen produced by 2,3-DPG, like the Bohr effect, permits greater oxygen delivery to the tissues at any partial pressure. After dephosphorylation 2,3-DPG is returned to the mainstream of glycolysis as 3-PG. The responsible phosphatase has proved to be a second catalytic activity of the same identical protein functioning as the mutase responsible for 2,3-DPG formation (22, 23). The Rapaport-Luebering shunt provides a reservoir of metabolizable triose; it helps modulate ATP production in accordance with body needs; and it serves important functions relative to tissue delivery of O_2 by hemoglobin.

Recently, autosomally transmitted, severe deficiency of the 2,3-DPG mutase has been documented (23). The proband had essentially undetectable activity of the mutase or the phosphatase in his red cells; the erythrocytes of his apparently heterozygous parents possessed about half-normal mutase activity. The 2,3-DPG was essentially undetectable in patient cells, although enough remained to support catalytically the interconversion of the monophosphate glycerates. In the proband, the nearly complete absence of 2,3-DPG in the red cells shifted the O_2 dissociation curve of hemoglobin unfavorably and hence, resulted in diminished O_2 delivery per unit of blood to the body tissues. There were no clinical manifestations and hemolysis was not present. Rather, there was a compensatory modest increase in the circulating red cell mass, or stated otherwise, moderate polycythemia was present in the proband. The syndrome graphically illustrates how an enzyme related to glycolysis secondarily exerts a regulatory influence on the function of a structural protein pigment, hemoglobin, and upon the magnitude of red cell production required of the bone marrow.

The human erythrocyte also has available another metabolic sequence, the aerobic hexosemonophosphate or pentose phosphate (PP) shunt. The importance of the PP shunt lies in the production of pentose from glucose, and in the generation of reduced nicotinamide adenine dinucleotide phosphate (NADPH). In functioning, the shunt relies not only on glucose-6-phosphate and 6-phosphogluconate dehydrogenases, but also upon the enzymatic mechanisms by which glutathione (GSH) is synthesized, by which it converts peroxides to harmless water, and by which its oxidized form (GSSG) is reduced (Figure 1).

Hemoglobin is subject to oxidative denaturation by peroxides, and this is accentuated when noxious oxidants are increased through the administration of certain drugs of the presence of other oxidant-producing stresses. Its chief protection against oxidative denaturation lies in the pentose phosphate pathway. The enzymatic detoxification of peroxides requires GSH and GSH peroxidase. The resultant GSSG must again be reduced by glutathione reductase, a process requiring NADPH which is produced in the human erythrocyte exclusively by the two dehydrogenases of the shunt. Under normal circumstances, probably about 10% of glucose traverses the pentose phosphate pathway. However, the amount varies widely, since it is strictly dependent upon the rate of oxidation of GSH, which in turn reflects the magnitude of oxidant stresses in the cell's environment of the moment. The products of the shunt are returned to the mainstream of glycolysis and travel the same final common pathway as glucose metabolized anaerobically.

Deficiency of G6PD was the first described and remains the commonest genetically-determined enzymopathy responsible for hemolytic anemia in man (24). It is X-chromosome linked, is associated with glutathione instability in the red cell, and in one form or another, has been estimated to affect some 100 million persons. It is not a single syndrome due to a single muted protein, but a host of disorders, varying widely in clinical severity and sharing the common feature of a mutant gene at the G6PD locus. Again, there is little evidence that any type of G6PD deficiency can be reasonably attributed to regulator or operator gene dysfunction. Rather, there is a profound polymorphism in man which results in a plethora of defective mutant structural genes coding for a large assortment of muted proteins. The latter differ electrophoretically, catalytically, in the stability of the enzyme protein, in the kinetics of their reactions with a variety of substrates, inhibitors, and pyridine cofactors, in pH optima and in other ways.

While G6PD deficiency is the commonest red cell enzymatic abnormality, other mutant enzymes affecting pentosephosphate shunt metabolism have been documented. Most of these are rare. Thus, hemolytic syndromes secondary to severe, genetically-determined deficiencies of both enzymes of glutathione synthesis (25-27), possibly, though less certainly, of glutathione peroxidase (28, 29), and of glutathione reductase (30, 31) have been defined (Figure 1). Mutant 6-phosphogluconate dehydrogenases have also been well documented (32), but it appears that abnormalities of the second dehydrogenase of the pentosephosphate shunt are unlikely to be associated with hemolysis. The pathogenesis of hemolysis in these cases is related predominantly to oxidative denaturation of hemoglobin, which triggers the hemolytic syndrome, and not primarily to metabolic depletion of ATP, as is believed to be the case in severe enzymopathies of the Embden-Meyerhof pathway. All shunt pathway enzymopathies then share with G6PD deficiency a propensity for oxidative denaturation of hemoglobin, and hemolysis is exacerbated by the same medications or other noxious oxidants which produce or exacerbate hemolysis in G6PD deficient subjects.

The possible consequences of structural gene mutations are altered catalytic activity, molecular instability, both, or absence of a detectable gene product. The latter may be the result of a "silent" deleted gene or of structural or, theoretically, regulatory gene mutations. In the case of structural gene mutants, "nonsense" codons or the insertion or deletion of single nucleotides resulting in a frameshift may result in absence of a detectable enzyme protein. Theoretically, synthesis of extremely unstable gene products would have a similar result. Also, structural mutations could be associated with production of highly labile messenger RNA. Mutated controller or operator genes, if these are present in man, could prevent the production of a gene product. However, virtually all the autosomally transmitted, genetically-induced, molecular diseases of man are associated with biochemically detectable, partial defects in the enzymatic machinery of heterozygotes. This is difficult to reconcile with the presence of diffusible substances elaborated by controller genes as described in bacteria (33). A mutated operator gene would still be compatible with the observations. To our knowledge the presence of controller or operator genes as regulator mechanisms in mammalian species has never been unequivocally documented, however.

In the case of red cell enzymopathies, the vast majority and nearly the totality of inborn errors appear associated with structurally altered proteins definable by immunologic cross reactivity, kinetic abnormalities, electrophoretic polymorphism, differences in isoelectric points, pH optima, behaviour toward substrates, cofactors and inhibitors, and by variations in catalytic activity and molecular stability. While the ultimate proof of structural abnormalities depends on amino acid sequencing, nonetheless the overwhelming burden of less direct evidence provides strong confirmation for a structural gene mutation in most instances. Absence of muscle type PFK, which usually constitutes about 50 percent of the red cell enzyme, characterizes the mild hemolytic syndrome associated with profound muscle dysfunction in the PFK deficiency syndrome described by Tarui (15, 16), and silent genes not coding for immunologically detectable gene products have been described in glucosephosphate isomerase (GPI) deficiency of the erythrocyte (34). Further, the evidence for structurally altered mutant proteins extends to their usual documentation in asymptomatic heterozygotes. Actually, in the absence of consanguinity, most severe autosomally transmitted red cell enzymatic defects are characterized by double heterozygosity for two separate mutant structural genes, and hence by the presence intracellularly of two separate mutant proteins (Figure 2).

Inherited enzymopathies may also result from failure to transform a proenzyme into an active form, either because the proenzyme is structurally abnormal, or because of a defect in

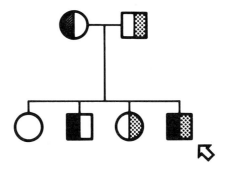

FIGURE 2. Genealogy of a typical family with PK deficiency. Cross-hatching indicates heterozygosity for an isozyme with impaired kinetics. Stippled areas indicate heterozygosity for an unstable isozyme. The proband, indicated by the arrow, is doubly heterozygous for both and is clinically affected, whereas the single heterozygotes are not.

the activator system. Both mechanisms have been documented in mammalian tissues other than the red cell, and, as discussed shortly, certain red cell pyruvate kinase variants may represent additional examples (34, 35). Finally, post-synthetic modifications of mutant enzyme proteins may be accompanied by abnormal enzyme maturation.

Hemolytic anemia associated with severe erythrocyte pyruvate kinase (PK) deficiency was described in our laboratory in 1961 (1, 2). Since then, PK deficiency has proved second in frequency to G6PD as a cause of hemolysis associated with mutant erythrocyte enzymes. The three non-equilibrium steps of glycolysis are catalyzed by hexokinase (HK), phosphofructokinase (PFK), and pyruvate kinase (PK), and in the human red cell these are rate controlling for lactate production. The fact that PK was the first Embden-Meyerhof pathway enzyme whose deficiency was demonstrated to cause hemolytic anemia, that it remains the best studied, and that it exercises one of the controlling roles in glycolysis render it suitable as a prototype, illustrating a battery of complex interactions triggered in a human tissue by variations in the activity of a single enzyme protein.

In the red cells of the first recognized PK deficient patients (1, 2), only a small fraction of the PK activity of normal erythrocytes was detectable irrespective of the substrate concentration employed (Figure 3). The erythrocytes of both patients and often certain siblings and other family members characteristically possessed PK activities averaging about half normal. The presence of a single defective gene

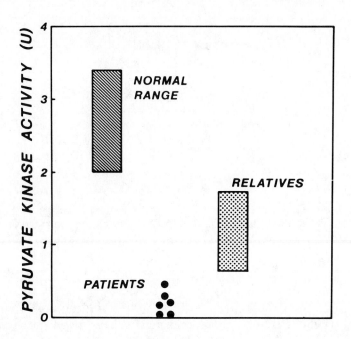

FIGURE 3. Erythrocyte PK activity in first subjects found to have PK deficiency.

in heterozygotes was usually unassociated with clinical or hematologic abnormalities; the inheritance of two defective mutant genes resulted in a variably severe hemolytic anemia. Initially, observations that the severity of hemolysis did not always correlate well with PK activity as assayed conventionally led to considerable puzzlement. We now know that enormous polymorphism for the PK gene exists (36-48). Mutant proteins, separable in terms of catalytic activity, aberrant kinetics, electrophoretic mobility, molecular stability, and behaviour toward allosteric activators, inhibitors, and nucleotide cofactors, abound. Defective mutant alleles of the PK gene remain concealed in phenotypically normal heterozygotes, surfacing at the clinical level only in subjects unfortunate enough to have inherited two such genes (Figure 2). Thus, in the absence of consanguinity, most hemolytic anemias associated with red cell PK deficiency reflect double heterozygosity for separate mutant genes, and are characterized by the presence of two separate mutant proteins within the same erythrocyte.

Figure 4 depicts the kinetics of PK in a kindred in which genetically determined hemolytic anemia was present. In the severely affected proband, significant PK activity

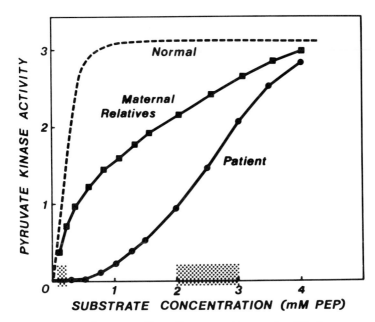

FIGURE 4. An example of an erythrocyte PK isozyme with severely defective kinetics.

was present in assays where substrate, purely for convenience, was present in high concentration. But kinetic studies demonstrated a $K_{0.5s}$ phosphoenolpyruvate (PEP) 10 times greater than normal, a fact that rendered proband PK virtually ineffective at PEP concentrations attainable in the circulating human red cell. The kinetics of PK in the cells of the clinically normal mother and certain maternal relatives revealed abnormalities intermediate between normal kinetics and those of proband PK. The kinetics of PK in the father's erythrocytes were normal, but maximal activity was only half that expected. The proband had inherited from the father a gene coding for a mutant PK protein virtually lacking in catalytic activity at any substrate concentration. A gene coding for a catalytically inefficient and kinetically grossly abnormal mutant PK was inherited from the mother. In proband erythrocytes, PK capable of effective catalysis under in vivo conditions was nearly totally lacking.

At least three isozymes of PK are postulated to exist (49, 54), and these have been often designated as M_1, M_2, and L. Liver possesses two isozymes, the M_2 and L protein, while the erythrocyte possesses for pactical purposes a single isozyme. This exists in two molecular forms closely related to the L protein of liver (51-53) and quite probably coded for by the same gene (51, 52). According to Marie, Kahn and Boivin (54), red cell PK purified 40,000 x to homogeneity showed complete immunologic identify with the L protein of liver. The L-type PK of liver was a homotetramer composed of 4 identical L subunits of molecular weight 58,000. In contrast, the PK of very young red cells and their nucleated precursors was composed of four identical but somewhat different subunits of slightly higher molecular weight (63,000) and designated L'. The major form of PK in circulating red cells was a heterotetramer designated as $L_2L'_2$. A mild tryptic attack could transform L'_4 into $L_2L'_2$, and a more aggressive proteolysis transforms $L_2L'_2$ into L_4. It was postulated that L type PK is initially synthesized as an L'_4 enzyme, secondarily partially proteolyzed by the erythrocyte into $L_2L'_2$, and more aggressively by the very active proteolytic systems of the liver into L_4 PK. Indeed a number of reports indicate that the liver has a defective mutant L protein when PK deficiency of the red cells is present. However, it has a redundant backup system in the form of the M_2 isozyme to compensate for single gene failure. If these findings stand the test of time and are confirmed, they represent an example of molecular evolution in which the maturation of a single gene product differs in accordance with metabolically different microenvironments in different mammalian tissues.

Further, the posttranslational alterations in peptide subunits influence the molecular behaviour of the PK enzyme. In the course of proteolytic maturation, the regulatory properties of PK appear to be improved (52). Both positive homotropic interactions at low PEP concentrations and negative interactions at intermediate concentrations appear, and ATP inhibition increases with transition from L'_4 to $L_2L'_2$ to L_4. The improvement of regulatory properties of L_4 as compared to L'_4 and $L_2L'_2$ theoretically favors the turning off of liver PK activity during gluconeogenic reversal of glycolysis, when futile cycling of PEP would be energetically wasteful. L'_4 is less stable than L_4, and has a lower affinity for PEP, both unfavorable characteristics. Cases are now reported of mutant PK associated with hemolytic syndromes in which the mutant protein is resistant to the favorable effects of proteolysis available to normal erythrocyte PK.

Human erythrocyte PK has been proposed to conform to the two-state model for allosteric enzymes as suggested by Monod et al (55). Experimental results support the validity of interconvertible conformations not dependent on alterations in peptide subunits (24, 26). The "R" conformation is favored by low pH and is characterized by hyperbolic kinetics, a decreased K_m for the positive allosteric effector, fructose-1,6-diphosphate ($F-1,6-P_2$), an increased affinity for substrate PEP, and a decreased affinity for the negative effectors ATP and alanine. The "T" conformation is favored by higher pH and characterized by decreased affinities for $F-1,6-P_2$ and PEP, and increased affinity for ATP and alanine (53, 56). Thus, R ⇌ T equilibrium of erythrocyte PK is modified in complex fashion by positive and negative allosteric modifiers. When PK activity is decreased to the point that obstruction to a smooth flow of glycolytic intermediates exists, the latter increase in concentration. $F-1,6-P_2$, hexose phosphates and particularly 2,3-DPG accumulate, and to an extent the increased concentration of $F-1,6-P_2$ favorably influences PK kinetics. A change in the ATP/ADP ratio within the cell likewise is capable of exerting allosteric effects. Thus, PK activity is secondarily influenced by the consequences of its own deficiency when this is severe, and its activity partially regulated by any cellular events which alter the concentration of $F-1,6-P_2$. For example, hexokinase deficiency and phosphofructosekinase deficiency, both of which are documented as inherited abnormalities of the human erythrocyte, depress the synthesis of $F-1,6-P_2$, and have the potential of influencing the behaviour of normal PK through alteration in feedforward regulatory phenomena. Triosephosphate isomerase (TPI) deficiency is also documented in association with hemolytic anemia (4, 5). Here, although knowledge of precise mechanisms is lacking, enormous accumulations of dihydroxyacetone phosphate (DHAP) have been demonstrated (57). This obviously alters the equilibrium involving $F-1,6-P_2$, DHAP, and 3-phosphoglyceraldehyde.

An interesting additional phenomenon is the marked increase in 2,3-DPG accompanying severe PK deficiency. This shifts the oxygen dissociation curve toward diminished oxygen affinity so that, in effect, the PK deficiency is associated with some improvement in oxygen delivery per unit of blood to the tissues. A conversely low level of 2,3-DPG has been documented in the cells of affected members of a kindred exhibiting increased red cell PK activity on an inherited basis (58). Irrespective of whether the Monod model fully characterizes the situation, there is no doubt that multiple interconvertible forms of normal PK exist. The study of mutant enzymes brought to light by the presence of associated hemolytic syndromes has amply documented a wide variety of PK variants having properties analogous to shifts in the postulated R ⇌ T equilibrium in one or the other direction as well as in molecular stability and susceptibility to postsynthetic maturation.

Human erythrocyte PK is known to have sulfhydryl groups whose alteration or oxidation can profoundly affect the kinetic properties of the enzyme. This has led to the postulate that levels of oxidized glutathione in the erythrocyte might exert regulatory influences on PK activity (59). In the case of PK mutant proteins, it is doubtful if this mechanism contributes significantly to deficient activity. However, acquired PK deficiency has now been documented in certain morbid states such as preleukemia and other dyserythropoietic syndromes (60-65). In a number of these, the PK deficiency was reversible by such procedures as prolonged dialysis, exposure of hemolysates to agents reducing sulfhydryl groups, or by partial purification of erythrocyte PK. Such reversibility points to posttranslational depression of PK activity and suggests that in certain pathologic states, abnormal accumulations of small molecules of undetermined nature can modify the behaviour of a normal enzyme protein (63, 64). Unfavorable modifications exert unwanted and pathologic regulatory effects on red cell metabolism.

Characterization of a mutant hexokinase (HK) with abnormal regulatory properties has been reported by Ryksen and Staal (66) in a patient with inherited hemolytic anemia. Since parents of the proband were cousins, homozygous expression of the defective gene was rendered highly likely. Of all the enzyme activities of glycolysis, the activity of HK is most dependent on cell age (11, 12) (Figure 5), and the HK activity of reticulocyte-rich, young erythrocyte populations is characteristically several fold greater than that of blood of normal mean red cell age. The HK activity of probrand erythrocytes was about 10 to 15 percent that expected when the degree of reticulocytosis was taken into account; the cells of both parents possessed HK activity about one-half that of normal blood. The mutant enzyme exhibited normal kinetics with respect to glucose and $MgATP^{2-}$, pH optimum, and heat stability at 40° C. Its behaviour in respect to regulation by glucose-1,6-diphosphate ($G-1,6-P_2$) and glucose-6-phosphate (G-6-P) which are normally strongly inhibitory, and in respect to inorganic phosphate (P_i) which is at least partially able to overcome this inhibition, was abnormal. The patient's HK showed a decreased affinity for $G-1,6-P_2$ and insensitivity to regulation by P_i. The altered electrophoretic pattern of proband red cell HK and other data were compatible with the hypothesis that band I_b normally increased in reticulocytes, and representing the $G-1,6-P_2$ inhibitable, P_i sensitive form of HK, was lacking, and that the P_i regulated form of the enzyme is modified as the normal erythrocyte ages to the insensitive form. A genetically determined lack of the P_i regulated HK was associated

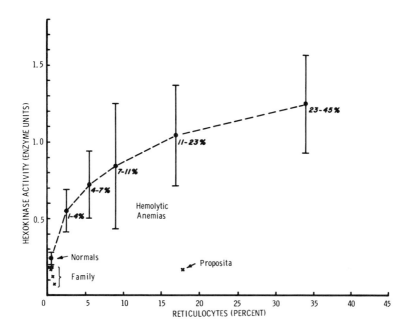

FIGURE 5. Dependence of erythrocyte hexokinase activity on a young mean cell age, as reflected by numbers of reticulocytes in patients with hemolytic anemias unrelated to hexokinase deficiency. Brackets enclose 1 S.D. above and below each mean.

with defective catalysis and hemolytic anemia. In effect, on the HK scale of aging, proband reticulocytes could already be regarded as senescent and incapable of prolonged survival.

In addition to its glycolytic machinery, certain enzymes of nucleotide metabolism are necessary for normal erythrocyte survival. As the reticulocyte matures, it loses its ribosomes and RNA disappears, its degradation being affected at least in part by ribonucleases (Figure 6). The end products of RNA breakdown include pyrimidine-containing ribonucleotides which are trapped by inability to diffuse from within the erythrocyte as long as they remain phosphorylated. The red cell contains, however, a unique, pyrimidine-specific 5'-nucleotidase through whose action uridine and cytidine-containing ribonucleotides are converted to diffusible nucleosides (67, 68). The latter freely leave the red cell to be metabolized or excreted elsewhere.

Severe, autosomal recessively transmitted pyrimidine-5'-nucleotidase deficiency has been described in several kindreds in our laboratory (67, 68) and confirmed elsewhere.

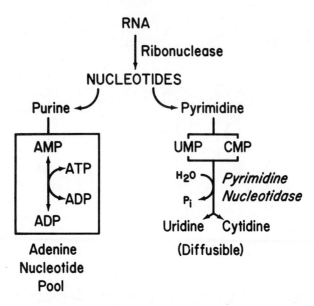

FIGURE 6. Outline of RNA degradation in the maturing reticulocyte.

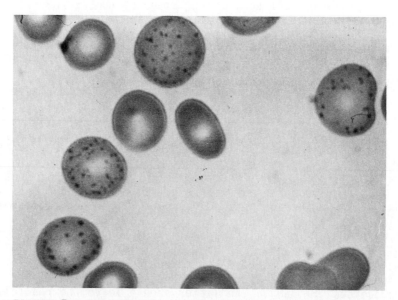

FIGURE 7. Peripheral red blood cells in patient with severe pyrimidine 5'-nucleotidase deficiency. The prominent basophilic stippling evident with Wright's stain represents aggregates of undegraded or partially degraded ribosomal RNA.

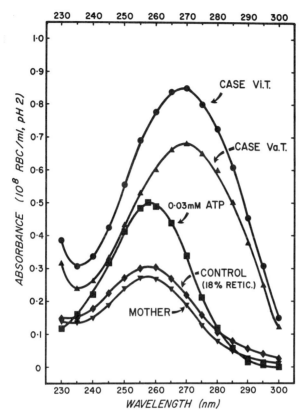

FIGURE 8. Ultraviolet absorption spectra of an aqueous solution of ATP compared to extracts of erythrocytes from two patients with severe pyrimidine 5'-nucleotidase deficiency (Vi. T. and Va. T.), their mother, and a subject with reticulocytosis due to sickle-cell anemia, all normalized to 10^8 erythrocytes. The wavelength peak shifts and amplitude differences reflect the presence of pyrimidine-containing nucleotides in cells from the two affected individuals (67).

Affected members of the kindred possess very little red cell nucleotidase activity and have a lifelong hemolytic anemia. Morphologically (Figure 7), there is evidence of retarded ribosomal degradation, which presumably is a consequence of negative feedback influences resulting from a unique and enormous accumulation of cytidine and uridine ribonucleotides. The latter result in severalfold increases in total nucleotide content of deficient erythrocytes in which as much as 80 percent of accumulated nucleotides may contain pyrimidines (Figure 8).

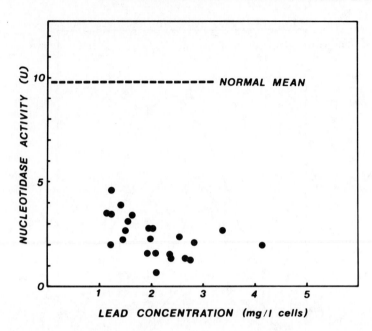

FIGURE 9. Pyrimidine 5'-nucleotidase activities in erythrocytes from subjects with lead overburden.

In the normal red cell, pyrimidine-containing nucleotides are present in vanishingly small amounts and are undetectable by most methods. The pyrimidine 5'-nucleotidase is exquisitely sensitive to inhibition by Pb^{++} in vitro and in vivo (68, 69). Nucleotidase deficiency of varying degrees is present in subjects with lead overburden (Figure 9), and in the very severe forms of lead poisoning this approximates the degree of deficiency observed in the genetically determined lesions (69-71). When this is the case, accumulations of red cell pyrimidine nucleotides become detectable. Thus far, the latter phenomena has been observed only in the lead induced and the inherited nucleotidase deficiency. The mechanism of hemolysis is complex and not fully understood. However, the fact that pyrimidine ribonucleotides can compete with far more effectual adenosine phosphates in crucial glycolytic reactions such as those catalyzed by HK, PGK, and PK offers one possible explanation for deleterious effects on red cell metabolism.

The maintenance of its adenine nucleotide pool poses special problems for the anucleated human erythrocyte. Unlike most tissues it does not possess the enzymatic machinery to synthesize adenosine phosphates de novo from small, precursor molecules (72, 73). While its AMP, ADP, and ATP are interconvertible as long as glycolysis is active, AMP is at risk to being lost through deamination or dephosphorylation, and firm evidence exists that such adenine nucleotide losses do occur. Replenishment of these losses is crucial to normal red cell survival and is dependent upon salvage pathways. Insight into control mechanisms governing maintenance of erythrocyte adenosine phosphates is provided by two experiments of nature involving the enzyme adenosine deaminase.

Adenosine, though toxic in larger amounts, is continously being formed in small quantities in various body tissues, from which it can diffuse into plasma and become a substrate for a salvage pathway available to the human red cell (74-79). But adenosine is the substrate for two competing enzymes of the erythrocyte. One, adenosine kinase, is capable of phosphorylating adenosine to AMP (80), and hence, is a viable mechanism for renewing the adenine nucleotide pool. The second, adenosine deaminase, converts adenosine to inosine. There is no mechanism by which inosine or inosine monophosphate can be utilized as a source of replenishment of adenosine phosphates (72, 73). In the normal red cell, the balance between the actions of the kinase and the deaminase are such that appropriate concentrations of adenine nucleotides are maintained.

In our laboratory, we have had the opportunity of investigating a kindred in which 12 of 24 members, spanning three generations, exhibit a dominantly transmitted, moderate hemolytic anemia characterized by inability to maintain the adenine nucleotide pool (76, 77). ATP concentrations in red cells of affected subjects are about half or less those expected, and this is believed to be the cause of diminished red cell survival. But, in addition, all affected subjects have red cell adenosine deaminase activities enormously increased from 45 to 75-fold above those found in normal subjects and in unaffected members of their own kindred (76, 77) (Figure 10). Under these circumstances, balance between adenosine kinase and adenosine deaminase activities is destroyed. The kinase is unable to compete with the greatly enhanced activity of the deaminase, and insufficient phosphorylation of adenosine results. Adenine nucleotide concentrations can not be maintained at normal levels, and hemolytic anemia is the result. The regulatory imbalance which permits the massive overproduction of adenosine deaminase clearly involves genetic mechanisms, but its precise nature remains entirely unclear. By all criteria thus far employed, the adenosine deaminase in affected family members appears to be a normal enzyme protein. Dr. William Osborne of the University of Washington in Seattle has purified the

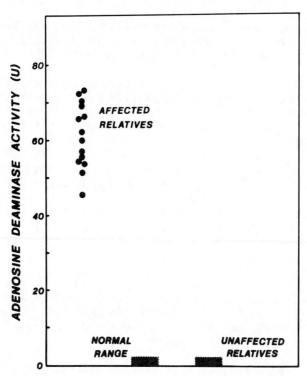

FIGURE 10. Adenosine deaminase activities in erythrocytes from subjects within a kindred affected by dominantly inherited hemolytic aniema associated with significantly lowered erythrocyte adenine nucleotide concentrations.

red cell deaminase in this kindred to homogeneity, and kinetically, electrophoretically, and in terms of specific activity, simple overproduction of a normal enzyme appears to be the result of the genetic lesion. Interestingly, blood leukocytes and cultured skin fibroblasts do not share the vastly increased activity of the erythrocyte deaminase in this hemolytic syndrome.

In support of the role of a balance between adenosine kinase and deaminase activities in maintaining normal concentrations of adenosine phosphates in the human red cell are observations made by other laboratories of another inherited syndrome, that of adenosine deaminase deficiency (81). The latter is unaccompanied by hemolytic anemia but, for reasons yet unclear, is associated with a syndrome involving lymphocyte dysfunction and severe immunoincompetence. In this disorder, adenosine and deoxyadenosine concentrations in red cells and lymphocytes are greater than normal (82-84). It is presumed that in this contrasting situation, the kinase acts virtually unopposed by deaminase, with the result that the adenine nucleotide pool is actually increased.

In conclusion, the anucleate human red cell, despite its comparative metabolic simplicity, provides a fertile source for genetically determined experiments of nature. These inborn errors resulting in molecular lesions of a single point in complex, interacting enzymatic machinery, in turn, often provide insights into regulatory mechanisms which modulate and fine-tune cellular metabolism.

REFERENCES

1. Valentine, W. N., Tanaka, K. R., and Miwa S. (1961). Trans. Assoc. Am. Physicians 74, 100.
2. Tanaka, K. R., Valentine, W. N., and Miwa S. (1962). Blood 19, 267.
3. Baughan, M. A., Valentine, W. N., Paglia, D. E., Ways, P. O., Simons, E. R., DeMarsch, Q. B. (1968). Blood 32, 236.
4. Schneider, A. S., Valentine, W. N., Hattori, M., and Heins, H. L., Jr. (1965). New Engl. J. Med. 272, 229.
5. Valentine, W. N., Schneider, A. S., Baughan, M. A., Paglia, D. E., and Heins, H. L., Jr. (1966). Am. J. Med. 41, 27.
6. Kahn, A., Etiemble, J., Meienhofer, M. C., and Boivin, P. (1975). Clin. Chim. Acta 61, 415.
7. Waterbury, L., and Frenkel, E. P. (1969). Clin. Res. 17, 347.
8. Valentine, W. N., Hsieh, H., Paglia, D. E., Anderson, H. M., Baughan, M. A., Jaffé, E. R., and Garson, O. M. (1969). New Engl. J. Med. 280, 528.
9. Beutler, E., Scott, S., Bishop, A., Margolis, N., Matsumoto, F., and Kuhl, W. (1973). Trans. Assoc. Am. Physicians 86, 154.
10. Kitamura, M., Iijima, N., Hashimoto, F., and Hiratsuka, A. (1971). Clin. Chim. Acta 34, 419.
11. Valentine, W. N., Oski, F. A., Paglia, D. E., Baughan, M. A., Schneider, A. S. and Naiman, J. L. (1967). New Engl. J. Med. 276, 1.
12. Valentine, W. N., Oski, F. A., Paglia, D. E., Baughan, M. A., Schneider, A. S., and Naiman, J. L. (1968). In "Hereditary Disorders of Erythrocyte Metabolism" (E. Beutler, ed.), p. 288. Grune and Stratton, New York.
13. Chen, S. H., Malcolm, L. A., Yoshida, A., and Giblett, E. R. (1971). Am. J. Hum. Genet. 23, 87.

14. Meera Kahn, S. P., Westerveld, A., Grzeschik, K. H. Deys, B. F., Garson, O. M., and Siniscalco, M. (1971). Am. J. Hum. Genet. 23, 614.
15. Tarui, S., Okuno, G., and Ikura, Y. (1965). Biochem. Biophys. Res. Commun. 19, 517.
16. Layzer, R. B., and Rasmussen, J. (1974). Arch. Neurol. 31, 411.
17. Lyon, M. F. (1961). Nature (Lond.). 190, 372.
18. Beutler, E., Yeh, M., and Fairbanks, V. F. (1962). Proc. Natl. Acad. Sci. U.S.A. 48, 9.
19. Rapaport, S., and Luebering, J. (1950). J. Biol. Chem. 183, 507.
20. Bartlett, G. R. (1959). J. Biol. Chem. 234, 449.
21. Benesch, R., Benesch, R. E., and Yu, C. I. (1968). Proc. Natl. Acad. Sci. U.S.A. 59, 526.
22. Rosa, R., Gaillardon, J., and Rosa, J. (1973). Biochem. Biophys. Res. Commun. 51, 536.
23. Rosa, R., Prehu, M-O., Beuzard, Y., and Rosa, J. (1978). J. Clin. Invest. 62, 907.
24. Beutler, E. In "The Metabolic Basis of Inherited Disease" (1978) (J. B. Stanbury, J. B. Wyngaarden, and D. S. Fredrickson, eds.), pp. 1430-1451. McGraw-Hill, Inc., 4th edition, New York.
25. Prins, H. K., Oort, M., Loos, J. A., Zürcher, C., and Beckers, T. (1966). Blood 27, 145.
26. Mohler, D. N., Majerus, P. W., Minnich, V., Hess, C. E., and Garrick, M. D. (1970). New Engl. J. Med. 283, 1253.
27. Konrad, P. N., Richards, F. II., Valentine, W. N., and Paglia, D. E. (1972). New Engl. J. Med. 286, 557.
28. Beutler, E., and Matsumoto, F. (1975). Blood 46, 103.
29. Necheles, T. F., Maldonado, N., Barquet-Chediak, A., and Allan, D. M. (1969). Blood 33, 164.
30. Beutler, E. (1969). J. Clin. Invest. 48, 1957.
31. Loos, H., Roos, D., Weening, R. and Houwerzijl, J. (1976). Blood 48, 53.
32. Parr, C. W., and Fitch, L. I. (1967). Ann. Hum. Genet. 30, 339.
33. Dreyfus, J-C. (1969). In "Progress in Medical Genetics" (A. G. Steinberg, and A. G. Bearn, eds.), Vol. 6, pp. 169-200. Grune and Stratton, New York.
34. Kahn, A. and Dreyfus, J-C. (1978). In "Metabolic Diseases". Proceedings of the 12th FEBS Meeting, Perganon Press, Oxford, London.
35. Kahn, A., and Marie, J. In "Model for the Study of Inborn Errors of Metabolism" (Hommes, Elsenier/North Holland, Amsterdam). In press.

36. Paglia, D. E., Valentine, W. N., Baughan, M. A., Miller, D. R., Reed, C. F., and McIntyre, O. R. (1968). J. Clin. Invest. 47, 1929.
37. Paglia, D. E., Konrad, P. N., Wolff, J. A., and Valentine, W. N. (1976). Clin. Chim. Acta 73, 395.
38. Nakashima, K., Miwa, S., Oda, S., Tanaka, T., Imamura, K., and Nishina, T. (1974). Blood 43, 537.
39. Nakashima, K. (1974). Clin. Chim. Acta. 55, 245.
40. Kahn, A., Marie, J., Galand, C., and Boivin, P. (1976). Scand. J. Haematol. 16, 250.
41. Miwa, S., Nakashima, K., Ariyoshi, K., Shinohara, K., Oda, E., and Tanaka, T. (1975). Brit. J. Haematol. 29, 157.
42. Shinohara, K., Miwa, S., Nakashima, K., Oda, E., Kageoka, T., and Tsujino, G. (1976). Am. J. Hum. Genet. 28, 474.
43. Paglia, D. E., and Valentine, W. N. (1971). Blood 37, 311.
44. Staal, G. E. J., Ceerdink, R. P., Vlug, A. M. C., and Hamelink, M. L. (1976). Clin. Chim. Acta 68, 11.
45. Tanaka, K. R. and Paglia, D. E. (1971). Semin. Hematol. 8, 367.
46. Paglia, D. E., and Valentine, W. N. (1970). J. Lab. Clin. Med. 76, 202.
47. Valentine, W. N., and Tanaka, K. R. (1978). In "The Metabolic Basis of Inherited Disease" (J. B. Stanbury, J. B. Wyngaarden, and D. S. Fredrickson, eds.), pp. 1410-1429, McGraw-Hill, Inc., New York. 4th edition.
48. Kahn, A., Marie, J., Galand, C., and Boivin, P. (1975). Humagenetik. 29, 271.
49. Ibsen, K. H. (1977). Cancer Research 37, 341.
50. Miwa, S., Nishima, T., Imamura, K., and Tanaka, T. (1972). Abstract 341. Abstracts of the XIV International Congress of Hematology, Sao Paulo, Brazil, July 16-21, 1972.
51. Kahn, A., Marie, J., and Boivin, P. (1976). Hum. Genet. 33, 35.
52. Kahn, A., Marie, J., Garreau, H., and Sprengers, E. D. (1978). Biochim. Biophys. Acta 523, 59.
53. Staal, G. E. J., Koster, J. F., Kamp, H., Van Milligen-Boersma, L., and Veeger, C. (1971). Biochim. Biophys. Acta 227, 86.
54. Marie, J., Kahn, A., and Boivin, P. (1977). Biochim. Biophys. Acta 481, 96.
55. Monod, J., Wyman, J., and Changeux, J-P. (1965). J. Mol. Biol. 12, 88.
56. Black, J. A., and Henderson, M. H. (1972). Biochim. Biophys. Acta 481, 96.

57. Schneider, A. S., Dunn, I., Ibsen, K. H., and Weinstein, I. M. (1968). In "Hereditary Disorders of Erythrocyte Metabolism" (E. Beutler, ed.) p. 273, Grune and Stratton, New York.
58. Zürcher, C., Loos, J. A., and Prins, H. K. (1965). Folia Haematol. 83, 366.
59. Van Berkel, J. C., Staal, G. E. J., Koster, J. F., and Nyessen, J. G. (1974). Biochim. Biophys. Acta 334, 361.
60. Valentine, W. N., Konrad, P. N., and Paglia, D. E. (1973). Blood 41, 857.
61. Boivin, P., Galand, C., and Dreyfus, B. (1969). Nouv. Rev. Franc. Hemat. 9, 105.
62. Arnold, H., Blume, K. G., Löhr, G. W., Boulard, M., and Najean, Y. (1974). Clin. Chim. Acta 57, 187.
63. Arnold, H., Blume, K. G., and Löhr, G. W. (1977). Blood 49, 1022.
64. Kahn, A., Marie, J., Bernard, J-F., Cottreau, D., and Boivin, P. (1976). Clin. Chim. Acta 71, 379.
65. Boivin, P., Galand, C., and Audollent, M. (1970). Path. Biol. 18, 175.
66. Rijksen, G., and Staal, G. E. J. (1978). J. Clin. Invest. 62, 294.
67. Valentine, W. N., Fink, K., Paglia, D. E., Harris, S. R., and Adams, W. S. (1974). J. Clin. Invest. 54, 866.
68. Paglia, D. E., and Valentine, W. N. (1975). J. Biol. Chem. 250, 7973.
69. Paglia, D. E., Valentine, W. N., and Dahlgren, J. G. (1975). J. Clin. Invest. 56, 1164.
70. Valentine, W. N., Paglia, D. E., Fink, K., and Madokoro, G. (1976). J. Clin. Invest. 58, 926.
71. Paglia, D. E., Valentine, W. N., and Fink, K. (1977). J. Clin. Invest. 60, 1362.
72. Lowy, B. A., Williams, M. K., and London, I. M. (1962). J. Biol. Chem. 237, 1622.
73. Bishop, C. (1960). J. Biol. Chem. 235, 3228.
74. Lerner, M. H., and Lowy, B. A. (1974). J. Biol. Chem. 249, 959.
75. Lerner, M. H., and Rubinstein, D. (1970). Biochim. Biophys. Acta 224, 301.
76. Paglia, D. E., Valentine, W. N., Tartaglia, A. P., Gilsanz, F., and Sparkes, R. S. (1978). In "The Red Cell" (G. J. Brewer, ed.), pp. 319-335, Alan R. Liss, Inc., New York.
77. Valentine, W. N., Paglia, D. E., Tartaglia, A. P., and Gilsanz, F. (1977). Science 195, 783.
78. Parker, J. C. (1970). Am. J. Physiol. 218, 1568.
79. Parks, R. E., Jr., and Brown, P. R. (1973). Biochem. 12, 3294.

80. Lowy, B. A., Jaffé, E. R., Vanderhoff, G. A., Crook, L., and London, I. M. (1958). J. Biol. Chem. 230, 409.
81. Giblett, E. R., Anderson, J. E., Cohen, F., Pollara, B., and Meuwissen, H. J. (1972). Lancet II. 1067.
82. Cohen, A., Hirschorn, R., Horowitz, S. D., Rubenstein, A., Polmar, S. H., Hong, R., and Martin, D. W. (1978). Proc. Natl. Acad. Sci. U.S.A. 75, 472.
83. Donofrio. J., Coleman, M. S., and Hutton, J. J. (1978). J. Clin. Invest. 62, 884.
84. Agarwal, R. P., Crabtree, G. W., Parks, R. E., Jr., Nelson, J. A., Kneightley, R., Parkman, R., Rosen, F. S., Stern, R. C., and Polmar, S.H. (1976). J. Clin. Invest. 57, 1025.

ENZYME REPLACEMENT THERAPY [1]

Ernest Beutler[2] and George L. Dale[2]

Division of Medicine, City of Hope Medical Center, Duarte, Ca. 91010 and Department of Clinical Research, Scripps Clinic and Research Foundation, La Jolla, Ca. 92037

ABSTRACT: Of the many genetic diseases which are now known to be due to an enzyme deficiency, the glycolipid storage diseases are among those which are most likely to yield to enzyme replacement therapy. To be successful, such therapy requires the availability of relatively pure enzyme from a human source, delivery of the enzyme to the storage cell, and contact with the storage material for a sufficiently long period of time to permit it to be catabolized. Various receptor sites on cells may divert the delivery of enzyme to its site of action.

Gaucher's disease is one of the most suitable disorders for enzyme replacement therapy. The results of administration of various amounts of enzyme by various routes to six patients with Gaucher's disease is reported. Little if any therapeutic effect has been observed to date and alterations in quantity of enzyme given or in the mode of delivery will presumably be required.

INTRODUCTION

Many diseases, particularly those of hereditary origin, may be ascribed to a clearly identified deficiency of an enzyme protein. Enzymes that perform a variety of functions may be affected. The deficient enzyme may be one that is important in catalyzing the complex transformations that are required to provide energy for the body's metabolic

---

[1]This work was supported by Grant No. AM 14755 from the National Institutes of Health.
[2]Present address: Department of Clinical Research, Scripps Clinic and Research Foundation, 10666 North Torrey Pines Road, La Jolla, Ca. 92037

activities. Galactosemia, due to a deficiency of galactose-1-phosphate uridyl transferase or galactokinase is such a disorder. The deficient enzyme alternatively may be one which plays a role in catalyzing one of the cascading reactions which characterize the complement or the coagulation system. An example is Cl esterase deficiency, which causes hereditary angioneurotic edema. A particularly interesting series of disorders results from deficiencies of enzymes which are required for the breakdown of complex molecules such as glycolipids into smaller components to be catabolized or reutilized by the body. Since it is this group of diseases which forms the primary target of enzyme replacement strategies being formulated today, it is perhaps appropriate to discuss these so-called lysosomal storage disorders in greater detail.

THE STORAGE DISEASES

The phagocytic vacuole of a cell may be likened to a small stomach. It is a digestive organelle lined initially by a portion of the outer membrane of the cell which has

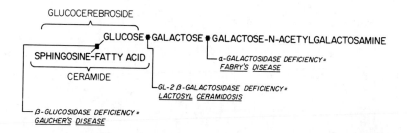

FIGURE 1. Globoside. Genetic defects in the ability to hydrolyze the linkages indicated cause various types of storage diseases.

invaginated, engulfing the particle to be phagocytosed. The primary phagocytic vacuole then apparently fuses with lysosomes which release their hydrolytic enzymes into the phagocytic vacuole, where they begin their digestive process (1).

When the ingested material contains complex glycolipids the ceramide core of the molecule remains resistant to digestion until the sugars attached to it have been removed, and these must be digested from the outer sugar inward. For example, in the degradation of globoside (fig. 1) the terminal N-acetylgalactosamine must be cleaved from the glycolipid, forming ceramide trihexoside, before the ceramide trihexoside formed can undergo enzymatic attack. The α-galactosidic linkage between the terminal galactose and the penultimate galactose of ceramide trihexoside must be removed by an α-galactosidase, α-galactosidase A, before the next galactose, held in a β-galactosidic bond can be removed by an appropriate β-galactosidase. Finally, the removal of glucose from ceramide is catalyzed by a β-glucosidase. The genetic absence of any of these enzymes results in accumulation of the compound which cannot be degraded (2). Thus, in Sandhoff's disease, when β-N-acetylgalactosaminide A and B are missing, globoside accumulates in the tissues. A deficiency of α-galactosidase A results in the accumulation of ceramide trihexoside in Fabry's disease, a rare sex-linked disorder characterized by the development of vascular abnormalities and usually renal failure leading to death in the third or fourth decade of life. Deficiency of β-galactosidase results in lactosyl ceramidosis, and Gaucher's disease is due to absence of the enzyme which cleaves glucose from ceramide. When this acid β-glucosidase, glucocerebrosidase, is absent, glucocerebroside (fig. 2) accumulates in the liver, spleen, and bone marrow. A rare infantile form of

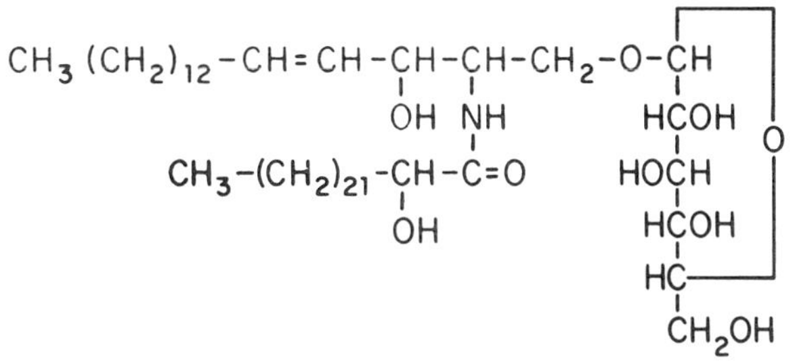

$$CH_3(CH_2)_{12}-CH=CH-\underset{OH}{CH}-\underset{NH}{CH}-CH_2-O-CH$$

$$CH_3-(CH_2)_{21}-\underset{OH}{CH}-C=O$$

FIGURE 2. The structure of glucocerebroside

the disorder is associated with the accumulation of glycolipid in the brain as well, and leads to the early demise of the unfortunate victim. In the more common, adult form of the disease, however, gradual accumulation of lipid deposits in the reticuloendothelial system leads to a reduced blood platelet count, anemia, bone pain, and bone fractures. Some of the changes in the blood are ameliorated by surgical removal of the spleen. Apart from this form of intervention, medical science today has little to offer to patients with Gaucher's disease.

Degradation of the neuraminic acid-containing complex glycolipids, the gangliosides, leads to a similar series of storage disorders which generally involve the central nervous system. Best known of these is a GM_2 ganglioside storage disorder, Tay-Sachs disease, caused by a deficiency of N-acetylgalactosaminidase A, and leading to blindness, neuromuscular degeneration and severe mental retardation beginning in infancy and terminating fatally in early childhood. Some of the disorders which result from hereditary absence of lysosomal acid hydrolases are extremely rare. Others, however, reach an appreciable frequency, particularly among Jews of Eastern European origin. The incidence of Tay-Sachs disease and of Gaucher's disease is about one in 5,000 in this ethnic group.

THE TREATMENT OF GENETIC DISEASES

General Considerations. A common misconception among scientists and laymen alike is that a hereditary disease is, by its very nature, untreatable. Nothing could be further from the truth. The successful management of patients with certain types of hereditary disease may be counted among the important successes of medicine in the 20th century. Diabetes mellitus and pernicious anemia are examples of diseases which are, at least in large measure, hereditary and which have yielded to treatment. The introduction of diets lacking the substrate which the genetically deficient person is unable to metabolize is another example of such treatment: a galactose-free diet allows a galactosemic child to grow into a normal adult; phenylalanine-poor diets ameliorate the consequences of phenylketonuria. In hereditary spherocytosis, a hereditary defect which affects the shape of the red cells, removal of the spleen, the graveyard of the abnormal red cells, is highly effective. In spite of their beneficial effects, however, these treatments represent relatively superficial approaches to basic problems. In the distant future, it is possible that genetic diseases may be treatable by engrafting a functioning, normal gene onto the genome of the affected individual. Yet, the technologic barriers to such gene-replacement therapy are so formidable that there is serious doubt whether this feat can ever be accomplished in any but a few very limited circumstances. A more practical approach to the treatment of disorders due to genetic absence of an enzyme is "enzyme replacement therapy".

Principles of Replacement Therapy. There is sometimes a tendency to view that which is old in medicine as new. This seems to be the case with enzyme replacement therapy. The transfusion of fresh-frozen plasma, a time-honored treatment for hemophilia, is surely a form of enzyme replacement therapy. The enzyme of blood coagulation, factor VIII, is infused into the circulation of the patient who lacks it. Similarly, transfusion of small amounts of plasma into patients who lack the plasma transglutaminase which firmly cross-links the fibrin clot, factor XIII, may be viewed as enzyme replacement therapy. Such standard forms of enzyme replacement are limited, however, to circulating proteins. We would like to be able to extend this form of management to those patients with hereditary disease in which the site of function of the missing enzyme is an intracellular one. The disorders which seem most amenable to such enzyme replacement treatment are the lysosomal storage diseases. The choice of the particular disease from this group is very important; efforts should be directed to those diseases in which there is reasonable hope of reversing the detrimental effects. If accumulation of the storage material has done irreparable harm, then its removal may stay the progress of the disease but will not produce any real benefit for the already severely damaged patient. From this point of view, treatment of diseases which severely damage the central nervous system seems futile. Thus, we would not expect that a patient with advanced Tay-Sachs disease or Niemann-Pick's disease would be helped by enzyme replacement therapy.

With these considerations, we have focused our own attention upon replacement therapy for the treatment of the adult form of Gaucher's disease. The central nervous system is spared in this disorder, and there is reason to hope that with the removal of the massive accumulations of glucocerebroside in the bone marrow, liver and spleen, the patient might be restored to a more normal state.

Method of Implementation of Replacement Therapy. Four problems must be overcome for this type of replacement therapy to be successful. First of all, it is necessary to purify the enzyme from a human source. Second, the enzyme must be delivered to its target, the storage cell. Third, it is necessary for the enzyme to survive in the target cell sufficiently long to be of practical value; if the enzyme were merely ingested by the cell and catabolized it would be no more useful in the treatment than is the protein hormone insulin when taken orally. Fourth, it is necessary for the intracellular enzyme to undergo productive contact with the storage material. If the storage material is segregated in such a way that the enzyme cannot reach it, one can hardly expect a beneficial effect.

Enzyme Purification. The limited availability of suitable enzymes has been a limiting factor in the initiation of replacement trials. In early clinical trials, enzymes from fungal and animal sources were utilized but often evoked life-threatening immunological reactions (3-5). For example, in 1968 α-galactosidase from Aspergillus niger was repeatedly administered to a patient with Type II glycogen storage disease; the patient subsequently died from immunological complications (4).

In recent clinical trials, only human enzymes have been used. Fresh plasma has served as an enzyme source for the treatment of various mucopolysaccharide storage disorders (6-8), Sandhoff's disease (9) and Fabry's disease (10). Leukocytes have been administered to patients with Fabry's disease (11) and Type II mucopolysaccharidosis (12). However, most recent trials have employed purified enzymes derived from human urine, plasma or placenta. Hexosaminidase A and B and arylsulfatase A have been purified from urine and used in the treatment of Sandhoff's disease (13) and metachromatic leukodystrophy patients (14), respectively. Human placenta has been a rich source for the purification of glucocerebrosidase (15,16) α-galactosidase A (11), β-glucuronidase (17) and α-glucosidase (18) for the treatment of Gaucher's disease, Fabry's disease, β-glucuronidase deficiency and Type II glycogen storage disease patients, respectively. Plasma has also been used as starting material for the purification of α-galactosidase A (19).

Our purification procedure for glucocerebrosidase (15) from placenta utilizes batch purification steps which are adaptable to industrial scale and subsequent affinity chromatography on concanavalin A-Sepharose and oleoyl-agarose. The net result is the purification of large amounts of glucocerebrosidase with a yield of 60%. Such large scale purification is desirable so that the availability of enzymes is not a limiting factor in evaluating the effectiveness of enzyme replacement therapy.

Enzyme Delivery. A key problem in enzyme replacement therapy is the delivery of enzymes to the correct tissue and to the correct subcellular location within that tissue. For example, in the treatment of Gaucher's disease, the accumulation site for the unmetabolized glucocerebroside is the lysosome of the reticuloendothelial system.

Some optimism with respect to our ability to introduce lysosomal enzymes into cells is engendered by the suggestion that cells may normally excrete lysosomal enzymes into their surroundings and subsequently package the enzymes in lysosomes after reincorporating them from the external medium (20). This theory is based upon observations made of cells of patients with a peculiar genetic abnormality which has been designated as I-cell disease. Fibroblasts from patients with I-cell disease synthesize and excrete lysosomal enzymes into the surrounding medium but the cells themselves are deficient in the enzymes. When exposed to lysosomal enzymes from normal fibroblasts, however, fibroblasts from patients with I-cell disease are able to incorporate these enzymes. Moreover, the lysosomal enzymes from I-cell fibroblasts are not incorporated into normal fibroblasts. These findings led Neufeld et al to propose that I-cell fibroblasts synthesize lysosomal enzymes which do not have the correct recognition signal (20). It must be emphasized that while many lysosomal acid hydrolases appear to participate in the excretion-reabsorption cycle, β-glucosidase and acid phosphatase are not among them. This is particularly important to recognize with respect to enzyme replacement therapy, since β-glucosidase is the glucocerebroside which is missing in Gaucher's disease. Recent studies suggest that the uptake signal of the lysosomal hydrolases may be a phosphomannose group (21).

Two studies (22,23) have demonstrated that lysosomal enzymes injected intravenously into rats are sequestered by the reticuloendothelial system. On the other hand, Pentchev et al (24) have demonstrated that intravenously administered glucocerebrosidase is predominantly localized in liver parenchymal cells rather than the reticuloendothelial system. This model system would predict that intravenous injection of glucocerebrosidase is not an efficient delivery mode. Although the liver cell type involved in the clearance of lysosomal enzymes is in question and may vary with different lysosomal enzymes, it is clear that the liver recognizes the carbohydrate portion of lysosomal glycoproteins. This uptake system has the greatest affinity for mannose residues in the clearance of these glycoproteins (23). It is apparently completely distinct from the recognition system which functions in the clearance of asialoglycoproteins, relying on the binding of galacto-terminal carbohydrate chains (23).

An alternative method for targeting of lysosomal enzymes to a specific tissue and organelle is to encapsulate the enzyme in either liposomes (25) or resealed erythrocytes (26). Both carriers provide efficient delivery of enzymes to the reticuloendothelial system. A major drawback to the use of liposomes as carriers is the poor incorporation of proteins into the vesicles, usually less than 8%. In contrast, proteins may be incorporated into resealed erythrocytes with yields of up to 40% (27). Another advantage of encapsulating proteins in resealed erythrocytes may be the possible attenuation of the immunological response compared to direct intravenous administration of proteins (28). Normally, human lysosomal enzymes would not be considered immunogenic upon reinfusion into humans, however new approaches to stabilizing enzymes for their prolonged intracellular survival may transform these proteins into immunogenic species (29). Therefore, it is necessary to consider the benefits of enzyme encapsulation in shielding proteins from immunological detection or inactivation.

EXPERIMENTAL ENZYME REPLACEMENT IN GAUCHER'S DISEASE

Attempts to treat Gaucher's disease by enzyme replacement were first reported by Brady and his colleagues (30,31). These investigators used a partially purified preparation of glucocerebrosidase that was precipitated on albumin prior to infusion into three patients with Gaucher's disease. The amount of enzyme administered was expressed in terms of the number of nanomoles of glucocerebroside hydrolyzed per hour. The units we employ (vide infra), in contrast, represent μmoles of fluorogenic substrate hydrolyzed per minute; thus, about 25,000 of the nanomolar units employed by Brady et al represent a single unit of activity as expressed in our investigations. A normal spleen contains approximately 10 U of enzyme as measured in our assay system. In their investigations, Brady and colleagues administered between 1.5 and 9.3 million units of enzyme, a quantity roughly equivalent to what might be found in between 6 and 35 normal spleens. Clinical observations were not reported by these investigators, who relied entirely on tissue glucocerebroside levels before and after enzyme infusion in appraising the results. They found that a liver biopsy sample obtained after treatment contained, in each case, less glucocerebroside than the biopsy taken prior to treatment. Multiplying the change in the glucocerebroside content by the weight of the liver, they concluded that a large amount of glucocerebroside had been hydrolyzed as a result of enzyme infusion. Interpretation of the data must be tempered, however, by the markedly heterogenous distribution of glucocerebroside in the liver of Gaucher's disease patients: in a study of one liver obtained at autopsy we found a range of 10-30 mg of glucocerebroside per gram wet weight in a very small series of samples (32).

Another very limited trial of glucocerebrosidase therapy of Gaucher's disease has been reported by Belchetz et al (33). These authors injected enzyme entrapped in liposomes to one patient with Gaucher's disease. Worrysome signs of embolization, including mental confusion and severe abdominal pain occurred following injections. Although hepatomegaly was thought to be somewhat decreased after a year's treatment, the changes reported by these authors are by no means convincing.

We initiated replacement trials in 1976. Our first subject was a young woman with very severe liver involvement, bone marrow failure, and a generally severely debilitated state. She was following a rapid, downhill course, and we considered it unlikely that she would survive more than a few months if her disease could not be arrested. Thus, she was an ideal subject for experimental therapy with a new modality, the risk of which could not be accurately predicted. The patient readily consented to participation in such a study. Several courses of glucocerebrosidase were given in various forms for a period of approximately one year. Several months after the last treatment she succumbed to an uncontrollable bleeding episode related to her liver and bone marrow involvement. The treatment of this patient has been described in some detail (32). She was given relatively small amounts of enzyme directly intravenously, encapsulated in red blood cells, and encapsulated in red blood cells coated with anti-Rh antibody. Some improvement in the appearance of her liver by technecium scan seemed to occur, but the improvement was by no means unequivocal. The level of leukocyte glucocerebroside seemed to decline somewhat during the one course of treatment in which measurements were made. Overall, it is difficult to state whether or not this patient benefited from treatment. Large amounts of glucocerebroside were present in her liver at autopsy, but since the pre-treatment levels were not known, no definitive conclusions can be drawn from this observation. Strangely, virtually no intact Gaucher's cells were found in the liver. Instead, there were large patches of necrotic cells. We presume that this represents an unusual agonal change that was unrelated to therapy, which had last been given several months before death.

A second patient was a child with severe thrombocytopenia. A single infusion of 2.3 Units of enzyme encapsulated in anti-Rh globulin coated erythrocytes was administered in the hope that this might produce a rise in her platelet count and obviate the need for splenectomy. After 4 months, however, no change in the platelet count had been observed and splenectomy was performed.

A third patient, a young woman with Gaucher's disease manifested marked bony involvement with collapse of thoracic vertebrae, hip and knee lesions, and marked splenomegaly. The results of some of the studies carried out on this patient have been reported previously (34). To date this patient has received 27 infusions of glucocerebrosidase. This included 9 direct intravenous infusions of 4 Units each, one series of six infusions of resealed red cells containing a total of 83 Units of enzyme, and subsequently another series of 11 infusions of red cell-entrapped enzyme totalling 27 Units. There has been no regression of liver size, followed in this patient by serial ultrasound scans. Her bony lesions appear to have stabilized, and one particularly troublesome lesion in the knee seems to have improved somewhat.

One patient whom we have treated received a total of only 2.2 Units of β-glucosidase in resealed red cells. No change in organomegaly was observed and a bony lesion which the patient had proved to be a manifestation of multiple myeloma, a relatively common complication in older patients with Gaucher's disease (35).

A 6-year-old child with Gaucher's disease was given a total of 27 Units of enzyme encapsulated in red cells and administered over a period of nine days in eight infusions. Although follow-up of this patient has been incomplete because he lives at a considerable distance, available data do not suggest any change in organomegaly.

Finally, we have administered two courses of enzyme encapsulated in gamma globulin-coated red cells to a 23-year-old girl. This patient suffers from far-advanced Gaucher's disease characterized by severe cirrhosis of the liver which necessitated a shunting operation during childhood. She received a total of 8 Units of enzyme in 3 injections in June of 1978 and in December of 1978 and in January of 1979 received an additional 7 infusions totalling 50 Units of enzyme. Again, no objective changes have been noted.

None of the six patients whom we have treated have experienced symptoms which may clearly be attributed to the infusion of enzyme preparations. One of the patients developed fever following treatment, but this patient had a marked proclivity to spontaneous febrile episodes, and we are reluctant to attribute the febrile episodes to enzyme infusion. Two of the patients complained of abdominal pain during the period of time that enzyme was being administered.

However, they also received enzyme infusions without developing pain and developed pain when they were not receiving enzyme. Thus, it is again doubtful that a cause-and-effect relationship exists between symptoms which these patients experienced and the enzyme infusion.

CONCLUSIONS

No clear-cut objective improvement has been documented in any of the six patients we treated by the infusion of varying amounts of glucocerebrosidase.

Most of the patients experienced a feeling of increased well-being following enzyme therapy, but we are reluctant to give much credence to such reports. On the other hand, the possibility that the patients may have benefited from therapy despite the absence of objective changes cannot be dismissed altogether. The accumulation of glycolipid in the organs of patients with Gaucher's disease is associated with extensive fibrosis, and it is possible that the removal of glycolipid may not be associated with alteration in organomegaly. In the absence of objective changes, however, it is our inclination either to deliver larger amounts of enzyme to the tissues of patients with Gaucher's disease or to attempt to devise means which more efficiently deliver enzyme to the target cells.

REFERENCES

1. Allison, A. C. (1967). Sci. Am. 217, 62.
2. Brady, R. O. (1969). Med. Clin. North Am. 53, 887.
3. Baudhuin, P., Hers, H. G., and Loeb, H. (1964). Lab. Invest. 13, 1139.
4. Hug, G., Schubert, W. K., and Chuck, G. (1968). Clinical Research 16, 345.
5. Greene, H. L., Hug, G., and Schubert, W. K. (1969). Arch. Neurol. 20, 147.
6. Dekaban, A. S., Holden, K. R., and Constantopoulos, G. (1972). Pediatrics 50, 688.
7. Erickson, R. P., Sandman, R. V. B., Robertson, W., and Epstein, C. J. (1972). Pediatrics 50, 693.
8. Dean, M. F., Muir, H., and Benson, P. F. (1973). Nature New Biol. 243, 143.
9. Desnick, R. J., Snyder, P. D., Desnick, S. J., Krivit, W., and Sharp, H. L. (1972). In "Sphingolipids, Sphingolipidoses and Allied Disorders" (B. W. Volk and S. M. Aronson, eds.), pp. 351-371. Plenum Publ. Corp., New York.

10. Mapes, C. A., Anderson, R. L., Sweeley, C. C., Desnick, R. J., and Krivit, W. (1970). Science 169, 987.
11. Brady, R. O., Tallman, J. F., Johnson, W. G., Gal, A. E., Leahy, W. R., Quirk, J. M., and Dekaban, A. S. (1973). N. Engl. J. Med. 289, 9.
12. Knudson Jr., A. G., Di Ferrante, N., and Curtis, J. E. (1971). Proc. Natl. Acad. Sci. USA 68, 1738.
13. Johnson, W. G., Desnick, R. J., Long, D. M., Sharp, H. L., Krivit, W., Brady, B., and Brady, R. O. (1973). In "Enzyme Therapy in Genetic Diseases, Birth Defects: Original Article Series" (D. Bergsma, ed.), Vol. IX, pp. 120-124. Williams and Wilkins, Baltimore.
14. Austin, J. H. (1967). In "Inborn Disorders of Sphingolipid Metabolism" (S. M. Aronson and B. W. Volk, eds.), pp. 359-387. Oxford, Pergamon.
15. Dale, G. L., and Beutler, E. (1976). Proc. Natl. Acad. Sci. USA 73, 4672.
16. Furbish, F. S., Blair, H. E., Shiloach, J., Pentchev, P.G., and Brady, R.O. (1977). Proc. Natl. Acad. Sci. USA 74, 3560.
17. Brot, F. E., Bell Jr., C. E., and Sly, W. S. (1978). Biochemistry 17, 385.
18. De Barsy, T. P., Jacquemin, P., Van Hoof, F., and Hers, H. G. (1973). In "Enzyme Therapy in Genetic Diseases, Birth Defects: Original Article Series" (D. Bergsma, ed.), Vol. IX, pp. 184-190. Williams and Wilkins, Baltimore.
19. Bishop, D. F., Wampler, D. E., Sgouris, J. T., Bonefeld, R.J., Anderson, D. K., Hawley, M. C., and Sweeley, C. C. (1978). Biochim. Biophys. Acta 524, 109.
20. Neufeld, E. F., Sando, G. N., Garvin, A. J., and Rome, L. H. (1977). J. Supramol. Struct. 6, 95.
21. Kaplan, A., Achord, D. T., and Sly, W. S. (1977). Proc. Natl. Acad. Sci. USA 74, 2026.
22. Achord, D. T., Brot, F. E., Bell, C. E., and Sly, W. S. (1978). Cell 15, 269.
23. Schlesinger, P. H., Doebber, T. W., Mandell, B. F., White, R., De Schryver, C., Rodman, J. S., Miller, M. J. and Stahl, P. (1978). Biochem. J. 176, 103.
24. Furbish, F. S., Steer, C. J., Barranger, J. A., Jones, E. A., and Brady, R. O. (1978). Biochem. Biophys. Res. Commun. 81, 1047.
25. Gregoriadis, G. (1976). N. Engl. J. Med. 295, 704.
26. Ihler, G. M., Glew, R. H., and Schnure, F. W. (1973). Proc. Natl. Acad. Sci. USA 70, 2663.
27. Dale, G. L., Villacorte, D., and Beutler, E. (1977). Biochem. Med. 18, 220.

28. Desnick, R. J., Fiddler, M. B., Douglas, S. D., Hudson, L. D. S. (1978). Adv. Exp. Med. Biol. 101, 753.
29. Desnick, R. J., Thorpe, S. R., and Fiddler, M. B. (1976). Physiol. Rev. 56, 57.
30. Brady, R. O., Pentchev, P. G., Gal, A. E., Hibbert, S. R., and Dekaban, A. S. (1974). N. Engl. J. Med. 291, 989.
31. Brady, R. O., Pentchev, P. G., Gal, A. E., Hibbert, S. R., Quirk, J. M., Mook, G. E., Kusiak, J. W., Tallman, J. F., and Dekaban, A. S. (1976). Adv. Exp. Med. Biol. 68, 523.
32. Beutler, E., Dale, G. L., Guinto, E., and Kuhl, W. (1977). Proc. Natl. Acad. Sci. USA 74, 4620.
33. Belchetz, P. E., Crawley, J. C. W., Braidman, I. P., and Gregoriadis, G. (1977). Lancet 2, 116.
34. Beutler, E. (1979). In "Genetic Diseases Among Ashkenazi Jews" (R. M. Goodman and A. G. Motulsky, eds.), pp. 157-167. Raven Press, New York.
35. Pratt, P. W., Estren, F., and Kochwa, S. (1968). Blood 31, 633.

Index

Numbers refer to the chapters in which the entries are discussed.

A

Activation
  complement, 13
  light-dependent, 7
Adenine nucleotides
  bacteria, growing and starving, 2
  Beneckea natriegens, 2
  energy transduction, 2
  Escherichia coli, 2
  eukaryotic organisms, 2
  hemolytic anemia, pool size, 26
  Klebsiella aerogenes, 2
  measurements, 2
  metabolic compartmentation, 2
  as metabolic intermediates, 2
  metabolic regulation, 2
  Peptococcus prévotii, 2
  phosphofructokinase, effect on, 1
  pool size and hemolytic anemia, 26
  turnover during growth, 2
Adenosine deaminase
  chemotaxis, leukocyte, 19
  eukaryotic chemotaxis, 18
Adenosine triphosphatase, effect of thioredoxin on, 7
Adenosine triphosphate
  measurements, 2
  photosynthesis, 8
  regeneration, 2
  turnover during growth, 2
  utilization, 2
S-Adenosyl-homocysteine, 18, 19
S-Adenosyl-methionine, 18
Adenylate cyclase, ADP-ribosylation, 4
Adenylate kinase, energy charge, 2

Adenylylation cycle, 12
Adenylyltransferase, 12
ADP-glucose pyrophosphorylase
  activation of, 10
  activator specificities of, 10
  bacterial glycogen mutants affected in the regulatory properties of, 10
  chemical and physical properties of, 10
  chemical modification of, 10
  E. coli B, 10
  evidence for *in vivo* regulation of, 10
  inhibition of, 10
  kinetic functions of activators and inhibitors, 10
  mutant CL1136, 10
  mutant JP51, 10
  mutant SG14, 10
  mutant SG5, 10
  primary sequence of allosteric activator binding site of, 10
  regulation of, 10
ADP-ribosyl histone hydrolase, 4
ADP-ribosyl transferase, 4
ADP-ribosylation
  adenylate cyclase, 4
  ADP-ribosyl histone hydrolase, 4
  ADP-ribosyl transferase, 4
  elongation factor 2, 4
  poly (ADP-ribose) glycohydrolase, 4
  poly (ADP-ribose) synthetase, 4
Adrenaline, 16
Allosteric effects
  heterotropic, 3
  homotropic, 3
Allosteric regulation, phosphofructokinase, 1

9-Aminoacridine, 8
Amplitude, 11
Aspartate transcarbamylase, 5
ATP, adenosine triphosphate
　flux, 11
　hydrolysis, in photosynthesis, 8
　synthesis, in photosynthesis, 8

B

Bacillus subtilis, enzyme inactivation and degradation in, 5
Bacteria
　adenine nucleotide content during growth and starvation, 2
　energy charge during growth and starvation, 2
　reserve polymers, 2
Bacterial, glycogen synthesis
　mutants altered in, 10
　regulation of, 10
Bacterial chemotaxis, 17
　Escherichia coli, 17
　methylation, 17
　Salmonella typhimurium, 17
Beneckea natriegens
　adenine nucleotides during growth and starvation, 2
　energy charge, 2
Blue-green algae, thioredoxin in, 7

C

Calcium, glycogen metabolism, 16
Calcium-dependent regulator protein, 16
Calmodulin, 16
Carbobenzoxy-phenylalanyl-methionine, 19
Carbon cycle
　photosynthesis, 8
　regulation, 8
Carboxyl groups, chemotaxis, leukocyte, 19
Carboxymethylation, 17, 18, 19
　chemotaxis
　　eukaryotic, 18
　　leukocyte, 19
Carboxymethyltransferase, chemotaxis, 17
Carboxypeptidase Y, proteolytic activation, 6
Cascade, glutamine synthetase, 12
Cascade system
　amplification, 11
　enzyme regulation, 11, 12
　kinetics of, 11
Casein kinase, 14
Catabolite inactivation, 6
　proteolytic, 6

Catabolite modification
　galactose uptake system, 6
　maltose uptake system, 6
Cell proliferation, 21
Chemical modification, ADP-glucose pyrophosphorylase, 10
Chemotaxis
　adenosine deaminase, 18, 19
　bacterial, 17
　divalent ions in, 18
　eukaryotic cells, 18, 19
　leukocyte, 19
　methylation and demethylation, 17
　methylation of phospholipids, 18
　mutants, 17
　receptors, 17
　Salmonella typhimurium, 17
Chitin synthase, 6
　activation of, 6
Chlorella pyrenoidosa, reductive pentose phosphate cycle in, 9
Chloroplasts
　pH gradient, 8
　photosynthesis, 7, 8, 9
　proton shuttles, 8
　reconstituted, 9
Cholera toxin, ADP-ribosylation, 4
Compartmentation, adenine nucleotides, 2
Complement
　activation
　　alternative pathway, 13
　　classical pathway, 13
　components, 13
　membrane-attack complex, 13
　physical properties, 13
　regulators, 13
Converter enzyme, 11
Cooperativity
　for binding, 3
　enzyme cascade, 11
　glutamine-synthetase cascade, 12
Cordycepin, 22
Covalently interconvertible enzyme, 11
Cyclic AMP
　cAMP-dependent protein kinase
　　from bovine cardiac muscle, 15
　　phosphorylation of, 15
　cell growth, inhibition of, 15
　Fc-mediated phagocytosis, 15
　glycogen metabolism, 16
　murine reticulum cell sarcoma, 15
　PEP carboxykinase synthesis, 22
　phosphodiesterase, 21

INDEX

phosphoprotein phosphatases, 15
plasminogen activator, effect on, 15
protein phosphorylation, 15
  insulin-dependent, 15
  3T3-L1 cell line, 15
Cyclic AMP-dependent protein kinase, 14, 15, 16
  bovine cardiac muscle, 15
  glycogen metabolism, 16
  reticulocytes, 14
Cyclic cascade
  bicyclic system, 11
  monocyclic system, 11
  multicyclic system, 11
  signal amplification, 11
Cyclic CMP, 20, 21
  liver, mice, 20
  phosphodiesterase, 21
    assay, 21
    regenerating liver, 21
Cyclic GMP phosphodiesterase, 21
Cyclic nucleotide-independent protein kinases
  hemin controlled repressor, 14
  reticulocytes, 14
Cyclic nucleotide phosphodiesterases, 21
  2'-deoxy cyclic nucleotide monophosphate, inhibition by, 21
  effectors of, 21
  Morris hepatoma 3924A,
  2'-Cyclocytidine, β-D-O^2, 20
Cytidine 3',5'-cyclic monophosphate, 20, 21
Cytidylate cyclase, 20
  effect on, 20
Cytosine arabinoside, 20

D

3-Deazaadenosine, 18, 19
Degradation, poly (ADP-ribose), 4
Degradation of enzymes, 5
  immunochemical methods for studying, 5
  sporulation, during, 5
2'-deoxy cyclic nucleotide monophosphate, effect on cyclic nucleotide phosphodiesterases, 21
Diabetes, peptide-chain initiation, 23
2,3-Diphosphoglycerate mutase, hemolytic anemia, 26
Diphtheria toxin
  ADP-ribosyl transferase, 4
  ADP-ribosylation, 4
Disallosterisms, 3

E

Ehrlich ascites tumor cell, eIF-2, 25
eIF-2, 24, 25
  cycling of, 25
  Ehrlich ascites tumor cell, 25
  phosphorylation of, 25
eIF-2 kinase, 25
Electron transport
  photosynthesis, regulation, 8
  vectorial, 8
Elongation factor 2, 4
  ADP-ribosylation, 4
Energy charge
  adenylate kinase, 2
  bacteria, growing and starving, 2
  Beneckea natriegens, 2
  Escherichia coli, 2
  eukaryotic organisms, 2
  Klebsiella aerogenes, 2
  Peptococcus prévotii, 2
  photosynthesis, 8
  problems and drawbacks, 2
  as a regulator of metabolic activity, 2
Enterotoxin LT, ADP-ribosylation, 4
Enzyme
  light-activation of, 7, 8
  replacement of, 27
Enzyme activation, by light, 7
Enzyme cascade, 11
  steady-state analysis, 11
Enzyme deficiencies
  Gaucher's disease, 27
  hemolytic anemia, 26
Enzyme inactivation
  Bacillus subtilis, 5
  glutamate dehydrogenase, NADP-dependent, 6
  proteinase, 6
  sporulation, during, 5
Enzyme regulation, 7, *see also* Regulation
  photochemical, 7
Enzyme replacement therapy, 27
  Gaucher's disease, 27
  genetic diseases, 27
Erythrocyte
  energy generation in, 26
  hemolytic anemia, 26
Escherichia coli
  adenine nucleotides during growth and starvation, 2
  ADP-glucose pyrophosphorylase, 10
  chemotaxis, 17
  energy charge, 2

thioredoxin in, 7
Eukaryotic cells, chemotaxis, 18, 19
Eukaryotic organisms
  adenine nucleotide contents, 2
  energy charge, 2
  metabolic compartmentation, 2
Evolution
  phosphofructokinase, 3
  of regulatory enzymes, 3

F

Fatty acids, 23
  peptide-chain initiation, 23
  protein synthesis, 23
Ferredoxin, 7, 9
  fructose 1, 6-bisphosphatase, 7
  glyceraldehyde 3-phosphate-dehydrogenase, NADP-linked, 7, 8
  NADP-linked, 7, 8
  light-activation, 7
  reductive pentose phosphate cycle, regulation, 9
  regulation of enzymes in photosynthesis, 7
Ferredoxin-thioredoxin reductase, 7
Ferredoxin–thioredoxin system, regulation of enzymes in photosynthesis, 7
Fetal tissues, cyclic nucleotide phosphodiesterases, 21
Formylmethyl peptides, leukocyte attractant, 19
Free energy change, reductive pentose phosphate cycle, 9
Fructose 1, 6-bisphosphatase
  enzyme evolution, 3
  ferredoxin–thioredoxin system in, 7
  photosynthesis, 7, 8, 9
Fructose 1, 6-bisphosphate, 7, 8, 9

G

Galactose uptake system, 6
  yeast, 6
Gaucher's disease, 27
  enzyme replacement therapy, 27
Genetic diseases, 27
  Gaucher's disease, 27
  globoside, accumulation of, 27
  lysosomal storage diseases, 27
  treatment of, 27
Globoside, accumulation of, 27
Glucagon, PEP carboxykinase synthesis, 22
Glucocorticoids, PEP carboxykinase synthesis, 22

Gluconeogenesis, hepatic, 22
Glucose-6-phosphate, peptide-chain initiation, 23
Glucose-6-phosphate dehydrogenase, hemolytic anemia, 26
Glutamate dehydrogenase, NADP-linked, inactivation of, 6
Glutamine phosphoribosylpyrophosphate amidotransferase, 5
Glutamine synthetase
  adenylylation, 12
  cascade, role of, 12
  enzyme multimodulation, 3
  metabolic control, 12
  uridylylation, 12
Glutamine synthetase cascade, 12
Glutathione, hemolytic anemia, 26
Glyceraldehyde 3-phosphate dehydrogenase, NADP-linked, ferredoxin in, 7
Glyceraldehydephosphate dehydrogenase (NADP), in photosynthesis, 8
Glycogen metabolism, 16
  cyclic AMP in, 16
  cyclic AMP-dependent protein kinase, 16
  insulin, 16
Glycogen synthase, 16
  phosphorylation of serine, 16
  purification, 16
Glycogen synthase kinase, 16
  assay, 16
Glycogen synthase kinase-2, 16
Glycogen synthesis, bacterial, 10
Glycolysis, 9
Growth, turnover of adenine nucleotides, during, 2
Guinea pigs, cyclic nucleotide phosphodiesterases, 21

H

HeLa cells, 4
  poly ADP-ribosylation, 4
Hemin controlled repressor (HCR), 14, 24
  cyclic nucleotide-independent protein kinase, 14
Hemoglobin
  denaturation of, 26
  hemolytic anemia, 26
Hemolytic anemia
  adenine nucleotide pool, 26
  alterations in metabolism, 26
  enzyme deficiencies in, 26
  hemoglobin, oxidative denaturation of, 26

INDEX

Hepatectomy, cyclic CMP- phosphodiesterase, effect on, 21
Hepatoma 3924A, Morris, cyclic nucleotide phosphodiesterases, 21
Hexokinase, hemolytic anemia, 26
Histone H1, poly ADP-ribosylation, 4
Histone H2B, poly ADP-ribosylation, 4
Homocysteine thiolactone, 18
Hormonal regulation, PEP carboxykinase synthesis, 22

I

Imidazole, effect on cyclic nucleotide phosphodiesterase, 21
Immunochemical methods, for studying enzyme degradation, 5
Immunohistochemistry, poly ADP-ribosylation in, 4
Inactivation of enzymes, 5, 6
   aspartate transcarbamylase, 5
   energy-dependent, 5
   glutamate dehydrogenase, NADP-dependent, 6
   glutamine phosphoribosylpyrophosphate amidotransferase, 5
   malate dehydrogenase, cytoplasmic, 6
   oxygen-dependent, 5
Initiation complex, protein synthesis, 24
Initiation factor, 14, 24, 25
   eIF-2, 24
      physical characteristics, 24
   phosphorylation of, 14
   protein synthesis, 24, 25
Insulin
   glycogen metabolism, 16
   PEP carboxykinase synthesis, 22
   peptide-chain initiation, 23
   protein phosphorylation, 15
   protein synthesis, 23
Interconvertible enzyme, 11
Iron, effect on cytidylate cyclase, 20
Iron-sulfur enzymes, 5
   degradation of enzymes, 5

K

Klebsiella aerogenes
   adenine nucleotides during continuous culture, 2
   energy charge, 2
Km, CO_2, ribulose 1,5-bisphosphate carboxylase, 9

L

Leukemia cells, poly ADP-ribosylation in, 4
Leukocyte
   chemotaxis, 18, 19
   formylmethyl peptides, in chemotaxis, 19
Light-activation
   of enzymes, 7, 8
   fructose 1,6-bisphosphatase, 7
   glyceraldehyde 3-phosphate dehydrogenase, NADP-linked, 7
   malate dehydrogenase, NADP-linked, 7, 8
   sedoheptulose 1,7-bisphosphatase, 7, 8, 9
Liver
   cyclic CMP, 20
   gluconeogenesis, 22
   regenerating, 21
Lysosomal storage diseases, treatment of, 27

M

Macrophages, 18
Malate dehydrogenase, NADP-linked
   dark deactivation of, 7
   inactivation of, 6, 7
   light-activation, 8
Maltose uptake system, 6
   yeast, 6
Membrane phospholipids, chemotaxis, leukocyte, 19
Mepacrine, 19
Mercaptoethanol, effect on cyclic CMP phosphodiesterase, 21
Metabolic compartmentation, adenine nucleotides, 2
Metabolic interconversion of enzymes, 3
Metabolic regulation, 2, 3, 12
   adenine nucleotides, 2
   energy charge, 2
   glutamine synthetase, 12
Metabolism, alterations in, hemolytic anemia, 26
Methylation
   chemotaxis, bacterial, 17
   chemotaxis, eukaryotic, 18
   leukocyte chemotaxis, 19
   phospholipids, 18, 19
Methylesterase, chemotaxis, 17
Methyltransferase, chemotaxis, 17
Microtubules, 18
Mitochondria adenine nucleotides, 2
Monocytes, 18

Multimodulation of enzymes, 3
  glutamine synthetase, 3
  phosphofructokinase, 3
  pyruvate kinase, 3
Murine reticulum cell sarcoma
  cAMP in, 15
  protein phosphorylation, 15
Muscle, bovine cardiac, cyclic AMP-dependent protein kinase, 15
Mutants
  ADP-glucose pyrophosphorylase, 10
  chemotaxis, 17
Myeloid leukemia, 20

N

NAD, *see* Nicotinamide adenine dinucleotide
N_4 phage, ADP-ribosylation, 4
Nicotinamide adenine dinucleotide, (NAD), role in ADP-ribosylation, 4
Nonhistone proteins, poly ADP-ribosylation, 4
Nucleotide biosynthesis, regulation of, 5
Nucleotides, effect on cyclic nucleotide phosphodiesterases, 21
Nutrient deprival, protein synthesis, effect on, 23, 25

O

Oxidative pentose phosphate cycle, 9
Oxygen-dependent, inactivation of enzymes, 5

P

Papaverine cyclic nucleotide phosphodiesterases, effect on, 21
Pentose-phosphate cycle
  Chlorella pyrenoidosa, 9
  free energy change, 9
  regulation, ferredoxin in, 9
Pentose-phosphate shunt, hemolytic anemia, 26
Peptide-chain initiation, 23
  diabetes, 23
  effect of fatty acids, 23
  effect of insulin, 23
  effect of noncarbohydrate substrates, 23
  heart, 23
  skeletal muscle, 23
  starvation, 23
Peptococcus prévotii
  adenine nucleotides during starvation, 2
  energy charge, 2
pH gradient
  chloroplast envelop, 8
  photosynthesis, 8

pH regulation, in photosynthesis, 8
Phagocytic cells, 18
Phagocytosis, cAMP in, 15
Phenylalanine, rates of protein synthesis, 23
Phosphatases, of phosphoprotein, 15
Phosphate
  anhydride bound in ATP, 2
  cyclic nucleotide phosphodiesterases, effect on, 21
  inorganic, in photosynthesis, 9
  turnover in ATP, 2
Phosphate translocator, 9
Phosphodiesterase, 4
Phosphodiesterase-cyclic CMP, 20
Phosphoenolpyruvate carboxykinase (GTP), 22
  diet, effect on, 22
  glucagon, effect on synthesis, 22
  glucocorticoids, effect on synthesis, 22
  *m*RNA levels of, 22
  regulation of synthesis, 22
Phosphoenolpyruvic acid (PEPA), 9
Phosphofructokinase
  adenine nucleotides, 1
  allosteric control, 1
  energy regulation, 1
  enzyme evolution, 3
  multimodulation, 3
  polyphosphate control, 1
  vanadate regulation, 1
Phosphoglucomutase, 9
6-Phosphogluconate, 9
Phospholipids
  leukocyte chemotaxis, 19
  methylation 18, 19
Phosphoprotein phosphatases, cyclic AMP, 15
Phosphoribulose kinase, 8, 9
Phosphorylase, 3, 16
  phosphorylation of serine, 16
Phosphorylase kinase, 16
  assay, 16
  deficient, mice (ICR/IAn), 16
Phosphorylation
  eIF-2, 24, 25
  protein, 15
  protein kinases, 14
    cAMP-dependent, 15
  protein synthesis, components of, 14
  ribosomal protein, S6, 25
  serine, in glycogen synthase, 16
Phosphorylation potential, 8
Phosphoserine, 16
Photophosphorylation, regulation of, 8

Photorespiration, 9
Photosynthesis, 7, 8, 9
  ATP synthesis and hydrolysis, 8
  carbon cycle, 8
  chloroplasts, 7, 8, 9
  electron transport in, 8
  energy charge, 8
  ferredoxin–thioredoxin system, 7
  pH gradient, 8
  regulation of, 7
  spinach, 9
Plasminogen activator effect of cyclic AMP on, 15
Poly(ADP-ribose)
  acceptor proteins, 4
  biosynthesis, 4
  degradation, 4
  immunohistochemistry, 4
  natural distribution, 4
Poly(ADP-ribose) glycohydrolase, 4
Poly(ADP-ribose) synthetase, 4
Polymers, as reserve in bacteria, 2
Polysomes, 23
Protease activated kinase, 14
Protein carboxyl groups, chemotaxis, leukocyte, 19
Protein degradation, 5
Protein kinase, 15, 24
  cAMP-dependent, from bovine cardiac muscle, 15
  eIF-2, 24
  phosphorylation of, 15
Protein phosphorylation, 15, 24
  eIF-2, 24
  insulin-dependent, 15
  murine reticulum cell sarcoma, 15
Protein synthesis
  diabetes, 23
  eIF-2, 24
  effect of fatty acids, 23
  effect of insulin, 23
  effect of noncarbohydrate substrates, 23
  effect of nutrient deprival, 25
  heart, 23
  inhibitors, 25
  initiation complex, 24
  initiation of, 25
  phosphorylation in, 14
  protein kinases in, 14
  rate determination, phenylalanine, 23
  skeletal muscle, 23
  starvation, 23
Proteinase, enzyme inactivation, 6

Proteinase-inhibitor, 6
Proteolytic inactivation
  Bacillus subtilis, 5
  yeast, 6
Proteolytic modulation, yeast, 6
Proton shuttles, chloroplast, 8
Pseudomonas exoenzyme S, ADP-ribosylation, 4
Pseudomonas toxin, ADP-ribosylation, 4
Pyrimidine 5'-nucleotidase, hemolytic anemia, 26
Pyrophosphate, in photosynthesis, 9
Pyruvate kinase
  hemolytic anemia, 26
  multimodulation of enzyme activity, 3
  in photosynthesis, 9

R

Rat tissues, cyclic CMP phosphodiesterase in, 21
Rate amplification, 11
Receptors, chemotaxis, 17
Reductive pentose phosphate cycle, 9
Regulation, 11
  covalent modification, 11
  nucleotide biosynthesis, 5
  photophosphorylation, 8
  starch synthesis, 10
Regulatory enzymes, evolution of, 3
Regulatory sites
  evolutionary origin, 3
  identification, 3
Replacement of enzyme as therapy, 27
Reticulocytes
  cyclic nucleotide-independent protein kinase, 14
  phosphorylation in, 14
  protein kinases, 14
Reversibility, reductive pentose-phosphate cycle, reactions in, 9
Ribonucleotide reductase, 3
Ribosomal cycle, 25
Ribosomal proteins, 25
  phosphorylation of, 25
  S6, 25
Ribosomal subunits, 23
Ribosomes, phosphorylation of subunits, 14
Ribulose 1,5-bisphosphate (RuBP), 9
Ribulose 1,5-bisphosphate carboxylase, 7, 8, 9
  K_m, CO_2, 9
RNA polymerase, multimodulation of, 3

S

Salmonella Typhimurium, chemotaxis, 17
SBPase, 9
Sedoheptulose 1,7-bisphosphatase
  ferredoxin-linked activation, 7, 8, 9
  light-activation of, 7, 8, 9
Sedoheptulose 1,7-bisphosphate, 9
Sensitivity, signal amplification in enzyme cascade, 11
Shapiro's regulatory protein, 12
Shuttles, of substrates, 8
Signal amplification, 11, 12
  glutamine synthetase cascade, 12
Skeletal muscle, glycogen metabolism, 16
Spinach, regulation of reductive pentose–phosphate cycle, 9
Sporulation, enzyme inactivation and degradation during, 5
Starch synthesis, regulation of, 10
Steady-state analysis
  of enzyme cascade, 11
  glutamine synthetase cascade, 12
Storage diseases, 27
Sucrose phosphate synthetase, 9

T

3T3-L1 cells, cAMP in, 15
T4 phage, ADP-ribosylation, 4

Thioredoxin
  ATPase activity, 7
  blue-green algae, 7
  Escherichia coli, 7
  fructose 1,6-bisphosphatase, 7
  light-dependent activation, 7
  multiple forms, 7
  regulation of enzymes in photosynthesis, 7
Translation, phosphorylation in, 14
Trifluoperazine, 16
Triose phosphate dehydrogenase, 9
Triose phosphates, 9
Troponin-I, 16

U

Unidirectional cascade, 11
Uridylyl removing enzyme, 12
Uridylyl transferase, 12
Uridylylation cycle, 12

V

Vanadate, regulation of phosphofructokinase, 1
Vectorial electron transport, 8

Y

Yeast
  galactose uptake system, 6
  inactivation of enzymes, 6
  maltose uptake system, 6